Th

Pock

Science

The **HUTCHINSON**

Pocket Dictionary of Science

Helicon Publishing Ltd
42 Hythe Bridge Street
Oxford OX1 2EP

Printed and bound in Great Britain by
Unwin Brothers Ltd, Old Woking, Surrey

ISBN 1–85986–229–2

British Cataloguing in Publication Data

A catalogue record for this book is available
from the British Library

Editorial director
Michael Upshall

Consultant editors
Yvonne Ryszkowski
Stephen Webster

Project editor
Sara Jenkins-Jones

Text editor
Catherine Thompson

Art editor
Terence Caven

Additional page make-up
Helen Bird

Production
Tony Ballsdon

abdomen the lower part of the body, containing the stomach, intestines, liver, and kidneys; in insects it is the third section of the body. In mammals, the abdomen is separated from the upper part of the body by the ◊diaphragm, a sheet of muscular tissue.

abiotic factor a nonliving feature of the environment, such as temperature, that has an effect on the life of organisms. Other examples are light, humidity, ◊soil structure and composition, and background radiation. Abiotic factors can be harmful; for instance, the acid rain produced as a result of sulphur-dioxide emissions from industry.

absolute zero the lowest temperature theoretically possible, zero kelvin, equivalent to –273.16°C, at which molecules are motionless. Near absolute zero, the physical properties of some materials change substantially; for example, some metals lose their electrical resistance and become superconducting.

absorption in biology, the method by which organisms take in the molecules they need for their cellular processes. For instance, the small molecules formed during digestion in the alimentary canal move across the gut wall into the blood stream – the molecules are then said to have been 'absorbed'.

The mechanism of absorption involves a simple semipermeable membrane separating two zones. If a particular molecule is in greater concentration on one side, then the molecule, provided it is small enough, will diffuse through to the other side. Thus glucose will move from the gut, where it is at a high concentration, into the blood system, where it is at a lower concentration. In biology, therefore, absorption is best understood as a process in which an organism obtains essential molecules by diffusion.

In physics and chemistry, absorption is more generally used to describe the taking-up of one substance by another, such as a liquid by a solid (ink by blotting paper) or a gas by a liquid (ammonia by water). The term can also refer to the phenomenon by which a substance retains radiation of particular wavelengths; for example, a piece of blue glass absorbs all visible light except the wavelengths in the blue part of the spectrum.

AC abbreviation for ◊alternating current.

acceleration the rate of change of the velocity of a moving body. It is measured in metres per second per second ($m\ s^{-2}$). Because velocity is a vector quantity (possessing both magnitude and direction), a body travelling at constant speed may be said to be accelerating if its direction of motion changes. According to Newton's second law of motion, a body will accelerate only if it is acted upon by an unbalanced or resultant ◊force.

The average acceleration a of an object travelling in a straight line may be calculated by using the formula

$$a = \Delta v / \Delta t$$

where Δv is the change in velocity and Δt is the time taken, or by

$$a = (v - u)/t$$

where u is the initial velocity of the object, v is its final velocity, and t is the time taken. A negative answer shows that the object is slowing down (decelerating).

Acceleration due to gravity is the acceleration of a body falling freely under the influence of the Earth's gravitational field; it varies slightly at different latitudes and altitudes. The value adopted internationally for gravitational acceleration is $9.806\ m\ s^{-2}$.

accommodation the ability of the vertebrate ◊eye to focus on near or far objects by changing the shape of the lens.

For something to be viewed clearly, the image must be precisely focused on the retina, the light-sensitive sheet of cells at the rear of the eye. Close objects can be seen when the lens takes up a more spherical shape. Distant objects become focused when the lens is stretched and made thinner. These changes in shape are produced by the contraction and relaxation of a ring of ciliary muscles lying beneath the iris. Accommodation is coordinated by the brain.

acetate common name for ◊ethanoate.

acetic acid common name for ◊ethanoic acid.

acetone common name for ◊propanone.

acetylene common name for ◊ethyne.

acid compound that releases hydrogen ions in the presence of water, and turns litmus paper red.

$$HCl_{(g)} + aq \leftrightarrow H^+_{(aq)} + Cl^-_{(aq)}$$

The reactions of acids are the reactions of the $H^+_{(aq)}$ ion. These are as follows:

with indicators They give a specific colour reaction with indicators; for example, litmus turns red.

with alkalis They react with alkalis to form a salt and water (neutralization).

$$HCl_{(aq)} + NaOH_{(aq)} \rightarrow NaCl_{(aq)} + H_2O_{(l)}$$

with carbonates With carbonates and hydrogencarbonates, acids form a salt and displace carbon dioxide.

$$HNO_3 + NaHCO_3 \rightarrow NaNO_3 + CO_2 + H_2O$$

with metals Acids react with metals to give off hydrogen and form a salt.

$$Mg + H_2SO_4 \rightarrow MgSO_4 + H_2$$

Acids react with many ◊bases, such as oxides and hydroxides, but the product is not always soluble in water so the reaction soon ceases, as when sulphuric acid reacts with calcium oxide, hydroxide, or carbonate.

Sulphuric, nitric, and hydrochloric acid are sometimes referred to as the mineral acids. The commonest naturally occurring acids are organic compounds, such as the ◊fatty acids.

acidic oxide oxide of a ◊non-metal. Acidic oxides are covalent compounds. Those that dissolve in water, such as sulphur dioxide, give acidic solutions.

$$SO_2 + H_2O \leftrightarrow H_2SO_{3(aq)} \leftrightarrow H^+_{(aq)} + HSO^-_{3(aq)}$$

All acidic oxides react with alkalis to form salts; for example, carbon dioxide reacts with sodium hydroxide to give sodium hydrogencarbonate.

$$CO_2 + NaOH \rightarrow NaHCO_3$$

acid rain acidic rainfall, thought to be caused principally by the release into the atmosphere of sulphur dioxide and oxides of nitrogen. Sulphur dioxide is formed from the burning of fossil fuels such as coal that contain high quantities of sulphur. Nitrogen oxides are produced by industrial activities and by cars.

Acid rain is linked with the damage and death of forests and lake organisms in Scandinavia, other parts of Europe, and eastern North America. It also results in corrosion of buildings.

Pollution by acid rain is difficult to control, because the acidic emissions drift in winds from one country to another, and even across continents. Much of the acid rain that falls in Europe, for instance, comes from emissions within E Europe, especially NE Germany, Czech Republic, and Poland. It is possible to remove sulphur dioxide from factory or power-station emissions, but the process is expensive.

acid salt compound formed by the partial neutralization of a dibasic or tribasic ◊acid. Although a salt, it contains replaceable hydrogen, so it may undergo the typical reactions of an acid. Examples are sodium hydrogen-sulphate ($NaHSO_4$) and disodium hydrogenphosphate (Na_2HPO_4).

acoustics in general, the experimental and theoretical science of sound and its transmission; in particular, the branch of the science that has to do with the phenomena of sound in a particular space such as a room or theatre.

actinide any of a series of 15 radioactive metallic chemical elements with atomic numbers 89 (actinium) to 103 (lawrencium).

activation energy the energy required to start a chemical reaction. Some elements and compounds will react together merely by bringing them into contact (spontaneous reaction). For others it is necessary to supply energy in order to start the reaction, even if there is ultimately a net output of energy. This initial energy is the activation energy. The lower the activation energy, the more likely it is that a reaction will take place. In biology, enzymes help certain reactions to occur by lowering the activation energy.

activity series alternative name for ◊reactivity series.

adaptation any feature of an organism that makes it 'suited' to its environment. Examples are the long neck of the giraffe (an adaptation that allows the giraffe to graze from trees) and the heavy coat of a polar bear (which keeps the bear warm). Adaptations are inherited, so that a well-adapted organism passes on favourable features (via the genetic material) to its offspring. The theory of evolution by natural selection holds that species become extinct when they are no longer adapted to their environment—for instance, if the climate becomes suddenly colder. It follows that all species that exist today must be reasonably well adapted.

adaption

The beaks of birds are adapted to suit their diet
oystercatcher
long, straight beak to prise open shells of bivalve molluscs, such as mussels
nightjar
wide gape and bristles around beak to trap flying insects
toucan
long beak to pick berries from thin twigs
eagle
sharp, hooked beak to tear flesh from prey
hummingbird
long, pointed beak to reach nectar deep inside flowers
skimmer
scoops up fish and traps them in its scissor-like beak
spoonbill
swings long, wide beak from side to side underwater to capture fish and aquatic insects

addition polymerization ◊polymerization reaction in which a single monomer gives rise to a single polymer, with no other reaction products.

addition reaction reaction in which the atoms of an element or compound react with a double or triple bond in an organic compound by opening up one of the bonds and becoming attached to it, as when hydrogen chloride reacts with ethene to give chloroethane.

$$CH_2{=}CH_2 + HCl \rightarrow CH_3CH_2Cl$$

An example is the addition of hydrogen atoms to ◊unsaturated compounds in vegetable oils to produce margarine.

additive in food technology, a chemical added to prolong shelf life (such as salt), alter colour, enhance flavour, or improve nutritional value (such as vitamins or minerals). Many chemical additives are used in the manufacture of food. They are subject to regulation because some people are affected by constant exposure to even small concentrations of certain additives, and suffer side effects such as hyperactivity. Within the European Community, approved additives are given an official ***E number***.

ADH (abbreviation for ***antidiuretic hormone***) hormone used by the body to maintain the correct salt/water balance in the blood. ADH is secreted by the brain and affects the kidneys, stimulating them to produce concentrated or dilute urine according to the concentration of salt in the blood. When water is in short supply and the blood becomes too salty, high levels of ADH are produced, which stimulate the kidneys to conserve water by producing concentrated urine. When the animal is able to drink plenty of water and the blood is in danger of becoming too dilute, low levels of ADH stimulate the kidneys to get rid of water from the body by producing dilute urine.

aerobe organism that can survive only if there is a reasonable supply of oxygen. Almost all living organisms (plants as well as animals) are aerobes, obtaining energy by respiring ('burning') food using oxygen. An organism that does not require oxygen is called an ◊anaerobe.

aerobic respiration form of respiration that requires the presence of oxygen for the efficient release of energy from food molecules.

Aerobic respiration occurs inside the cell, and, unlike ◊anaerobic respiration, involves the complete breakdown of glucose to give carbon dioxide, water, and large amounts of energy. The carbon dioxide is a waste gas and passes from the organism; the energy is used to drive a wide variety of life processes, including growth and movement. The chemical reactions also produce heat as a waste product; in endotherms, or warm-blooded animals, this heat is used to keep the animal warm and therefore more active.

$$C_6H_{12}O_6 + O_2 \rightarrow 6CO_2 + 6H_2O + 2{,}880\ \mathrm{kJ}$$

affinity force of attraction (see ◊bond) between chemical elements that helps to keep them in combination in a molecule. A given element may have a greater affinity for one particular element than for another (for example, hydrogen has a great affinity for chlorine, with which it easily and rapidly combines to form hydrochloric acid, but has little or no affinity for argon).

AIDS (acronym for ***a****cquired* ***i****mmune* ***d****eficiency* ***s****yndrome*) the newest and gravest of the sexually transmitted diseases (◊STDs). It is caused by the ***human immunodeficiency virus*** (HIV), which is transmitted in body fluids – mainly blood and sexual secretions.

Sexual transmission of the AIDS virus endangers heterosexual men and women as well as high-risk groups, such as homosexual and bisexual men, prostitutes, intravenous drug-users sharing needles, and haemophiliacs and surgical patients treated with contaminated blood products. The virus has a short life outside the body, which makes its transmission by methods other than sexual contact, blood transfusion, and shared syringes extremely unlikely.

Infection with HIV does not necessarily mean that a person has AIDS; many people who have the virus in their blood are not ill; others suffer AIDS-related illnesses but not the full-blown disease. The effect of the virus in those who become ill is the devastation of the immune system, leaving the victim susceptible to diseases that would not otherwise develop. Some AIDS sufferers die within a few months of the outbreak of symptoms, some survive for several years; roughly 50% are dead within three years. There is no cure for the disease, and the search continues for an effective vaccine.

air see ◊atmosphere.

alcohol any member of a group of organic chemical compounds characterized by the presence of one or more OH (hydroxyl) groups in the molecule, and forming ◊esters with acids. The main uses of alcohols are as solvents for gums, resins, lacquers, and varnishes; in the making of dyes;

for essential oils in perfumery; and for medical substances in pharmacy. Alcohol (ethanol) is produced naturally in the ◊fermentation process and is consumed as part of alcoholic beverages.

Alcohols may be liquids or solids, according to the size and complexity of the molecule. The five simplest alcohols form a series in which the number of carbon and hydrogen atoms increases progressively, each one having an extra CH_2 (methylene) group in the molecule: methanol or wood spirit (methyl alcohol, CH_3OH); ethanol (ethyl alcohol, C_2H_5OH); propanol (propyl alcohol, C_3H_7OH); butanol (butyl alcohol, C_4H_9OH); and pentanol (amyl alcohol, $C_5H_{11}OH$). The lower alcohols are liquids that mix with water; the higher alcohols, such as pentanol, are oily liquids not miscible with water, and the highest are waxy solids—for example, hexadecanol (cetyl alcohol, $C_{16}H_{33}OH$) and melissyl alcohol ($C_{30}H_{61}OH$), which occur in sperm-whale oil and beeswax respectively.

alcoholic solution solution produced when a solute is dissolved in ethanol.

aldehyde any of a group of organic chemical compounds prepared by oxidation of primary alcohols, so that the OH (hydroxyl) group loses its hydrogen to give an oxygen joined by a double bond to a carbon atom (the aldehyde group, –CHO).

The name is made up from ***al***cohol ***dehyd***rogenation; that is, alcohol from which hydrogen has been removed. Aldehydes are usually liquids and include methanal, ethanal, benzaldehyde, formaldehyde, and citral.

alimentary canal another word for ◊gut, the tube through which food passes and in which it is processed, digested, and absorbed.

aliphatic compound any organic chemical compound that is made up of chains of carbon atoms, rather than rings, as in ◊cyclic compounds. The chains may be linear, as in hexane (C_6H_{14}), or branched, as in 2-propanol (isopropanol) $(CH_3)_2CHOCH$.

alkali water-soluble ◊base. The four main alkalis are sodium hydroxide (caustic soda, NaOH); potassium hydroxide (caustic potash, KOH); calcium hydroxide (slaked lime or limewater, $Ca(OH)_2$); and aqueous ammonia ($NH_{3\,(aq)}$). Their solutions all contain the hydroxide ion OH^-, which gives them a characteristic set of properties.

with indicators Alkalis give a specific colour reaction with indicators; for example, litmus turns blue.

with acids They react with acids to form a salt and water (neutralization).

$$KOH + HNO_3 \rightarrow KNNO_3 + H_2O$$

$$OH^- + H^+ \rightarrow H_2O \text{ (ionic equation)}$$

with ammonium salts Alkalis displace ammonia gas from ammonium salts.

$$NH_4Cl + NaOH \rightarrow NaCl + NH_3 + H_2O$$

$$NH^+_{4(s)} + OH^-_{(aq)} \rightarrow NH_{3\,(g)} + H_2O$$

with soluble salts Alkalis precipitate the insoluble hydroxides of most other methods from soluble salts.

$$FeCl_2 + 2NaOH \rightarrow Fe(OH)_2 + 2NaCl$$

$$Fe^{2+}_{(aq)} + 2OH^-_{(aq)} \rightarrow Fe(OH)_{2\,(s)}$$

alkali metal any of a group of six metallic elements with similar properties: lithium, sodium, potassium, rubidium, caesium, and francium. They form a linked group (group I) in the ◊periodic table of the elements. They are univalent and of very low density (lithium, sodium, and potassium float on water); in general they are reactive, soft, low-melting-point metals. Because of their reactivity they are found only as compounds in nature, and are used as chemical reactants rather than as structural metals.

alkaline-earth metal any of a group of six metallic elements with similar properties: beryllium, magnesium, calcium, strontium, barium, and radium. They form a linked group (group II) in the ◊periodic table of the elements. They are strongly basic, bivalent, and occur in nature only in compounds. They and their compounds are used to make alloys, oxidizers, and drying agents.

alkane member of the group of ◊hydrocarbons having the general formula C_nH_{2n+2} (common name ***paraffins***). Lighter alkanes are colourless gases (for example methane, ethane, propane, and butane); in nature they are found dissolved in petroleum. Heavier ones are liquids or solids. Because alkanes contain only single ◊covalent bonds, they are said to be saturated.

Their principal reactions are combustion and bromination.

$$CH_4 + 2O_2 \rightarrow CO_2 + H_2O$$

$$C_3H_8 + 5O_2 \rightarrow 3CO_2 + 4H_2O$$

$$C_2H_6 + Br_2 \xrightarrow{200°C} C_2H_5Br + HBr$$

alkanes

name	*molecular formula*	*structural formula*
methane	CH_4	H above; $\mathrm{H{-}C{-}H}$; H below
uses: domestic fuel (natural gas)		
ethane	C_2H_6	$\mathrm{H\ H}$ above; $\mathrm{H{-}C{-}C{-}H}$; $\mathrm{H\ H}$ below
uses: industrial fuel and chemical feedstock		
propane	C_3H_8	$\mathrm{H\ H\ H}$ above; $\mathrm{H{-}C{-}C{-}C{-}H}$; $\mathrm{H\ H\ H}$ below
uses: bottled gas (camping gas)		
butane	C_4H_{10}	$\mathrm{H\ H\ H\ H}$ above; $\mathrm{H{-}C{-}C{-}C{-}C{-}H}$; $\mathrm{H\ H\ H\ H}$ below
uses: bottled gas (lighter fuel, camping gas)		

alkene member of the group of ◊hydrocarbons having the general formula C_nH_{2n} (commonly known as ***olefins***). Lighter alkenes, such as ethene and propene ($CH_3CH=CH_2$), are gases, obtained from the ◊cracking of oil fractions. Alkenes are unsaturated compounds, characterized by one or more double bonds between adjacent carbon atoms. They react by addition, and many useful compounds, such as polyethene and ethanol, are made from them.

alkyne member of the group of ◊hydrocarbons with the general formula C_nH_{2n-2} (commonly known as ***acetylenes***). They are unsaturated compounds, characterized by one or more triple bonds between adjacent carbon atoms. Lighter alkynes are gases (for example, ethyne); heavier ones are liquids or solids.

allele one of two or more alternative forms of a ◊gene at a given position (locus) on a chromosome, caused by a difference in the ◊DNA.

Organisms with two sets of chromosomes (diploids) will have two copies of each gene. If the two alleles are identical the individual is said to be ◊homozygous at that locus; if different, the individual is ◊heterozygous at that locus.

allotropy property whereby certain elements exist in different forms (allotropes) in the same physical state. The allotropes of carbon are diamond and graphite; those of sulphur are rhombic and monoclinic sulphur. These have different crystal structures when solids, as do the the white and grey forms of tin. Oxygen also exists in two different forms: 'normal' oxygen (O_2) and ozone (O_3), which have different molecular configurations.

alloy metal blended with some other metallic or non-metallic substance to give it special qualities, such as resistance to corrosion, greater hardness, or tensile strength. Useful alloys include bronze, brass, cupronickel, gunmetal, pewter, solder, steel, and stainless steel.

The most recent alloys include the superplastics: alloys that can stretch to double their length at specific temperatures, permitting, for example, their injection into moulds as easily as plastic.

alpha particle positively charged, high-energy particle emitted from the nucleus of a radioactive ◊atom. It is one of the products of the spontaneous disintegration of radioactive elements such as radium and thorium, and is identical with the nucleus of a helium atom – that is, it consists of two protons and two neutrons. The process of emission, ***alpha decay***, transforms one element into another, decreasing the atomic (or proton) number by two and the mass (or nucleon) number by four. See ◊radioactivity.

Because of their large mass, alpha particles have a short range of only a few centimetres in air, and can be stopped by a sheet of paper. They are capable of damaging living cells.

alternating current (AC) electric current that flows for an interval of time in one direction and then in the opposite direction for an equal interval; that is, a current that flows in alternately reversed directions through or around a circuit. Electric energy is usually generated as alternating current in a power station, and alternating currents may be used for both power and lighting.

The advantage of alternating current over direct current (DC), as from a battery, is that its voltage can be raised or lowered economically by a transformer: high voltage for generation and transmission, and low voltage for safe utilization. Railways, factories, and domestic appliances, for example, use alternating current.

alternation of generations the life cycle of land plants and some seaweeds in which there are two distinct forms occurring alternately: ***diploid*** (having two sets of chromosomes) and ***haploid*** (one set of chromosomes). The diploid generation produces haploid spores by ◊meiosis, and is called the sporophyte, while the haploid generation produces gametes (sex cells), and is called the gametophyte. The gametes fuse to form a diploid ◊zygote which develops into a new sporophyte; thus the sporophyte and gametophyte alternate.

The life cycles of certain animals (such as the jellyfish) are sometimes said to show alternation of generations, but this is rarely as regular and clearly defined as in plants.

alternative energy energy from sources that are renewable and ecologically safe, as opposed to sources that are nonrenewable and that often have toxic by-products, such as coal, oil, or gas (fossil fuels), and uranium (for nuclear power). The most important alternative energy source is flowing water, harnessed as ◊hydroelectric power. Other sources include the ocean's tides and waves, the wind (harnessed by windmills and wind turbines), the Sun (solar energy), and the heat trapped in the Earth's crust (geothermal energy).

alternator electricity ◊generator that produces an alternating current.

alum double sulphate of a monovalent metal or radical (such as sodium, potassium, or ammonium) and a trivalent metal (such as aluminium or iron). The commonest is the double sulphate of potassium and aluminium,

$$K_2Al_2(SO_4)_4.24H_2O$$

a white crystalline powder that is readily soluble in water. Alums are used in papermaking and to fix dye in textiles.

alumina or ***corundum*** Al_2O_3 oxide of aluminium that is widely distributed in clays, slates, and shales. It is formed by the decomposition of the feldspars in granite and used as an abrasive.

Typically it is a white powder, soluble in most strong acids or caustic alkalis but not in water. Impure alumina is called 'emery'. Rubies and sapphires are corundum gemstones.

aluminium lightweight, silver-white, ductile and malleable, metallic element, symbol Al, atomic number 13, relative atomic mass 26.9815. It is the third most abundant element (and the most abundant metal) in the Earth's

aluminium

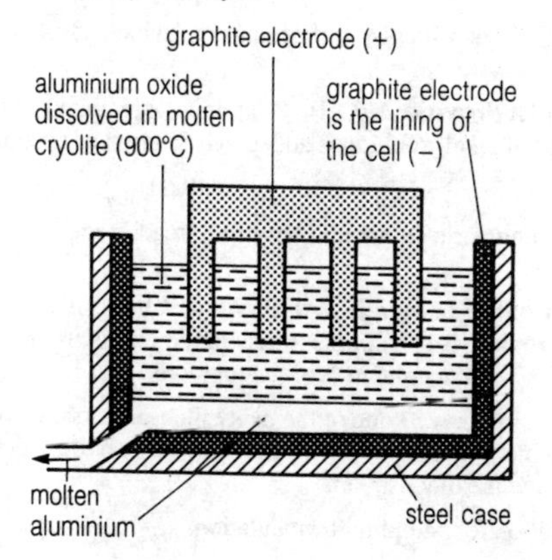

crust, of which it makes up about 8.1% by mass. It oxidizes easily, the layer of oxide on its surface making it highly resistant to tarnish. It is an excellent conductor of electricity. In its pure state it is a weak metal, but when combined with elements such as copper, silicon, or magnesium it forms alloys of great strength. In nature it is found only in the combined state in many minerals, and it is prepared commercially from the ore bauxite. The pure metal was not readily obtained until the middle of the 19th century.

Because of its light weight (specific gravity 2.70), aluminium is widely used in the shipbuilding and aircraft industries. Consumer uses include food and beverage packing, foil, outdoor furniture, and homebuilding materials. It is also much used in steel-cored overhead cables and for canning uranium slugs for nuclear reactors. Aluminium is an essential constituent in some magnetic materials and, as a good conductor of electricity, is used as foil in electrical capacitors.

The metal is extracted from purified bauxite (aluminium oxide, Al_2O_3) dissolved in molten cryolite at 900°C by electrolysis in a Hall cell. The reactions that occur at each electrode are as follows:

negative electrode:

$$2Al^{3+} + 6e^- \rightarrow 2Al$$

positive electrode:

$$3O^{2-} - 6e^- \rightarrow 1\,{}^1/_2\,O_2$$

The oxygen reacts with the carbon anodes, which must be replaced at intervals.

aluminium hydroxide $Al(OH)_3$ gelatinous ◊precipitate formed when a small amount of alkali solution is added to a solution of an aluminium salt.

$$Al^{3+}_{(aq)} + 3OH_{(aq)} \rightarrow Al(OH)_{3(s)}$$

It is an ◊amphoteric compound because it readily reacts with both acids and alkalis.

aluminium oxide or ***alumina*** Al_2O_3 white solid formed by heating aluminium hydroxide. It is an ◊amphoteric oxide and is used as a refractory (furnace lining) and in column ◊chromatography.

alveolus (plural *alveoli*) one of the many thousands of tiny air sacs in the ◊lungs in which exchange of oxygen and carbon dioxide takes place between air and the bloodstream.

AM abbreviation for ◊amplitude modulation.

alveolus

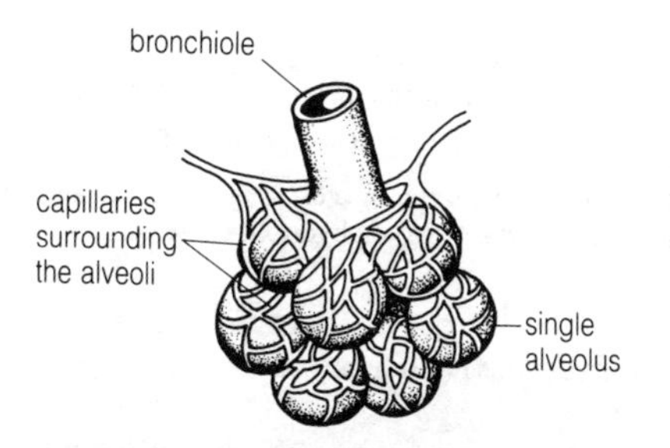

amino acid water-soluble organic ◊molecule, mainly composed of carbon, oxygen, hydrogen, and nitrogen, containing both a basic amine group (– NH_2) and an acidic carboxyl (–COOH) group. When two or more amino acids are joined together, they are known as ◊peptides; ◊proteins are made up of interacting polypeptides (peptide chains consisting of more than three amino acids) and are folded or twisted in characteristic shapes.

Many different proteins are found in the cells of living organisms, but they are all made up of the same 20 amino acids, joined together in varying combinations, (although other types of amino acid do occur infrequently in nature). Eight of these, the ***essential amino acids***, cannot be synthesized by humans and must be obtained from the diet. Children need a further two amino acids that are not essential for adults. Other animals also need some preformed amino acids in their diet, but green plants can manufacture all the amino acids they need from simpler molecules, relying on energy from the Sun and minerals (including nitrates) from the soil.

ammeter instrument that measures electric current, usually in ◊amperes. The ammeter is placed in series with the component through which current is to be measured, and is constructed with a low internal resistance in order to prevent the reduction of that current as it flows through the instrument itself. A common type is the moving-coil meter, which measures direct current (DC), but can, in the presence of a rectifier, measure alternating current (AC) also.

ammonia NH_3 colourless pungent- smelling gas, lighter than air and very soluble in water. It is made on an industrial scale by the ◊Haber process, and used mainly to produce nitrogenous fertilizers, some explosives, and nitric acid.

The gas has several typical reactions.

with metal oxides It will reduce some metal oxides when heated.

$$3CuO + 2NH_3 \rightarrow 3Cu + N_2 + 3H_2O$$

with water It dissolves readily in water to give an alkaline solution. Because ammonia is a weak base, its hydroxide-ion concentration is not high (pH 10).

$$NH_3 + H_2O \leftrightarrow NH_4^+ + OH^-$$

with indicators The gas turns moist litmus blue; this is used as a test for the presence of ammonia.

with hydrogen chloride Ammonia gas combines with hydrogen chloride gas to form white clouds of ammonium chloride.

$$NH_{3\,(g)} + HCl_{(g)} \leftrightarrow NH_4Cl_{(s)}$$

ammoniacal solution in chemistry, a solution produced by dissolving a solute in aqueous ammonia.

ammonium NH_4^+ ion formed when ammonia accepts a proton (H^+) from an acid. It is the only positive ion that is not a metal and that forms a series of salts. When heated with an alkali it produces ammonia gas.

$$NH_4^+ + OH^- \rightarrow NH_3 + H_2O$$

This reaction is used to detect the presence of the ion. Ammonium is a useful source of nitrogen for plants and is present in many fertilizer preparations.

amoeba (plural ***amoebae***) one of the simplest living animals, genus *Amoeba*, consisting of a single cell and belonging to the ◊protozoa group. The body consists of granular and colourless cytoplasm. Its activities are controlled by its nucleus, and it feeds by flowing round and engulfing organic debris. It reproduces by ◊binary fission. Some species of amoeba are harmful parasites.

amp abbreviation for ◊***ampere***, a unit of electrical current.

ampere SI unit (abbreviation amp, symbol A) of electrical current. Electrical current is measured in a similar way to water current, in terms of an amount per unit time; one ampere represents a flow of about 6.28×10^{18}

◊electrons per second, or a rate of flow of charge of one coulomb per second.

amphibian vertebrate of a type that generally spends its larval (tadpole) stage in fresh water before undergoing ◊metamorphosis. Having developed limbs, amphibians move onto land and become sexually mature; breeding, however, takes place in water, which explains why amphibians are usually found near pools and lakes. Like fish and reptiles, they continue to grow throughout life and cannot maintain a temperature greatly differing from that of their environment. The class includes salamanders, newts, frogs, and toads.

amphoteric (of elements) showing the properties of both ◊metals and ◊non-metals. For example, aluminium and zinc react with both acids and alkalis.

$$Zn + 2HCl \rightarrow ZnCl_2 + H_2$$

$$ZnO + 2NaOH \rightarrow Na_2ZnO_2 + H_2$$

The term also applies to the oxides and hydroxides of these elements, which also react with both acids and alkalis.

amplifier electronic device that magnifies the strength of a signal, such as a radio signal. The ratio of the amplitude of the output signal to that of the input signal is called the ◊gain of the amplifier (to a specific ratio in the ◊voltage amplifier). As well as achieving high gain, an amplifier should be free from distortion and able to operate over a range of frequencies. Practical amplifiers are usually complex circuits, although simple amplifiers can be built from single transistors or valves.

amplitude maximum displacement of an oscillation from its equilibrium position. For a transverse wave motion, it is the height of a crest (or the depth of a trough). The amplitude of a sound wave corresponds to the intensity (loudness) of the sound.

amylase one of a group of ◊enzymes that break down ◊starches into their component molecules (sugars) for use in the body. It occurs widely in both plants and animals. In humans, it is found in saliva and in pancreatic juices.

Simple experiments with amylase involve mixing starch with amylase, and monitoring the reaction. The starch will be broken down to glucose, which can be tested for using Benedict's solution (see ◊food tests). The efficiency with which amylase works, and therefore the amount of glucose produced, will vary according to such factors as temperature and pH.

anaemia condition caused by a shortage of haemoglobin, the oxygen-carrying component of red blood cells. The main symptoms are tiredness, paleness of the skin, breathlessness, palpitations, and poor resistance to infection.

Anaemia arises either from abnormal loss or poor production of haemoglobin. Excessive loss occurs, for instance, with chronic slow bleeding or with accelerated destruction of red blood cells. Causes of defective production include iron or cyanocobalamine (vitamin B_{12}) deficiency, malnutrition, and blood diseases such as sickle-cell disease. Untreated anaemia puts a strain on the heart and may prove fatal.

anaerobe organism that can obtain energy from food without using oxygen (by ◊anaerobic respiration). Anaerobes include many bacteria, yeasts, and internal parasites. Many of the organisms that live in the mud at the bottom of a river are anaerobic. Some anaerobes cannot survive if oxygen is present; others are able to respire anaerobically if necessary, but will use oxygen if present.

anaerobic respiration form of respiration that does not require the presence of oxygen for the release of energy from food molecules, such as glucose. Because the food molecule is only partly broken down in the process, anaerobic respiration has only one-twentieth the efficiency of aerobic respiration.

In yeasts and bacteria, anaerobic respiration involves the breakdown of glucose to give ethanol (alcohol), carbon dioxide, and energy, a process also known as ◊fermentation.

$$C_6H_{12}O_6 \rightarrow 2C_2H_5OH + 2CO_2 + 210\ \text{kJ}$$

The process is exploited by both the brewing and the baking industries.

Normally aerobic animal cells will respire anaerobically for short periods of time when oxygen levels are low, breaking down glucose and producing energy and lactic acid. However, the build-up of lactic acid poisons the cells temporarily and causes fatigue.

$$C_6H_{12}O_6 \rightarrow 2C_3H_6O_3 + 150\ \text{kJ}$$

This form of anaerobic respiration is seen particularly in muscle cells during intense activity (such as a 100-metre sprint), when the demand for oxygen can outstrip supply (see ◊oxygen debt). At the end of the activity, when the need for oxygen is reduced, the muscle cells return to fully aerobic respiration.

Although anaerobic respiration is a primitive and inefficient form of energy release, deriving from the period when oxygen was missing from the atmosphere, it can also be seen as an ◊adaptation. To survive in some habitats, such as the muddy bottom of a polluted river, an organism must be to a large extent independent of oxygen.

analogue (of a quantity or device) changing continuously; by contrast a ◊digital quantity or device varies in series of distinct steps. For example, an analogue clock measures time by means of a continuous movement of hands around a dial, whereas a digital clock measures time with a numerical display that changes in a series of discrete steps.

Most computers are digital devices. Therefore, any signals and data from an analogue device must be passed through a suitable ◊analogue-to-digital converter before they can be received and processed by computer. Similarly, output signals from digital computers must be passed through a digital-to-analogue converter before they can be received by an analogue device.

analogue-to-digital converter (ADC) in electronics, a circuit that converts an analogue signal into a digital one. Such a circuit is used to convert the signal from an analogue device into a digital signal for input into a computer. For example, many sensors designed to measure physical quantities, such as temperature and pressure, produce an analogue voltage signal and this must be passed through an ADC before computer input and processing. A digital-to-analogue converter performs the opposite process.

analysis the determination of the composition or properties of substances.

AND gate in electronics, a type of ◊logic gate.

anemometer device for measuring wind speed. A ***cup-type anemometer*** consists of cups at the ends of arms, which rotate when the wind blows. The speed of rotation indicates the wind speed in kilometres per hour or knots.

angiosperm flowering plant in which the seeds are enclosed within an ovary, which ripens to a fruit. Angiosperms are divided into monocotyledons (single seed leaf in the embryo) and dicotyledons (two seed leaves in the embryo). They include most flowers, herbs, grasses, and trees (except conifers).

angle of incidence

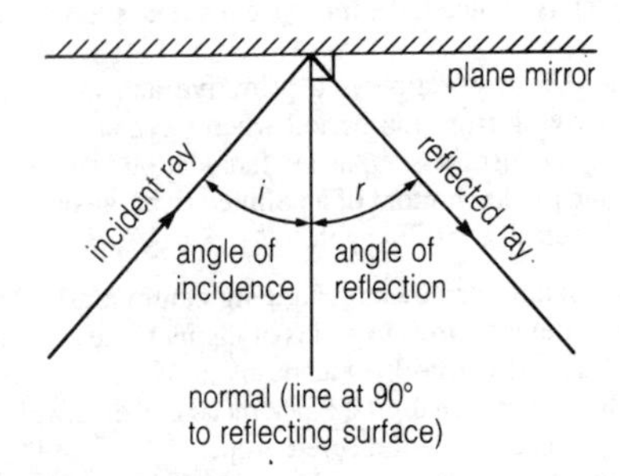

angle of incidence angle between a ray of light striking a mirror and the perpendicular to that mirror. It is equal to the ◊angle of reflection.

angle of reflection angle between a ray of light reflected from a mirror and the perpendicular to that mirror. It is equal to the ◊angle of incidence.

anhydride compound obtained by the removal of water from another compound, usually an acid. For example, sulphur trioxide (SO_3) is the anhydride of sulphuric acid (H_2SO_4). For monobasic acids, such as organic fatty acids, the formation of an anhydride involves the loss of a molecule of water from two molecules of acid.

anhydrous (of a substance) having no water content.

If the water of crystallization is removed from blue crystals of copper(II) sulphate ($CuSO_4$), a white powder (anhydrous copper(II) sulphate) results. Liquids from which all traces of water have been removed are also described as anhydrous.

annual plant plant that completes its life cycle within one year, during which time it germinates, grows to maturity, bears flowers, produces seed, and then dies. Examples include the common poppy and groundsel. Plants that live more than one year are ◊biennial plants or ◊perennial plants.

annual ring or ***growth ring*** one of a set of concentric rings visible on the wood of a cut tree trunk. Each ring represents a period of growth when new ◊xylem is laid down to replace tissue being converted into wood (secondary xylem). The wood formed from xylem produced in the spring and early

annual ring

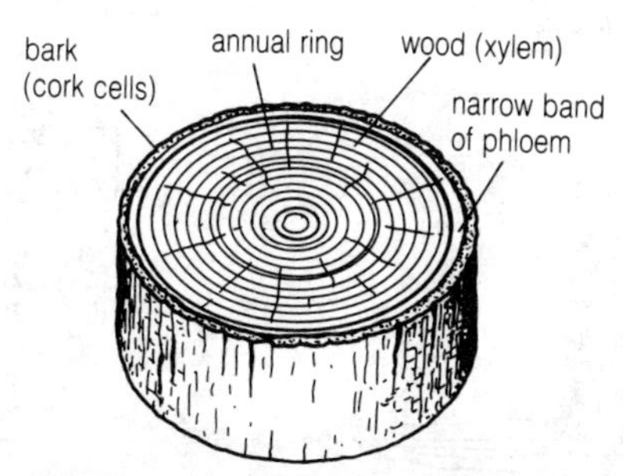

summer has larger and more numerous vessels than the wood formed from xylem produced in autumn when growth is slowing down. The result is a clear boundary between the pale spring wood and the denser, darker autumn wood. Annual rings may be used to estimate the age of the plant, although occasionally more than one growth ring is produced in a given year.

anode the positively charged electrode, or terminal, towards which negative particles (anions or electrons) move within a device such as an electrolysis cell, a cathode-ray tube, or a diode.

antagonistic muscles pair of muscles allowing coordinated movement of skeletal joints. The extension of the arm, for example, requires one set of muscles to relax, while another set contracts. The individual components of antagonistic pairs can be classified into ◊extensors and ◊flexors.

anther the place where pollen grains are produced. When mature, anthers are visible as dusty, sometimes coloured bodies at the tip of the stamens. They are the male part of the reproductive apparatus of the flowering plant.

antibiotic drug that kills or inhibits the growth of disease-causing microorganisms (pathogens). It is derived from other living microorganisms such as fungi or bacteria. Examples of antibiotics include penicillin (the first to be discovered) and streptomycin.

Each type of antibiotic acts in a different way and may be effective against either a broad range or a specific type of disease. Bacteria are constantly evolving and quickly become resistant to antibiotics; therefore new antibiotics are continually required in order to overcome them.

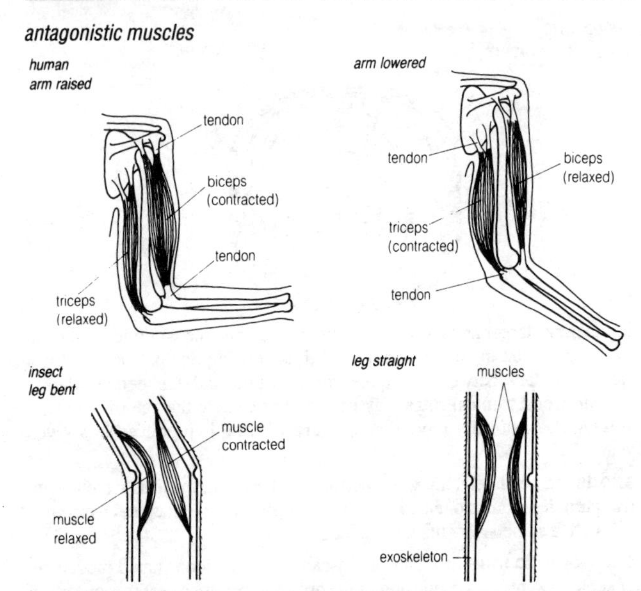

antibody proteins produced in the body in response to the presence of foreign or invading substances (◊antigens), which include the proteins carried on the surface of infecting microorganisms. Antibody production is an essential part of the immune system (see ◊immunity).

Each antibody acts against only one kind of antigen, and combines with it to form a 'complex'. This action may render the antigens harmless, or it may destroy microorganisms by setting off chemical changes that cause them to self-destruct. In other cases, the formation of a complex will cause antigens to form clumps that can then be detected and engulfed by white blood cells.

Each bacterial or viral infection will bring about the manufacture of a specific antibody, which will then fight the disease. Many diseases can only be contracted once because antibodies remain in the blood after the infection has passed, preventing any further invasion. Vaccination boosts a person's resistance by causing the production of antibodies specific to particular infections.

antifreeze substance added to a water-cooling system (for example, that of a car) to prevent it freezing in cold weather. The most common types of antifreeze contain the chemical ◊ethylene glycol ($HOCH_2CH_2OH$), an organic alcohol with a freezing point of about –15°C.

The addition of this chemical depresses the freezing point of water significantly. A solution containing 33.5% by volume of ethylene glycol will not freeze until about –20°C. A 50% solution will not freeze until –35°C.

antigen any substance that causes the production of ◊antibodies. Common antigens include the proteins carried on the surface of bacteria, viruses, and pollen grains. The proteins of incompatible blood groups or tissues also act as antigens, which has to be taken into account in medical procedures such as blood transfusions and organ transplants.

antioxidant type of food ◊additive, used to prevent fats and oils from becoming rancid, and thus extend their shelf life.

antiseptic any chemical that kills or inhibits the growth of microorganisms.

appendix area of the mammalian gut associated with the digestion of ◊cellulose. No mammal is able to produce the enzyme necessary for digesting cellulose. Some herbivores—rabbits, for instance—have evolved a system where millions of bacteria live in an enlarged appendix, and these bacteria produce the enzyme. The human appendix is tiny and serves no digestive function.

aquatic living in water. The aquatic environment, where life originated, has many advantages for organisms: dehydration is almost impossible, temperatures usually remain stable, and the heaviness of water provides physical support.

aqueous humour watery fluid found in the space between the cornea and lens of the ◊eye.

aqueous solution solution in which the solvent is water.

argon colourless, odourless, gaseous element, symbol Ar, atomic number 18, relative atomic mass 39.948. It is grouped with the ◊noble gases, since it was believed not to react with other substances; observations now indicate that it can be made to combine with boron fluoride to form compounds.

It constitutes almost 1% of the Earth's atmosphere. Its main industrial use is in the manufacture of light bulbs, where it forms the gaseous interior.

aromatic compound any organic chemical that incorporates a ◊benzene ring in its structure (see also ◊cyclic compounds). Aromatic compounds undergo ◊substitution reactions.

arsenic brittle, greyish-white, weakly metallic element (a metalloid), symbol As, atomic number 33, relative atomic mass 74.92. It occurs in many ores and occasionally in its elemental state, and is widely distributed, being present in minute quantities in the soil and the sea. Arsenic is a well-known poison.

arteriosclerosis hardening and thickening of the arteries due to the deposition of substances such as cholesterol, fatty acids, and calcium. The condition lowers the elasticity of the artery walls, and can lead to high blood pressure, loss of circulation, heart disease, and death. It is associated with smoking, ageing, and a diet high in saturated fats (see ◊polyunsaturate).

artery vessel that carries blood from the heart to the rest of the body. Arteries are built to withstand considerable pressure, having thick walls that are impregnated with muscle and elastic fibres. During contraction of the heart muscles, arteries expand in diameter to allow for the sudden increase in pressure that occurs; the resulting ◊pulse, or pressure wave, can be felt at the wrist. Not all arteries carry oxygenated (oxygen-rich) blood; the pulmonary arteries convey deoxygenated (oxygen-poor) blood from the heart to the lungs.

Arteries are vulnerable to damage by the build-up of deposits of such substances as fatty acids and cholesterol—a condition known as ◊arteriosclerosis.

articulating surface surface of a bone that rubs against or glides over the surface of another in a joint. Such surfaces are covered with a protective layer of ◊cartilage.

artificial radioactivity radioactivity arising from human-made radioisotopes (radioactive isotopes or elements that are formed when other elements are bombarded with subatomic particles—protons, neutrons, or electrons—or small nuclei).

artificial selection selective breeding of individuals that exhibit the particular characteristics that a plant or animal breeder wishes to develop. In plants, desirable features might include resistance to disease, high yield (in crop plants), or attractive appearance. In animal breeding, selection has led to the development of particular breeds of cattle for improved meat production (such as the Aberdeen Angus) or milk production (such as Jerseys).

ascorbic acid or ***vitamin C*** relatively simple organic acid found in fresh fruits and vegetables. It is soluble in water and destroyed by prolonged boiling, so soaking or overcooking of vegetables reduces their vitamin C content.

In the human body, ascorbic acid is necessary for the correct synthesis of ◊collagen, an important skin protein. Lack of it causes skin sores or ulcers, tooth and gum problems, and burst capillaries (all symptoms of the deficiency disease scurvy). See ◊vitamin.

asexual reproduction reproduction that does not involve the manufacture and fusion of sex cells, nor the necessity for two parents. The process carries a clear advantage in that there is no need to search for a mate or develop complex pollinating mechanisms; every asexual organism can reproduce on its own. Asexual reproduction can therefore lead to a rapid population build-up.

In evolutionary terms, the disadvantage of asexual reproduction arises from the fact that only identical individuals, or clones, are produced—there is no variation between individuals. In the field of horticulture, where standardized production is needed, this is useful, but in the wild, an asexual population of identical individuals may be at risk of extinction if the environment changes.

Asexual processes include ◊binary fission, in which the parent organism splits into two or more 'daughter' organisms, and ◊budding, in which a new organism is formed initially as an outgrowth of the parent organism. Many plants reproduce asexually, or vegetatively, by means of ◊runners, ◊rhizomes, and ◊bulbs.

asphalt semisolid brown or black ◊bitumen, used in the construction industry. Asphalt is mixed with rock chips to form paving material, and the purer varieties are used for insulating material and for waterproofing masonry. It can be produced artificially by the distillation of ◊petroleum.

assimilation the process by which absorbed food molecules, circulating in the blood, pass into the cells and are used for growth, tissue repair, and other metabolic activities. The destiny of each food molecule depends not only on its type, but also on the body requirements at that time.

asexual reproduction

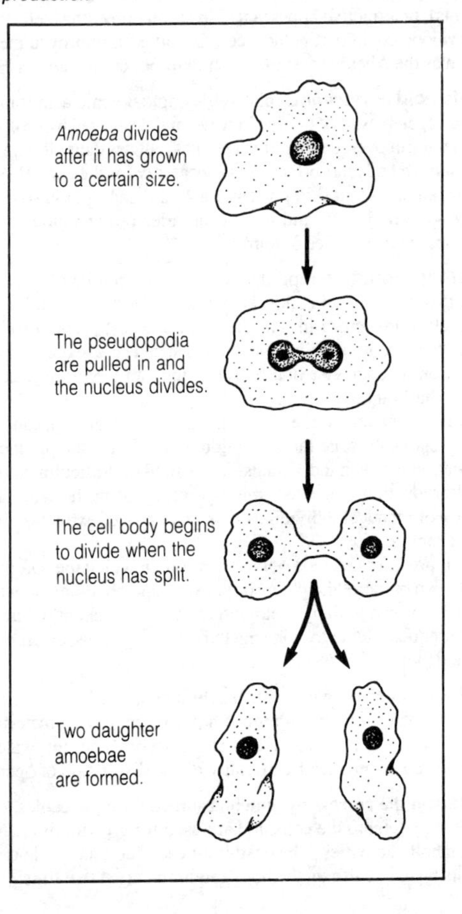

astatine non-metallic, radioactive element, symbol At, atomic number 85, relative atomic mass 210. It is a member of the ◊halogen group, and is very rare in nature. Astatine is highly unstable, with many isotopes; the longest-lived has a half-life of about eight hours.

atmosphere the mixture of gases that surrounds the Earth; it is prevented from escaping by the pull of the Earth's gravity. ◊Atmospheric pressure decreases with height in the atmosphere.

The lowest level of the atmosphere, the ***troposphere***, is heated by the Earth, which is warmed in turn by infrared and visible radiation from the Sun. Warm air cools as it rises in the troposphere, causing rain and most other weather phenomena. The upper levels of the atmosphere, particularly the ***ozone layer***, absorb almost all the ultraviolet light radiated by the Sun, and prevent lethal amounts from reaching the Earth's surface.

atmospheric pressure the ◊pressure at any point in the atmosphere that is due to the weight of air above it; it therefore decreases with height. At sea level the pressure is about 101 kilopascals; however, the exact value varies according to temperature and weather. Changes in atmospheric pressure, measured with a ◊barometer, are used in weather forecasting.

atom the smallest unit of matter that can take part in a chemical reaction, and that cannot be broken down chemically into anything simpler. An atom is made up of protons and neutrons in a central nucleus surrounded by electrons (see ◊atomic structure). The atoms of the various elements differ in atomic number, relative atomic mass, and behaviour. There are 109 different types of atom, corresponding to the 109 known elements.

Atoms are much too small to be seen even by the microscope (the largest, caesium, has a diameter of 0.0000005 mm), and they are in constant motion.

atom, electronic structure the arrangement of electrons around the nucleus of an atom, in distinct energy levels, also called orbitals or shells. These shells can be regarded as a series of concentric spheres, each of which can contain a certain maximum number of electrons. The energy levels are usually numbered beginning with the shell nearest to the nucleus. The outermost shell is known as the ◊valency shell because it contains the valence electrons.

The atomic number of an element indicates the number of electrons in a neutral atom. From this it is possible to deduce its electronic structure. For example, sodium has atomic number 11 ($Z = 11$) and its electronic arrangement (configuration) is two electrons in the first energy level, eight electrons in the second energy level and one electron in the third energy level—generally written as 2.8.1. Similarly for sulphur (Z = 16), the electron arrangement will be 2.8.6. The electronic structure dictates whether two elements will combine by ionic or covalent bonding (see ◊bond) or not at all.

atomic mass see ◊relative atomic mass.

atomic number or ***proton number*** the number (symbol Z) of protons in the nucleus of an atom. It is equal to the positive charge on the nucleus. The 109 elements are numbered 1 (hydrogen) to 109 (unnilennium) in the periodic table of elements. See also ◊nuclear notation.

atomic radiation or ***nuclear radiation*** energy given out by disintegrating atoms during ◊radioactive decay. The energy may be in the form of fast-moving particles, known as ◊alpha particles and ◊beta particles, or in the form of high-energy electromagnetic waves known as ◊gamma radiation. Overlong exposure to atomic radiation can lead to radiation sickness. Radiation biology studies the effect of radiation on living organisms.

atomic structure the internal structure of an ◊atom. The core of the atom is the ***nucleus***, a particle only one ten-thousandth the diameter of the atom itself. The simplest nucleus, that of hydrogen, comprises a single positively charged particle, the ***proton***. Nuclei of other elements contain more protons and additional particles of about the same mass as the proton but with no electrical charge, ***neutrons***. Each element has its own characteristic nucleus with a unique number of protons, the atomic number. The number of neutrons may vary: where atoms of a single element have different numbers of neutrons, they are called ◊isotopes. Although some isotopes tend to be unstable and exhibit ◊radioactivity, they all have identical chemical properties.

The nucleus is surrounded by a number of ***electrons***, each of which has a negative charge equal to the positive charge on a proton, but which weighs only 1/1,839 times as much. A neutral atom has exactly the same number of protons as electrons. The chemical properties of an element are determined by the ease with which its atoms can gain or lose electrons. This is dependent on both the number of electrons associated with the nucleus and the force exerted on them by its positive charge.

Atoms are held together by the electrical forces of attraction between each negative electron and the positive protons within the nucleus. The latter repel one another with relatively enormous forces; a nucleus holds together only because other forces, not of a simple electrical character, attract the protons and neutrons to one another. These additional forces act only so long as the protons and neutrons are virtually in contact with one another. If, therefore, a fragment of a complex nucleus, containing some protons, becomes only slightly loosened from the main group of neutrons and protons, the strong natural repulsion between the protons will cause this fragment to fly apart from the rest of the nucleus at high speed. It is by such fragmentation of atomic nuclei (◊nuclear fission) that nuclear energy is released.

atomic weight another name for ◊relative atomic mass.

atrium (plural ***atria***) one of the upper chambers of the heart, receiving blood under low pressure as it returns from the body. Atrium walls are thin and stretch easily to allow blood into the heart. On contraction, the atria force blood into the thick-walled lower chambers (◊ventricles), which then give a second, more powerful beat.

auditory canal tube leading from the outer ◊ear opening to the eardrum.

auxin plant ◊hormone that promotes stem and root growth in plants. Auxins influence many aspects of plant growth and development, including cell enlargement, inhibition of development of axillary buds, ◊tropisms, and the initiation of roots. ***Synthetic auxins*** are used in rooting powders for cuttings and in some weedkillers, where high auxin concentrations cause such rapid growth that the plants die. They are also used to prevent premature fruitdrop in orchards. The most common naturally occuring auxin is indoleacetic acid (IAA). It is synthesized in the shoot tip and transported to other parts of the plant.

Avogadro's hypothesis law stating that equal volumes of all gases, when at the same temperature and pressure, have the same numbers of molecules.

axon the long threadlike extension of a ◊neuron (nerve cell) that conducts electrochemical impulses away from the cell body towards other neurons, or towards an effector organ such as a muscle. Axons terminate in ◊synapses with other neurons, muscles, or glands.

B

bacteria (singular ***bacterium***) microscopic unicellular organisms. They differ from ordinary cells in not containing a nucleus. They usually reproduce by ◊binary fission, and since this may occur approximately every 20 minutes, a single bacterium is potentially capable of producing 16 million copies of itself in a day.

Some bacteria are parasitic, causing disease; others, if not checked by refrigeration or other preservation measures, can spoil food and cause food poisoning. However, most bacteria perform essential roles in the ecosystem—digesting ◊cellulose in the intestines of herbivores, bringing about the decomposition of waste or dead matter, and improving soil fertility by processing nitrogen compounds (see ◊nitrogen cycle). Others perform useful functions in the dairy industry—fermenting cheeses and yoghurt—and in the treatment of sewage.

baking powder mixture of ◊bicarbonate of soda (sodium hydrogencarbonate, $NaHCO_3$) and solid tartaric acid, used in cooking as a raising agent. When added to flour, the presence of water and heat causes carbon dioxide to be released, which makes the dough rise.

balance state of equilibrium achieved by an object when the forces acting on it (or the moments of those forces) cancel each other. The resultant force acting on a balanced object is zero.

balanced diet diet that includes carbohydrate, protein, fat, vitamins, water, minerals, and roughage. Although it is agreed that all these substances are needed if a person is to be healthy, argument occurs over how much of each type a person needs.

A very active person will have a higher carbohydrate requirement than someone who works in an office. Similarly, a person who is growing, or pregnant, may need a high intake of minerals. When a person does not have a balanced diet, a whole range of problems can arise, depending on which foods are lacking or in excess. Obesity, starvation, vitamin-deficiency diseases, cancer of the lower intestine, heart disease, and ◊arteriosclerosis are all caused by dietary mismanagement. Poverty is one of the biggest causes of bad diet, because some elements of the balanced meal—for instance, foods rich in protein and vitamins—are relatively expensive. Cheap diets tend to be high in carbohydrate.

ball-and-socket joint ◊joint allowing considerable free movement in three dimensions; for instance, the hip joint between the pelvis and the femur.

barium soft, silver-white, metallic element, symbol Ba, atomic number 56, relative atomic mass 137.33. It is one of the alkaline-earth metals, found in nature as barium carbonate and barium sulphate. Barium sulphate is important in medicine; it is mixed with liquid (a 'barium meal') and its progress followed using X-rays. This can reveal abnormalities of the alimentary canal. Barium is also used in alloys, pigments, and safety matches.

barium chloride $BaCl_2$ white, crystalline solid, used in aqueous solution to test for the presence of the sulphate ion. When a solution of barium chloride is mixed with a solution of a sulphate, a white precipitate of barium sulphate is produced.

$$Ba^{2+}_{(aq)} + SO^{2-}_{4(aq)} \leftrightarrow BaSO_{4(a)}$$

bark protective outer layer on the stems and roots of woody plants, composed mainly of dead cells. To allow for expansion of the stem, the bark is continually added to from within, and the outer surface often becomes cracked or is shed as scales.

Bark technically includes the tissues external to the vascular ◊cambium, for instance the ◊phloem.

barometer instrument that measures ◊atmospheric pressure as an indication of weather. Most often used are the ***mercury barometer*** and the ***aneroid barometer***.

In a mercury barometer, a column of mercury in a glass tube roughly 0.75 m high (closed at one end, curved upwards at the other) is balanced by the pressure of the atmosphere on the open end; any change in the height of the column reflects a change in pressure. An aneroid barometer achieves a similar result by changes in the distance between the faces of a shallow cylindrical metal box, from which most of the air has been removed.

base substance that accepts protons. Bases can contain negative ions such as the hydroxide ion (OH^-), which is the strongest base, or be molecules

barometer

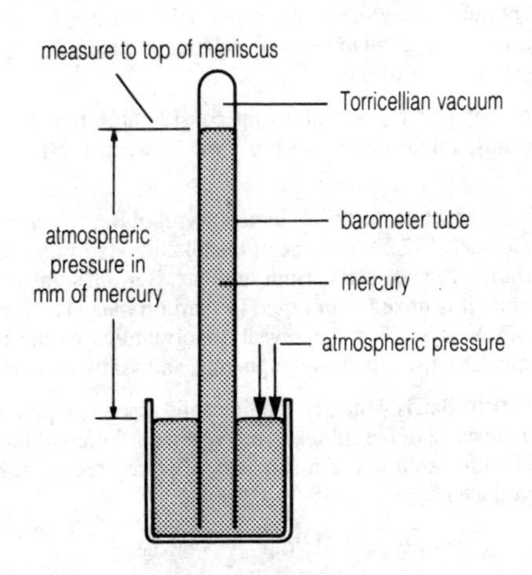

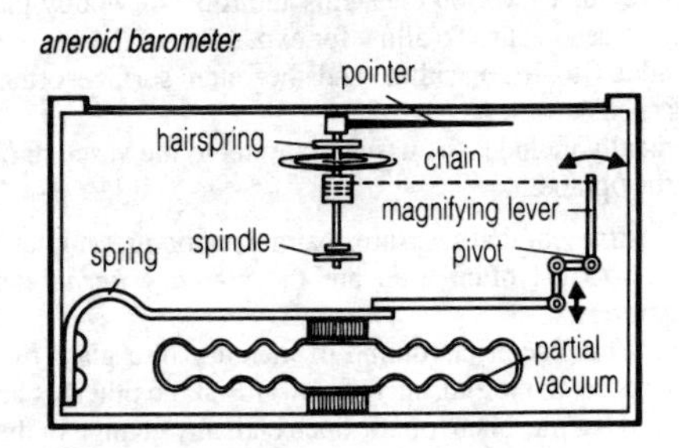

such as ammonia (NH_3). Ammonia is a weak base, as only some of its molecules accept protons.

$$OH^- + H^+_{(aq)} \rightarrow H_2O_{(l)} \; NH_3 + H_2O \leftrightarrow NH_4^+ + OH^-$$

Bases that dissolve in water are called ◊alkalis.

Inorganic bases are usually oxides or hydroxides of metals, which react with dilute acids to form a salt and water. Many carbonates also react with dilute acids, additionally giving off carbon dioxide.

basicity the number of replaceable hydrogen atoms in an acid. Nitric acid (HNO_3) is monobasic, sulphuric acid (H_2SO_4) is dibasic, and phosphoric acid (H_3PO_4) is tribasic.

basic oxide compound formed by a metal and oxygen, containing the O^{2-} ion. If, like sodium oxide (Na_2O), it is soluble, it forms the metal hydroxide when added to water.

$$Na_2O_{(s)} + H_2O_{(l)} \rightarrow 2NaOH_{(aq)}$$

All basic oxides react with acids to form a salt and water.

$$CaO + 2HNO_3 \rightarrow Ca(NO_3)_2 + H_2O$$

battery any energy-storage device allowing release of electricity on demand. A battery is made up of one or more cells, each containing two conducting electrodes (one positive, one negative) immersed in an ◊electrolyte, in a container. When an outside connection (such as through a light bulb) is made between the electrodes, a current flows through the circuit, and chemical reactions releasing energy take place within the cells.

Primary-cell batteries are disposable; secondary-cell batteries are rechargeable. The common ***dry cell*** is a primary-cell battery that consists of

battery

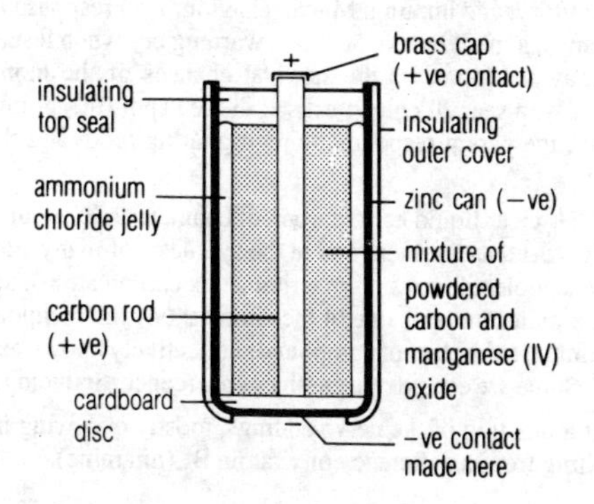

a central carbon electrode immersed in a paste of manganese dioxide and ammonium chloride as the electrolyte. The zinc casing forms the other electrode. It is dangerous to try to recharge a primary-cell battery.

The introduction of rechargeable nickel–cadmium batteries has revolutionized portable electronic newsgathering (sound recording, video) and information processing (computing). These batteries offer a stable, short-term source of power free of noise and other hazards associated with mains electricity.

The lead–acid ***car battery*** is a secondary-cell battery, or accumulator. The car's generator continually recharges the battery. It consists of sets of lead (positive) and lead peroxide (negative) plates in an electrolyte of sulphuric acid.

battery acid ◊sulphuric acid of 70% concentration used in lead-cell batteries (as in motor vehicles).

bauxite the principal ore of ◊aluminium, consisting of a mixture of hydrated aluminium oxides and hydroxides, generally contaminated with compounds of iron, which give it a red colour. Chief producers of bauxite are Australia, Guinea, Jamaica, Kazakhstan, Russia, Surinam, and Brazil.

becquerel SI unit (symbol Bq) of ◊radioactivity, equal to one radioactive disintegration (change in the nucleus of an atom when a particle or ray is given off) per second.

behaviour animal activity that has survival value; for instance, courting, defending territory, and hunting. Much behaviour is in response to a stimulus. For instance, a monkey may let out a warning cry when it sees a predator. This behaviour increases the survival chances of the monkey or its group. Similarly, a seagull chick pecks at the red spot (the stimulus) on its parent's beak; the parent responds by regurgitating food, and the chick is able to feed.

benzene C_6H_6 clear liquid hydrocarbon of characteristic odour, occurring in coal tar. It is used as a solvent and in the synthesis of many chemicals.

The benzene molecule consists of a ring of six carbon atoms, all of which are in a single plane, and it is one of the simplest ◊cyclic compounds. Benzene is the simplest of a class of compounds collectively known as ***aromatic compounds***. Some are considered carcinogenic (cancer-inducing).

beriberi inflammation of the nerve endings, mostly occurring in the tropics and resulting from a deficiency of vitamin B_1 (thiamine).

beryllium hard, lightweight, silver-white, metallic element, symbol Be, atomic number 4, relative atomic mass 9.012. It is one of the ◊alkaline-earth metals, with chemical properties similar to those of magnesium; in nature it is found only in combination with other elements. It is used to make sturdy, light alloys and to control the speed of neutrons in nuclear reactors. It was discovered in 1798 by French chemist Louis-Nicolas Vauquelin (1763–1829).

beta particle electron ejected with great velocity from a radioactive atom that is undergoing spontaneous disintegration. Beta particles do not exist in the nucleus but are created on disintegration of neutrons.

Beta particles are more penetrating than ◊alpha particles, but less so than gamma radiation; they can travel several metres in air, but are stopped by 2–3 mm of aluminium. A beam of beta particles is easily deflected by magnetic and electric fields.

bicarbonate common name for ◊hydrogencarbonate.

bicarbonate indicator pH indicator used to detect levels of carbon dioxide. Although CO_2 is not a particularly acidic gas, the weak acid it forms when dissolved in water will cause this sensitive indicator to change colour, from red to yellow, as the pH lowers. The indicator is used in photosynthesis and respiration experiments to find out whether carbon dioxide is being liberated. For instance, if indicator is placed at the bottom of a test tube, and germinating seeds are suspended above, using netting or wire gauze, then the CO_2 from the seeds will produce a colour change.

bicarbonate of soda $NaHCO_3$ (technical name ***sodium hydrogencarbonate***) white crystalline solid that neutralizes acids and is used in medicine to treat acid indigestion. It is also used in baking powders and fizzy drinks.

biennial plant a plant that completes its life cycle in two years. During the first year it grows vegetatively and the surplus food produced is stored in its ◊perennating organ, usually the root. In the following year these food reserves are used for the production of leaves, flowers, and seeds, after which the plant dies. Many root vegetables are biennials, including the carrot and the parsnip.

bile brownish fluid produced by the liver. It is stored in the gall bladder, but passes into the intestine, where it helps in the digestive processes. The enzymes in the small intestine work best in slightly alkaline conditions;

food emerging from the stomach is highly acidic, however. Chemicals within the bile neutralize the stomach contents, and therefore allow the intestinal enzymes to function. Bile is also important for fat digestion, because it breaks up fatty foods into tiny droplets, which can then be more easily attacked by enzymes.

bimetallic strip strip made from two metals that expand by different amounts when the temperature rises. The strip therefore bends when subjected to a change in temperature. Such strips are used widely for temperature measurement and control.

binary fission form of ◊asexual reproduction in which a single-celled organism divides into two smaller 'daughter' cells. It can also occur in a few simple multicellular organisms, such as sea anemones, producing two smaller sea anemones of equal size.

binoculars magnifying optical instrument for viewing an object with both eyes. Binoculars consist of two telescopes containing convex lenses and prisms, which together produce a stereoscopic, seemingly three-dimensional effect as well as magnifying the image. The use of prisms has the effect of 'folding' the light path, allowing for a compact design.

binocular vision or ***stereoscopic vision*** vision in which the eyes face forwards and can both focus on an object at the same time. It is characteristic of predatory animals, including humans.

Although the eyes provide two slightly different images of the world, these are coordinated by the brain to give a three-dimensional perception that allows the animal to judge accurately the position and speed of prey. See also ◊monocular vision.

biodegradable capable of being broken down by living organisms, principally bacteria and fungi (◊decomposers). Biodegradable substances, such as food and sewage, can therefore be rendered harmless by natural processes. The process of decay leads to the release of nutrients that are then recycled by the ecosystem. Nonbiodegradable substances, such as glass, heavy metals, and most types of plastic, present serious problems of disposal.

biological control the control of pests such as insects and fungi through biological means, rather than the use of chemicals. This can include: breeding resistant crop strains; inducing sterility in the pest; infecting the pest species with disease organisms; or introducing the pest's natural predator. Biological control tends to be naturally self-regulating, but as ecosystems are so complex, it is difficult to predict all the consequences of introducing a biological controlling agent.

biomass the total mass of living organisms present in a given area. It may be specified for a particular species (such as earthworm biomass) or for a general category (such as herbivore biomass). Estimates also exist for the entire global plant biomass. Measurements of biomass can be used to study interactions between organisms, the stability of those interactions, and variations in population numbers.

biosphere or ***ecosphere*** the region of the Earth's surface (both land and water), together with the atmosphere above it, that can be occupied by living organisms.

biotechnology the industrial use of living organisms to manufacture food, drugs, or other products. The brewing and baking industries have long relied on the yeast microorganism for ◊fermentation purposes, whereas the dairy industry employs a range of bacteria and fungi to convert milk into cheeses and yoghurts. Recent advances include ◊genetic engineering, in which single-celled organisms with modified ◊DNA are used to produce insulin and other drugs. ◊Enzymes, whether extracted from cells or produced artificially, are central to most biotechnological applications.

biotic factor living variable affecting an ◊ecosystem; for example, the changing population of elephants and its effect on the African savanna.

bird member of the vertebrate class Aves, the biggest group of land vertebrates, characterized by warm blood, feathers, forelimbs modified as wings, breathing through lungs, and egg-laying by the female.

birth the act of producing live young from within the body of female animals.

birth rate the number of births in a population per unit time; it is usually expressed as the number per year per thousand of the population.

In the 20th century, the UK's birth rate has fallen from 28 to less than 10 owing to increased use of contraception, better living standards, and falling infant mortality. The average household now contains 1.8 children. The population growth rate remains high in developing countries. While it is now below replacement level in the UK, in Brazil it stands at 23, in Bangladesh at 33, and in Nigeria at 34 per thousand people per year.

bismuth hard, brittle, pinkish-white, metallic element, symbol Bi, atomic number 83, relative atomic mass 208.98. It is the last of the stable elements; all from atomic number 84 up are radioactive. Bismuth occurs in ores and occasionally as a free metal. It is a poor conductor of heat and electricity, and is used in alloys of low melting point and in medical compounds to soothe gastric ulcers.

bistable circuit or *flip-flop* in electronics, a simple circuit that remains in one of two stable states until it receives a pulse (logic 1 signal) through one of its inputs, upon which it switches, or 'flips', over to the other state. Because it is a two-state device, it can be used to store binary digits and is widely used in the ◊integrated circuits used to build computers.

bit (contraction of ***bi**nary dig**it***) in computing, a single binary digit, either 0 or 1. A bit is the smallest unit of data stored in a computer; all other data must be coded into a pattern of individual bits. A ◊byte represents sufficient computer memory to store a single character of data (for example, a letter of the alphabet); it usually consists of eight bits.

bitumen impure mixture of hydrocarbons, including such deposits as petroleum, asphalt, and natural gas, although sometimes the term is restricted to a soft kind of pitch resembling asphalt.

Solid bitumen may have arisen as a residue from the evaporation of petroleum. If evaporation took place from a pool or lake of petroleum, the residue might form a pitch or asphalt lake, such as Pitch Lake in Trinidad. Bitumen was used in ancient times as a mortar, and by the Egyptians for embalming.

bladder hollow elastic-walled organ in which urine is stored. Urine enters the bladder through two ureters, one leading from each kidney, and leaves it through the urethra.

blast furnace smelting furnace in which the temperature is raised by the injection of an air blast. It is used in the extraction of metals from their ores, chiefly pig iron from iron ore. The principle has been known for thousands of years, but the present blast furnace is a heavy engineering development combining a number of special techniques.

In the extraction of iron, the ingredients of the furnace are iron ore (iron(III) oxide), coke (carbon), and limestone (calcium carbonate). The coke is the fuel and provides the agent for the ◊reduction of the iron ore, carbon monoxide. The carbon monoxide may be produced and added separately in the direct-reduction process. This is useful because it does not

blast furnace

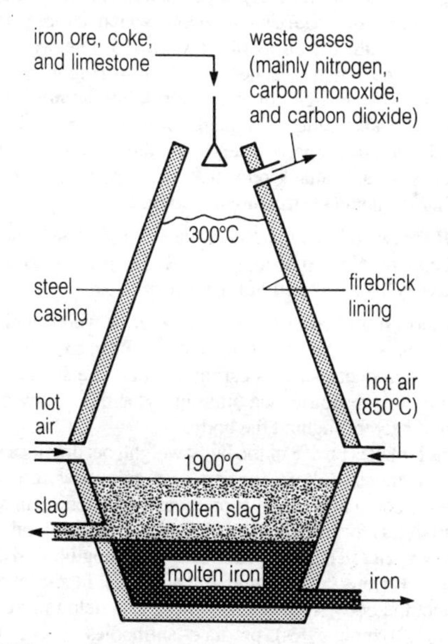

require such high temperatures.

$$C + O_2 \rightarrow CO_2$$

$$CO_2 + C \rightarrow 2CO$$

At the high temperature of the furnace, iron oxide is reduced to iron.

$$Fe_2O_3 + 3CO \rightarrow 2Fe + 3CO_2$$

The limestone is decomposed to quicklime (calcium oxide), which combines with the acidic impurities of the ore to form a molten mass known as slag (calcium silicate).

$$CaCO_3 \rightarrow CaO + CO_2$$

$$CaO + SiO_2 \rightarrow CaSiO_3$$

bleaching decolorization of coloured materials. The two main types of bleaching agent are the ***oxidizing bleaches***, which add oxygen and remove hydrogen, and include the ultraviolet rays in sunshine, hydrogen peroxide, and chlorine in household bleaches; and the ***reducing bleaches***, which add hydrogen or remove oxygen, for example sulphur dioxide.

Both natural and synthetic pigments usually possess highly complex molecules, the colour property often being due to only a part of the molecule. Bleaches usually attack only that small part, yielding another substance similar in chemical structure but colourless.

blind spot area where the optic nerve and blood vessels pass through the retina of the ◊eye. No visual image can be formed because there are no light-sensitive cells in this part of the retina.

blood transport liquid circulating in the arteries, veins, and capillaries of vertebrate animals. It carries nutrients and oxygen to individual cells and removes waste products, such as carbon dioxide. It also plays an important role in the immune response (see ◊immunity) and, in many animals, in the distribution of heat throughout the body.

In humans it makes up 5% of the body weight, occupying a volume of 5.5 litres in the average adult. It consists of a colourless, transparent liquid called ***plasma***, containing microscopic cells of three main varieties. ***Red cells*** (erythrocytes) form nearly half the volume of the blood, with 5 billion cells per litre. Their red colour is caused by ◊haemoglobin. ***White cells*** (leucocytes) are of various kinds. Some (◊phagocytes) ingest invading bacteria and so protect the body from disease; these also help to repair injured tissues. Others (◊lymphocytes) produce antibodies, which help provide immunity. Cell fragments called ***platelets*** (thrombocytes) assist in ◊blood clotting.

Blood cells constantly wear out and die, and are replaced from the bone marrow. Dissolved in the plasma are salts, proteins, sugars, fats, hormones, and fibrinogen, which are transported around the body. Fibrinogen is important in clotting.

blood clotting complex series of events that prevents excessive bleeding and the entry of disease-causing microorganisms after injury. It results in the formation of a meshwork of protein fibres that seals the wound.

When platelets (cell fragments) in the bloodstream come into contact with a damaged blood vessel, they and the vessel wall itself release the enzyme ***thrombokinase***, which brings about the conversion of the inactive

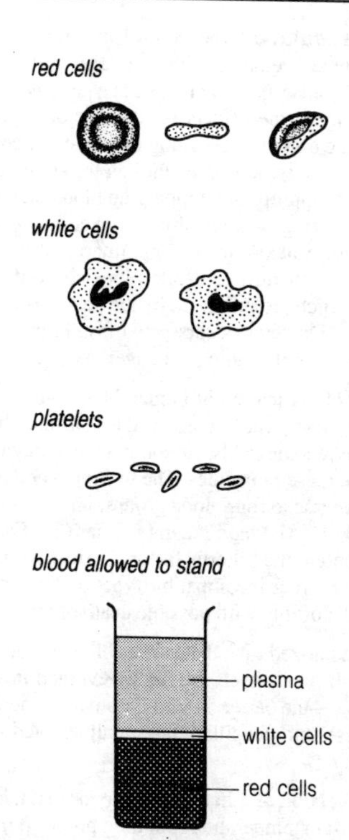

enzyme ***prothrombin*** into the active ***thrombin***. Thrombin in turn catalyses the conversion of the soluble protein ***fibrinogen***, present in blood plasma, to the insoluble ***fibrin***. This fibrous protein forms a net over the wound that traps red blood cells and seals the wound; the resulting jellylike clot hardens on exposure to air to form a hard scab. Calcium, vitamin K, and a variety of enzymes called factors are also necessary for efficient blood clotting.

blood-glucose regulation the control of sugar concentration in the blood, so that it always remains within certain limits. Blood-sugar levels are likely to vary, because an animal may have just eaten or have been physically active. However, the effects on the nervous system of an erratic blood-sugar level are disastrous, leading in humans to coma or fits.

The maintenance of a steady state relies on the ability of the liver to store or release glucose as appropriate, topping up blood-sugar levels when needed (for example, during or after physical exercise) and removing sugar from the blood after a heavy meal. This important homeostatic function (see ◊homeostasis) is in turn controlled by two hormones, ◊insulin and glucagon, both of which are produced by the pancreas and have a powerful effect on the liver. ◊Diabetes occurs when insulin production by the pancreas becomes insufficient, leading to dangerously high blood-sugar levels.

blood group the classification of human blood types. Red blood cells of one individual may carry molecules on their surface that act as antigens (molecules that are recognized by antibodies) in another individual whose red blood cells lack these molecules. The two main antigens are designated A and B. These give rise to four blood groups: having A only (A), having B only (B), having both (AB), and having neither (O). Each of these groups may or may not contain the ◊rhesus factor. Correct typing of blood groups is vital in transfusion, since incompatible types of donor and recipient blood will result in blood clotting, with possible death of the recipient.

blood vessel specialized tube that carries blood around the body of multicellular animals. Blood vessels are highly evolved in vertebrates, where the three main types—the ◊arteries (and the smaller arterioles), ◊veins (and the smaller venules), and ◊capillaries—are all adapted for their particular role within the body.

boiling rapid conversion of a liquid into vapour that takes place when the liquid reaches a certain temperature (◊boiling point). It involves the formation of vapour bubbles within the body of a liquid, whereas ◊evaporation occurs only at the surface.

boiling point for any given liquid, the temperature at which the application of heat raises the temperature of the liquid no further, but converts it to vapour. The boiling point of water under normal pressure is 100°C/212°F. The lower the pressure, the lower the boiling point. At high altitudes, the boiling point of water is less than 100°C.

bond in chemistry, any of the forces of attraction that hold atoms or ions together. The principal types of bonding are ◊ionic, ◊covalent, and ◊metallic. The type of bond formed depends on the elements concerned and their electronic structure.

bone the hard material making up the skeleton of most vertebrate animals. Bones are important for support, movement, and protection. They are alive, containing cells that need a supply of food and oxygen. Bones are constructed from protein hardened with calcium. It is this combination that gives the bone great strength, comparable in some cases with that of reinforced concrete.

bone marrow soft tissue in the centre of some limb bones that manufactures red and white blood cells.

boron non-metallic element, symbol B, atomic number 5, relative atomic mass 10.811. In nature it is found only in compounds; for instance, with sodium and oxygen in borax. It exists in two forms: a brown amorphous powder and very hard, brilliant crystals. Its compounds are used in the preparation of boric acid, water softeners, soaps, enamels, glass, and pottery glazes. In alloys it is used to harden steel. Because it absorbs slow neutrons, it is used to make boron carbide control rods for nuclear reactors. It is a necessary trace element in the human diet.

brain The controlling centre of the nervous system, constructed of ◊neurons and with excellent connections to specialist cells detecting light, sound, and smell. The brain is especially well developed in vertebrates, but simple brains are found in simpler animals such as worms and insects. In vertebrates it is composed of three main regions. An enlarged portion of the upper spinal cord, the medulla, contains centres for the control of involuntary processes such as respiration, heart rate, and blood pressure. Overlying this is the cerebellum, which is concerned with coordinating balance and movement. This, too, is automatic. The cerebral hemispheres are often considered the most 'intelligent' part of the brain, because they are involved in thought, memory, decision-making, and voluntary movement.

braking distance distance travelled by a vehicle while the brakes are being applied at maximum efficiency. The ◊stopping distance travelled by a vehicle is the sum of the braking distance and the ◊thinking distance.

brass metal ◊alloy of copper and zinc, with not more than 5–6% of other metals. The zinc content ranges from 20% to 45%, and the colour of brass varies accordingly from coppery to whitish yellow. Brasses are characterized by the ease with which they may be shaped and machined; they are strong and ductile, resist many forms of corrosion, and are used for electrical fittings, ammunition cases, screws and household fittings.

breathing in land animals, the muscular movements by which air is taken into the ◊lungs and then expelled. These movements are very noticeable among the vertebrates because they are large animals, and diffusion on its own would not supply oxygen quickly enough to the body cells. Breathing movements, supplying fresh air to a specialist ◊respiratory surface, are important if a large animal is to be active. Most of the larger, more active insects, such as locusts, show breathing movements of the abdomen.

breathing

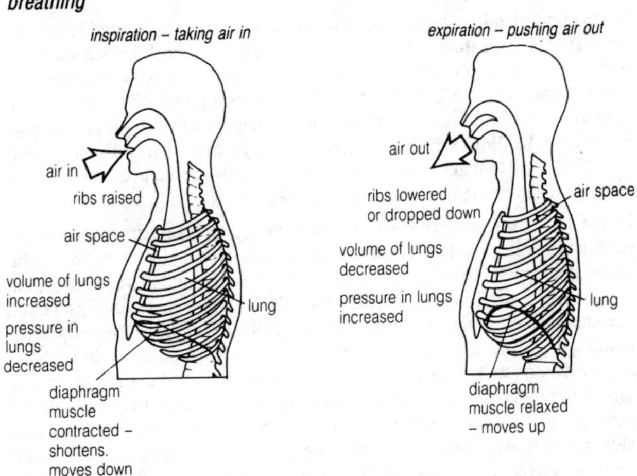

breathing rate the number of times a minute that the lungs inhale and exhale. It increases during exercise when the body needs more oxygen.

breeding the crossing and selection of animals and plants to change the characteristics of an existing breed (variety), or to produce a new one, usually by ◊artificial selection.

Cattle may be bred for increased meat or milk yield, sheep for thicker or finer wool, and horses for speed or stamina. Plants, such as wheat or maize, may be bred for disease resistance, heavier and more rapid cropping, and hardiness to adverse weather.

brewing the manufacture of beer by natural fermenting techniques. Barley is mashed and allowed to germinate so that the starch within the seeds turns into the sugar maltose. Yeast is then added, which respires anaerobically, feeding on the maltose, fermenting it and producing alcohol and the waste gas carbon dioxide. This crude liquid is filtered, and then flavoured with hops. As in wine-making, the enzymes of yeast are crucial to the process. Brewing differs from wine-making in that the sugar is not naturally present at the beginning of the process, but is released when the barley is allowed to germinate.

brine common name for a solution of sodium chloride (NaCl) in water. Brines are used extensively in the food-manufacturing industry for canning vegetables, pickling vegetables (sauerkraut manufacture), and curing meat. Industrially, brine is the source from which chlorine, caustic soda (sodium hydroxide), and sodium carbonate are made.

brittle material inelastic material that breaks suddenly under stress. Brittle materials may also break suddenly when given a sharp knock. Pottery, glass, and cast iron are examples of brittle materials. A ◊ductile material is one that does not break.

bromine dark reddish-brown, non-metallic element, a volatile liquid at room temperature, symbol Br, atomic number 35, relative atomic mass 79.904. It is a member of the ◊halogen group, has an unpleasant odour, and is very irritating to mucous membranes. Its salts are known as bromides.

Bromine was formerly extracted from salt beds, but is now mostly obtained from sea water, where it occurs in small quantities. Its compounds are used in photography and in the chemical and pharmaceutical industries.

bromine water red-brown solution of bromine in water. It is used to test for unsaturation in organic compounds, as such compounds decolorize the solution as they react with the bromine.

bronchiole air tube found in the lung. Bronchioles lead off from the larger ◊bronchus and branch many times before ending in the many thousands of air sacs or alveoli that form the bulk of the lung tissue. Bronchioles are damaged by infection and by smoking.

bronchus (plural ***bronchi***) one of a pair of large tubes splitting off from the trachea (windpipe) and passing into the lung. Apart from their size, bronchi differ from the ◊bronchioles in possessing cartilaginous rings, which give rigidity and prevent collapse during breathing movements.

Numerous glands secrete a slimy mucus, which traps dust and other particles and is constantly being propelled upwards to the mouth by thousands of tiny hairs or cilia. The bronchus is damaged by several respiratory diseases and by smoking, which damages the cilia and therefore the lung-cleaning mechanism.

Brownian motion the continuous random movement of microscopic particles (such as smoke particles) when suspended in a fluid medium (liquid or gas). It occurs because the suspended particles are small enough to be jostled when bombarded by surrounding molecules. Brownian motion provides evidence for the ◊kinetic theory of matter.

bud in plants, an undeveloped shoot usually enclosed by protective scales; inside is a very short stem and numerous undeveloped leaves, or flower parts, or both.

budding ◊asexual reproduction in which an outgrowth develops from a cell to form a new individual. Most yeasts reproduce in this way.

In a suitable environment, yeasts grow rapidly, forming long chains of cells as the buds themselves produce further buds before being separated from the parent. Simple invertebrates, such as hydra, can also reproduce by budding.

buffer mixture of chemical compounds chosen to maintain a steady ◊pH.

The commonest buffers consist of a mixture of a weak organic acid and one of its salts or a mixture of ◊acid salts of phosphoric acid. The addition of either an acid or a base causes a shift in the ◊chemical equilibrium, thus keeping the pH constant.

bulb in plants, an underground bud with fleshy leaves containing a reserve food supply and with roots growing from its base.

bunsen burner gas burner used in laboratories, consisting of a vertical metal tube through which a fine jet of fuel gas is directed. Air is drawn in through airholes near the base of the tube and the mixture is ignited and burns at the tube's upper opening.

The heat of the flame may be adjusted by turning a metal collar to close or partially close the airholes.

bur or ***burr*** fruit or seed covered with hooks. Burs catch in the feathers or fur of passing animals and can therefore be dispersed over wide areas.

burette apparatus used in ◊titrations for the controlled delivery of measured variable quantities of a liquid. It consists of a long, narrow, calibrated glass tube, with a tap at the bottom, leading to a narrow-bore exit.

burning common name for ◊combustion.

butane C_4H_{10} one of two gaseous alkanes (paraffin hydrocarbons) having the same formula but differing in structure. Normal butane ($CH_3CH_2CH_2CH_3$) is derived from natural gas; isobutane (2-methylpropane, $CH_3CH(CH_3)CH_3$) is a by-product of petroleum manufacture. Liquefied under pressure, it is used as a fuel for industrial and domestic purposes (for example, in portable cookers).

butene C_4H_8 fourth member of the ◊alkene series of hydrocarbons. It is an unsaturated compound, containing one double bond.

by-product substance formed incidentally during the manufacture of some other substance; for example, slag is a by-product of the production of iron in the ◊blast furnace. For industrial processes to be economical, by-products must be recycled or used in other ways as far as possible; in this example, slag is used for making roads.

byte in computing, enough computer memory to store a single character of data (such as a number or a letter of the alphabet). The character is stored in the byte of memory as a pattern of ◊bits (binary digits), using a particular code—for example, the letter F might be stored as the bit pattern 11000110. A byte usually contains eight bits.

A single byte can specify 256 values, such as the decimal numbers from 0 to 255; three bytes (24 bits) can specify 16,777,216 values. Computer memory size is measured in ***kilobytes*** (1,024 bytes) or ***megabytes*** (1,024 kilobytes).

C

cadmium soft, silver-white, ductile and malleable, metallic element, symbol Cd, atomic number 48, relative atomic mass 112.40. Cadmium occurs in nature as a sulphide or carbonate in zinc ores. It is a toxic metal that, because of industrial dumping, has become an environmental pollutant. Its uses include batteries, electroplating, and as a constituent of alloys used for bearings with low coefficients of friction; it is also a constituent of an alloy with a very low melting point.

caecum in the ◊digestive system, a blind-ending tube branching off from the first part of the large intestine, terminating in the appendix. It is used for the digestion of cellulose by some grass-eating mammals; the rabbit caecum and appendix, for example, contain millions of bacteria that produce cellulase, the enzyme necessary for the breakdown of cellulose to glucose. In humans, the caecum and appendix are ◊vestigial organs, which means they serve no function.

caesium (Latin *caesius* 'bluish-grey') soft, silvery-white, ductile, metallic element, symbol Cs, atomic number 55, relative atomic mass 132.905. It is one of the ◊alkali metals, and is the most electropositive of all the elements. In air it ignites spontaneously, and it reacts vigorously with water. It is used in the manufacture of photoelectric cells. The name comes from the blueness of its spectral line.

The rate of vibration of caesium atoms is used as the standard of measuring time. Its radioactive isotope Cs-137 (half-life 30.17 years) is one of the most dangerous waste products of the nuclear industry; it is a highly radioactive biological analogue for potassium, produced as a fission product of nuclear explosions and in the reactors of nuclear power plants.

calcium soft, silvery-white, metallic element, symbol Ca, atomic number 20, relative atomic mass 40.08. It is one of the ◊alkaline-earth metals. One of the most widely distributed elements, it is the fifth most abundant element (the third most abundant metal) in the Earth's crust. It is found mainly as its carbonate $CaCO_3$, which occurs in a fairly pure condition as chalk and limestone. Calcium is an essential component of bones, teeth, shells, milk and leaves, and it forms 1.5% of the human body by mass. Calcium ions in animal cells are involved in regulating muscle contraction, hormone secretion, digestion, and glycogen metabolism in the liver.

calcium carbonate $CaCO_3$ white solid, found in nature as limestone, marble, and chalk. In its reactions it is a typical ◊carbonate. It is a valuable resource, used in the making of iron, steel, cement, glass, slaked lime, bleaching powder, sodium carbonate and bicarbonate, and many other industrially useful substances. However, its quarrying causes scarring of the landscape, creating dust, noise, and additional traffic in what are often the most attractive areas of the countryside.

calcium hydrogencarbonate $Ca(HCO_3)_2$ substance found in ◊hard water, formed when rainwater passes over limestone rock.

$$CaCO_{3(s)} + CO_{2(g)} + H_2O_{(l)} \rightarrow Ca(HCO_3)_{2(aq)}$$

When this water is boiled it reforms calcium carbonate, removing the hardness; this type of hardness is therefore known as temporary hardness.

calcium hydroxide $Ca(OH)_2$ or ***slaked lime*** white solid, slightly soluble in water. A solution of calcium hydroxide is called ◊limewater and is used in the laboratory to test for the presence of carbon dioxide. It is manufactured industrially by adding water to calcium oxide (quicklime) in a strongly exothermic reaction.

$$CaO + H_2O \rightarrow Ca(OH)_2$$

It is used to reduce soil acidity and as a cheap alkali in many industrial processes.

calcium oxide or ***quicklime*** CaO white solid compound, formed by heating ◊calcium carbonate.

$$CaCO_3 \rightarrow CaO + CO_2$$

When water is added it forms calcium hydroxide (slaked lime) in an exothermic reaction.

$$CaO + H_2O \rightarrow Ca(OH)_2$$

It is a typical basic oxide, turning litmus blue.

calcium sulphate $CaSO_4$ white, solid compound, found in nature as gypsum. It dissolves slightly in water to form ◊hard water; this hardness is not removed by boiling, and is therefore sometimes called permanent hardness.

calorific value the amount of heat energy generated by a given mass of food when it is completely burned, or when it is completely oxidized by aerobic respiration. It is measured in joules (or kilojoules).

The calorific value of a food item can be measured experimentally by burning that item and using the heat released to raise the temperature of a known mass of water.

$$\text{calorific value (J)} = \text{mass of water (g)} \times \text{temperature rise (°C)} \times 4.2$$

cambium a layer of actively dividing cells found within stems and roots, responsible for manufacturing new xylem and phloem.

camera an optical instrument used to make a permanent image of a scene. The simplest type is the ◊***pinhole camera***.

A ***lens camera*** uses a converging lens (or a combination of several lenses) to produce a real, inverted image on a light-sensitive film at the back of the camera. The image can be focused by moving the lens nearer to or further from the film. Light is allowed onto the film when a spring-loaded shutter is opened.

The amount of light reaching the film is controlled in two ways. One is by varying the exposure time; this is done by changing the width of the gap in the shutter through which the light passes, so that each section of film is exposed to light for a greater or lesser time. The shorter the exposure time, the less light shines on the film. The second method is by adjusting the aperture in the centre of the iris diaphragm, a multi-leaved structure between the lens and the shutter that can open or close to produce a larger or smaller opening through which light passes. The smaller the aperture, the less light reaches the film.

cancer group of diseases characterized by abnormal proliferation of cells. Cancer (malignant) cells are usually degenerate, capable only of reproducing themselves (forming ◊tumours) until they outnumber the surrounding healthy cells. Malignant cells tend to spread from their site of origin by travelling through the bloodstream or lymphatic system.

There are more than 100 types of cancer. Some, like lung or bowel cancer, are common; others are rare. The cause remains unexplained. Triggering agents (carcinogens) include chemicals such as those found in cigarette smoke, other forms of smoke, asbestos dust, exhaust fumes, and many industrial chemicals. Some viruses can also trigger the cancerous growth of cells, as can X-rays and radioactivity. Dietary factors are important in some cancers; for example, lack of fibre in the diet may predispose people to bowel cancer.

canine in mammalian carnivores, long, often pointed tooth found at the front of the mouth between the incisors and premolars. Canines are used for catching prey, for killing, and for tearing flesh. They are absent in herbivores such as rabbits and sheep, and are much reduced in humans.

capacitor or ***condenser*** device for storing electric charge, used in electronic circuits; it consists of two or more metal plates separated by an insulating layer called a dielectric.

Its ***capacitance*** is the ratio of the charge stored on either plate to the potential difference between the plates. The SI unit of capacitance is the farad (symbol F), but most capacitors have much smaller capacitances, and the microfarad (a millionth of a farad) is the commonly used practical unit.

capillary the narrowest blood vessel in vertebrates, between 8 and 20 micrometres in diameter, barely wider than a red blood cell. Capillaries are distributed as ***beds***, complex networks connecting arteries and veins.

Capillary walls are extremely thin, consisting of a single layer of cells, and so nutrients, dissolved gases, and waste products can easily pass through them. This makes the capillaries the main area of exchange between the fluid (◊lymph) bathing body tissues and the blood.

carbide compound of carbon and one other chemical element, usually a metal, silicon, or boron.

carbohydrate organic molecule containing carbon, hydrogen, and oxygen, and possessing the general formula $C_x(H_2O)_y$.

The basic unit of all carbohydrates is a simple sugar molecule (monosaccharide), such as glucose. These units may be combined into disaccharides (two sugars), such as sucrose, or into polysaccharides (many sugars linked together in a chain), such as starch, glycogen, and cellulose.

Carbohydrates are an important part of a balanced human diet, providing energy for life processes including growth and movement. Excess carbohydrate intake can be converted into fat and stored in the body. In digestion, complex carbohydrates, for instance the starches, are broken down into simple sugars which are easily absorbed across the gut wall. ◊Starch itself is too large a molecule to be absorbed. It is found particularly in seeds and tubers, where it acts as an energy store for plants.

carbon non-metallic element, symbol C, atomic number 6, relative atomic mass 12.011. It occurs on its own as diamond and graphite, in carbonaceous rocks such as chalk and limestone, as carbon dioxide in the atmosphere, as hydrocarbons in petroleum, coal, and natural gas, and as a constituent of all organic substances. Life on Earth is often described as 'carbon-based', because of the importance of carbon chemistry (see ◊organic chemistry) in cell construction and metabolism.

Carbon is most familiar as coal, charcoal, and soot. Of the inorganic carbon compounds, the chief one is ***carbon dioxide*** (CO_2), a colourless gas formed when carbon is burned in an adequate supply of air, and ***carbon monoxide*** (CO), formed when carbon is oxidized in a limited supply of air. ***Carbon disulphide*** (CS_2) is a dense liquid with a sweetish odour. Another group of compounds is the ***carbon halides***, including ◊carbon tetrachloride (CCl_4).

When added to steel, carbon forms a wide range of alloys with useful properties. In pure form, it is used to control the chain reactions in nuclear reactors; as colloidal graphite it is a good lubricant and, when deposited on

carbon

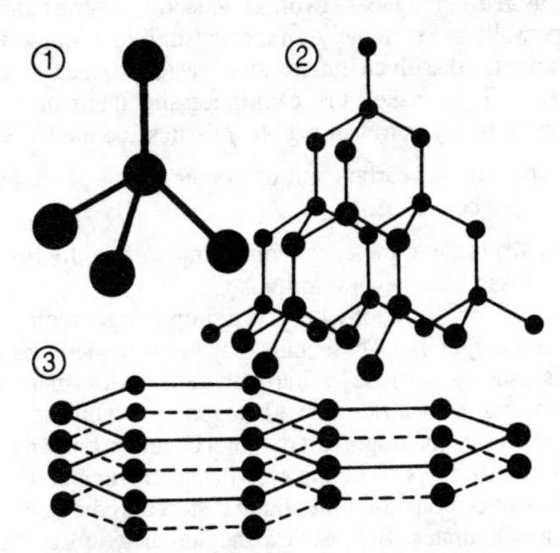

1. the basic unit of the diamond structure
2. *diamond*, a giant three-dimensional structure
3. *graphite*, a two-dimensional structure

a surface in a vacuum, reduces photoelectric and secondary emission of electrons. Carbon is used as a fuel in the form of coal or coke. The radioactive isotope carbon-14 (half-life 5,730 years) is used as a tracer in biological research.

The element has the following characteristic reactions.

with air or oxygen It burns on heating to form carbon dioxide in excess air, or carbon monoxide in a limited supply of air.

$$C + O_2 \rightarrow CO_2$$
$$(\Delta H = -394 \text{ kJ mol}^{-1})$$
$$2C + O_2 \rightarrow 2CO$$

with metal oxides It reduces many metal oxides at high temperatures.

$$Fe_2O_3 + 3C \rightarrow 2Fe + 3CO$$

with steam It forms water gas (a cheap, useful, industrial fuel) when steam is passed over white-hot coke.

$$C + H_2O \rightarrow CO + H_2$$

with concentrated acids With hot, concentrated sulphuric or nitric acids it forms carbon dioxide.

carbonate CO_3^{2-} ion formed when carbon dioxide dissolves in water, and any salt formed by this ion and another chemical element, usually a metal.

Carbon dioxide (CO_2) dissolves in water to form the weakly acidic carbonic acid (H_2CO_3), which unites with various substances to form carbonates. Calcium carbonate ($CaCO_3$) (chalk, limestone, and marble) is one of the most abundant carbonates known, being a constituent of mollusc shells and the hard outer skeletons of crustaceans.

Carbonates give off carbon dioxide when heated or treated with dilute acids. This reaction is used as the laboratory test for the presence of the ion, as it gives an immediate effervescence, with the gas turning limewater (a solution of calcium hydroxide, $Ca(OH)_2$) milky. See ◊sodium carbonate and ◊calcium carbonate.

carbon cycle the sequence by which carbon circulates and is recycled through the natural world. A massive amount of carbon is stored in oceans and in rocks, as well as in the atmosphere. The cycle involves the movement of carbon dioxide between these three zones, or 'sinks'. Carbon dioxide, released into the atmosphere by animals as a result of ◊respiration, is

carbon cycle

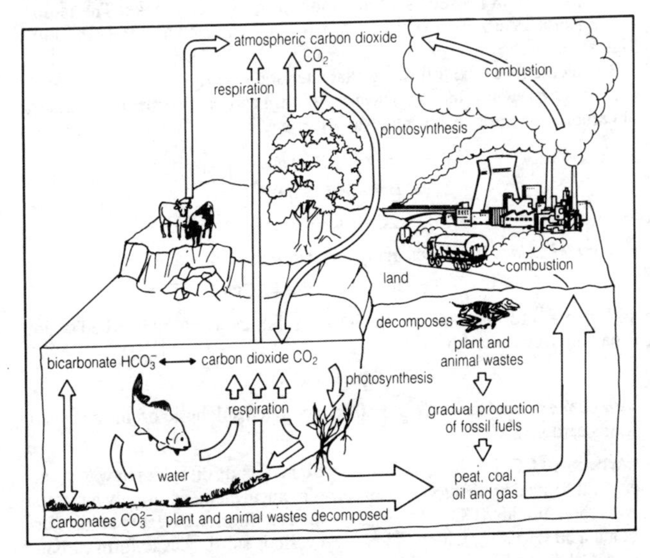

taken up by plants during ◊photosynthesis and converted into carbohydrates; the oxygen component is released back into the atmosphere. Similarly, carbon dioxide from other sources (for instance the burning of fossil fuels), may be taken up by plants. Another link in the carbon cycle occurs when an animal eats a plant and carbon is transferred from, say, a leaf cell to the animal body. Finally, organisms also produce carbon dioxide when they decay.

Today, the carbon cycle is being distorted by the increased burning of fuels and by the burning of large tracts of tropical forests, as a result of which levels of carbon dioxide are building up in the atmosphere and probably contributing to climatic change. There is speculation that increased carbon dioxide in the atmosphere will be removed by the oceans (by dissolving it) or by plants (by increased growth or photosynthesis rates).

However, it seems unlikely that these processes can keep up with the steadily increasing carbon dioxide emissions from world industry. See ◊greenhouse effect.

carbon dioxide CO_2 colourless gas, slightly soluble in water and denser than air. It is formed when carbon and carbon-containing compounds are fully oxidized, as when they are burnt in an excess of air. It is also produced when acids are added to carbonates or hydrogencarbonates, and when these salts are heated. It is a typical acidic oxide, dissolving in water to give a solution of the weak dibasic acid carbonic acid, and forming salts with alkalis.

$$H_2O + CO_2 \leftrightarrow H_2CO_3 \leftrightarrow H^+_{(aq)} + HCO^-_{3(aq)}$$

$$NaOH + CO_2 \rightarrow NaHCO_3$$

With a solution of calcium hydroxide (limewater) the gas forms a white (milky) precipitate of calcium carbonate. This reaction is used as the confirmatory test for carbon dioxide.

$$Ca(OH)_{2(aq)} + CO_{2(g)} \rightarrow CaCO_{3(s)} + H_2O_{(l)}$$

The gas supports the combustion of burning magnesium, but extinguishes lower-temperature flames; it is used in fire extinguishers.

Carbon dioxide plays a vital role in the ◊carbon cycle and in the formation of ◊hard water. It forms 0.03% of the atmosphere, but this proportion is increasing year by year as a result of increased industrial activity. See ◊greenhouse effect.

carbonic acid H_2CO_3 weak, dibasic acid formed by dissolving carbon dioxide in water.

$$H_2O + CO_2 \leftrightarrow H_2CO_3$$

It forms two series of salts: ◊carbonates and ◊hydrogencarbonates. Fizzy drinks are made by dissolving carbon dioxide in water under pressure; soda water is a solution of carbonic acid.

carbon monoxide CO colourless, odourless gas formed when carbon is oxidized in a limited supply of air. It is poisonous because, when inhaled, it binds irreversibly with haemoglobin in the blood, preventing the haemoglobin from being used to transport oxygen to the body tissues. Carbon monoxide from car exhaust fumes is a major constituent of air pollution.

In industry carbon monoxide is used as a reducing agent in metallurgical processes, such as the extraction of iron in a ◊blast furnace. It is a component of cheap industrial fuels such as water gas (see ◊carbon). In air it burns with a luminous blue flame to form carbon dioxide.

$$2CO + O_2 \rightarrow 2CO_2$$

carbon tetrachloride common name for ◊tetrachloromethane.

Carborundum trademark for a very hard, black abrasive, consisting of silicon carbide (SiC), an artificial compound of carbon and silicon. First produced in 1891, it is harder than corundum (see ◊alumina) but not as hard as ◊diamond.

carboxylic acid R–COOH organic acid containing the carboxyl group (COOH) attached to another group (R), which can be hydrogen (giving methanoic acid, HCOOH) or a larger molecule (up to 24 carbon atoms). The smaller carboxylic acids form a homologous series, with all the names ending -oic (methanoic acid, HCOOH; ethanoic acid, CH_3COOH; propanoic acid, C_2H_5COOH, and so on). Larger ones are often found as ◊esters in fats, often with glycerine, and so are called ◊fatty acids.

carcinogen any agent that increases the chance of a cell becoming cancerous, including various chemicals, some viruses, X-rays, and forms of nuclear radiation. In the past, testing for carcinogens has often involved the use of animals.

carnivore in a food chain, a type of ◊consumer that eats other animals. Carnivores therefore include not only the large vertebrates such as sharks and tigers but also many invertebrates, including microscopic ones.

carotene a naturally occurring pigment responsible for the orange, yellow, and red colours of carrots, tomatoes, oranges, and crustaceans. In vertebrates, carotene is converted to retinol (vitamin A) and to pigments important in vision. See ◊vitamin.

carotid artery one of a pair of major blood vessels, one on each side of the neck, supplying blood to the head.

carpel the term given to the entire assembly of female parts within the flower. It therefore consists of an ovary, where eggs are stored; a style, which allows pollen tubes to grow down towards the eggs; and a stigma, on which pollen grains land and germinate. The fine organization of the carpel, and the number of carpels within a flower, varies greatly according to species.

cartilage stiff bluish-white material found in the body, an important component of the skeleton. It is made of protein, but differs from bone in lacking mineral strengthening. In cartilaginous fish (the sharks) cartilage in fact forms the entire skeleton. Among other vertebrates cartilage forms the skeleton of the developing foetus, but is gradually 'mineralized', transforming into bone as the animal develops. However, even an adult vertebrate has zones where cartilage remains, for instance the ends of bones and the cushioning discs in the backbone. Other cartilaginous areas include the nose and the ear.

casein main protein of milk, from which it can be separated by the action of acid, the enzyme rennin, or bacteria (souring); it is also the main component of cheese.

cast iron cheap but invaluable constructional material, most commonly used for car engine blocks. Cast iron is partly refined pig (crude) ◊iron, which is very fluid when molten and highly suitable for shaping by casting, as it contains too many impurities, such as carbon, to be readily shaped in any other way. Solid cast iron is heavy and can absorb great shock but is very brittle.

catalyst substance that alters the speed of (or makes possible) a chemical or biological reaction but that remains unchanged at the end of the reaction. ◊Enzymes are natural biological catalysts.

Catalysts in industrial use include vanadium(V) oxide (see ◊contact process) and finely divided iron (see Haber process). A commonly used laboratory catalyst is ◊manganese(IV) oxide.

catalytic converter device fitted to the exhaust system of a motor car, aimed at reducing toxic emissions from the internal-combustion engine. It converts harmful exhaust products to relatively harmless ones by passing exhaust gases over a mixture of catalysts, coated on a metal or ceramic honeycomb. The nitrous oxides are converted to nitrogen, while carbon monoxide and the hydrocarbons are converted to carbon dioxide. Catalytic converters do not remove carbon dioxide, and so no car can be considered 'non-polluting'. Over the next ten years catalytic converters will become a feature on most cars. Catalytic converters become ineffective if the petrol contains lead.

catalytic cracking a form of ◊cracking.

cathode in electrolysis, the negatively charged electrode towards which positive particles (cations) move; in electronics, it is the part of an electronic device that emits electrons.

The cathode of an electronic device such as a cathode-ray tube or diode is kept at a negative potential with respect to the device's other electrodes (the anodes) so that the freed electrons will stream away from the cathode (as ◊cathode rays) and towards the anodes.

cathode-ray oscilloscope (CRO) instrument used to measure electrical potentials or voltages that vary over time and to display the waveforms of electrical oscillations or signals. Readings are displayed graphically on the screen of a ◊cathode-ray tube.

cathode rays streams of fast-moving electrons that travel from a cathode (negative electrode) towards an anode (positive electrode) in a vacuum tube. They carry a negative charge and can be deflected by electric and magnetic fields. Cathode rays focused into fine beams of fast electrons are used in cathode-ray tubes, the electrons' ◊kinetic energy being converted into light energy as they collide with the tube's fluorescent screen.

cathode-ray tube (CRT) vacuum tube in which a beam of electrons is produced and focused onto a fluorescent screen. CRTs are essential components of television receivers, computer VDUs, and cathode-ray oscilloscopes.

cell discrete, membrane-bound portion of living matter, the smallest unit capable of an independent existence. Each is formed by the cell division (◊mitosis or ◊meiosis) of another cell.

All living organisms consist of one or more cells, with the exception of viruses. Bacteria, protozoa, and many other microorganisms consist of single cells, whereas a human is made up of billions of cells.

Essential features of a cell are the ◊cell membrane, which encloses it and restricts the flow of substances in and out; the jellylike material, or ◊cytoplasm, within; the ribosomes, structures that carry out protein synthesis; and the ◊DNA, which forms the hereditary material. The cells of plants, bacteria, and fungi possess a rigid outer cell wall that protects the cell and maintains its shape. Many plant cells also contain a large sac, or vacuole, containing a watery sap. This sap is usually under pressure, and helps maintain rigidity.

In the cells of animals and plants, the DNA is organized into ◊chromosomes and contained within a ◊nucleus. The only cells of the human body that have no nucleus are the red blood cells. The nuclei of some cells contain a denser spot called the ◊nucleolus. Animal and plant cells also possess structures such as ◊mitochondria, ◊chloroplasts, endoplasmic reticulum, and Golgi apparatus, which perform specialized tasks.

Bacteria have a different, simpler, cell organization. Instead of being arranged as chromosomes, the DNA forms a simple loop and there is no nucleus.

cell, chemical or ***electrical cell*** apparatus that produces electrical energy as a result of a chemical reaction; the popular name is 'battery', but this actually refers to a collection of cells in one unit. A ***primary*** electric cell cannot be replenished, whereas in a ***secondary*** cell or accumulator, the action is reversible and the original condition can be restored by an electric current.

The reactions that take place in a simple cell depend on the fact that some metals are more reactive than others. If two different metals are joined by an ◊electrolyte and a wire, the more reactive metal loses electrons to form ions. The ions pass into solution in the electrolyte, while the electrons flow down the wire to the less reactive metal. At the less reactive metal the electrons are taken up by the positive ions in the electrolyte, which completes the circuit. If the two metals are zinc and copper and the electrolyte is dilute sulphuric acid, the following cell reactions occur.

The zinc atoms dissolve as they lose electrons (oxidation).

$$Zn - 2e^- \rightarrow Zn^{2+}$$

The two electrons travel down the wire and are taken up by the hydrogen ions in the electrolyte (reduction).

$$2H^+ + 2e^- \rightarrow H_2$$

The overall cell reaction is obtained by combining these two reactions; the zinc rod slowly dissolves and bubbles of hydrogen appear at the copper rod.

$$Zn + 2H^+ \rightarrow Zn^{2+} + H_2$$

If each rod is immersed in an electrolyte containing ions of that metal, and the two electrolytes are joined by a salt bridge, metallic copper deposits on the copper rod as the zinc rod dissolves in a ◊redox reaction, just as if zinc had been added to a copper salt solution.

$$Zn - 2e^- \rightarrow Zn^{2+}$$

$$Cu^{2+} + 2e^- \rightarrow Cu$$

$$Zn_{(s)} + Cu^{2+}_{(aq)} \rightarrow Zn^{2+}_{(aq)} + Cu_{(s)}$$

cell, chemical

the basic principle of a chemical cell

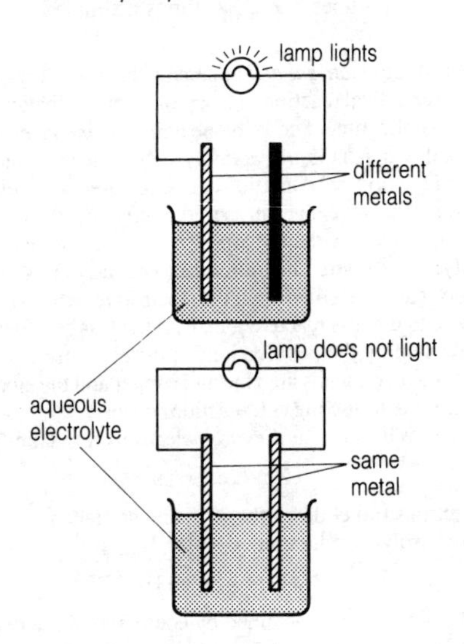

a simple cell

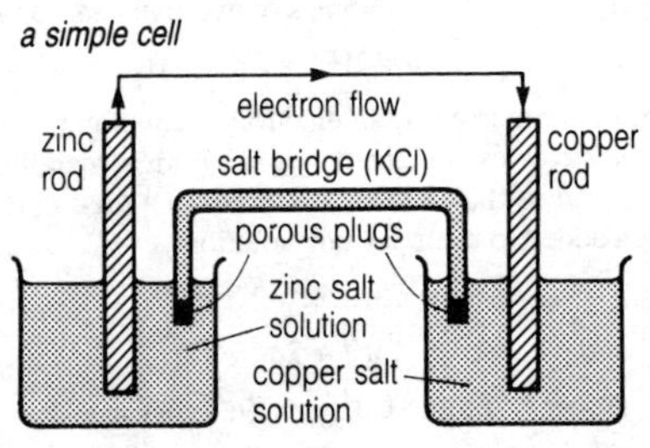

This mechanism is the basis of the sacrificial protection of metals (see ◊rust prevention).

cell division the process by which a cell divides, either by ◊meiosis, associated with sexual reproduction, or by ◊mitosis, associated with growth, cell replacement or repair. Both forms involve the duplication of DNA and the splitting of the nucleus.

cell, electrical alternative name for ◊cell, chemical.

cell membrane or ***plasma membrane*** thin semipermeable layer of protein and fat that encloses cells and controls the flow of substances into and out of the cytoplasm. Generally, small molecules such as water, glucose, and amino acids can penetrate the membrane, while large molecules such as starch cannot. Membranes also play a part in hormone function, and cell metabolism.

cell sap dilute fluid found in the large central ◊vacuole of many plant cells. It is made up of water, amino acids, glucose, and salts. The sap has many functions, including storage of useful materials, and provides mechanical support for non-woody plants.

celluloid transparent or translucent, highly inflammable, plastic material (a ◊thermosoftening plastic made from nitrocellulose and camphor) once used for toilet articles, novelties, and photographic film. It has been replaced by the non-inflammable substance cellulose acetate.

cellulose ◊polysaccharide composed of long chains of glucose units. It is the most abundant substance found in the plant kingdom, being the principal constituent of the cell wall of higher plants, and is a vital ingredient in the diet of many ◊herbivores. Molecules of cellulose are organized into long, unbranched fibres (microfibrils) that give support to the cell wall.

No vertebrate is able to produce cellulase, the enzyme necessary for the breakdown of cellulose into sugar. Yet most mammalian herbivores (such as rabbits and cows) rely on cellulose, using secretions from microorganisms living in the gut to break it down. Humans cannot digest the cellulose of the cell walls; they possess neither the correct gut microorganisms nor the necessary grinding teeth. However, cellulose still forms a necessary part of the human diet as ◊roughage.

cell wall in plants, bacteria, and fungi, the tough outer surface of the cell. The cell wall of plants is constructed from a mesh of ◊cellulose and is very strong and relatively inelastic. Most living plant cells are turgid (swollen with water; see ◊turgor) and develop an internal hydrostatic pressure (wall pressure) that acts against the cellulose wall. The result of this turgor pressure is to give the cell, and therefore the plant, rigidity. Plants that are not woody are particularly reliant on this form of support.

Celsius scale temperature scale in which one division or degree is taken as one hundredth part of the interval between the freezing point (0°C) and the boiling point (100°C) of water at standard atmospheric pressure.

The degree centigrade (°C) was officially renamed Celsius in 1948 to avoid confusion with the angular measure known as the centigrade (one hundredth of a grade). The Celsius scale is named after the Swedish astronomer Anders Celsius (1701—44), who devised it in 1742.

centigrade scale common name for the ◊Celsius temperature scale.

central nervous system the part of the nervous system with a concentration of ◊neurons (nerve cells) that coordinates body functions. In vertebrates, the central nervous system consists of a brain and a dorsal nerve cord (the spinal cord) within the spinal column.

centre of mass or ***centre of gravity*** the point in or near an object from which its total weight appears to originate and can be assumed to act. A symmetrical homogeneous object such as a sphere or cube has its centre of mass at its physical centre; a hollow shape (such as a cup) may have its centre of mass in space inside the hollow. See also ◊stability.

centrifugal force useful concept based on an apparent (but not real) force. It may be regarded as a force that acts radially outwards from a spinning or orbiting object, thus balancing the centripetal force (which is real and acts inwards). For an object of mass m moving with velocity v in a circle of radius r, the centrifugal force F is given by the formula:

$$F = mv^2r$$

centrifuge apparatus that rotates containers at high speeds, creating centrifugal forces. One use is for separating mixtures of substances of different densities. The mixtures are placed in the containers and the rotation sets up centrifugal forces, causing them to separate according to their densities. A common example is the separation of the lighter plasma from the heavier blood corpuscles in certain blood tests.

cerebellum in vertebrates, the part of the brain that controls muscular movements, balance, and coordination. The human cerebellum is well developed, because of the need for balance when walking or running, and for coordinated hand movements.

cerebral hemisphere one of the two halves of the ◊cerebrum.

cerebrum part of the vertebrate ◊brain, formed from two paired cerebral hemispheres. In birds and mammals it is the largest part of the brain. It is covered with an infolded layer of grey matter, the ***cerebral cortex***, which integrates brain functions. The cerebrum coordinates the senses, and is responsible for learning and other higher mental faculties.

cervix the neck of the ◊uterus; it is a narrow passage that opens into the ◊vagina.

CFC abbreviation for ◊chlorofluorocarbon.

chain reaction mechanism that produces very fast, ◊exothermic reactions, as in the formation of flames and explosions.

The reaction begins with the formation of a single reactive molecule. This combines with an inactive molecule to form two reactive molecules. These two produce four (or more) reactive molecules; very quickly, very many reactive molecules are produced, so the reaction rate accelerates dramatically. The reactive molecules contain an unpaired electron and are called ◊free radicals; they last only a short time because they are so reactive.

In nuclear physics, the chain reaction occurs when ◊neutrons, released by the splitting of some atomic nuclei themselves go on to split others, releasing even more neutrons. Such a reaction can be controlled (as in a nuclear reactor) by using moderators to absorb excess neutrons. Uncontrolled, a chain reaction produces a nuclear explosion (as in an atom bomb).

chalk soft, fine-grained, whitish rock composed of calcium carbonate $CaCO_3$, extensively quarried for use in cement, lime, and mortar, and in the manufacture of cosmetics and toothpaste. ***Blackboard chalk*** in fact consists of gypsum (calcium sulphate, $CaSO_4$).

change of state change in the physical state (solid, liquid, or gas) of a material. For instance, melting, boiling, evaporation, and their opposites, solidification and condensation, are changes of state. The former set of changes are brought about by heating or decreased pressure; the latter by cooling or increased pressure.

In the unusual change of state called ***sublimation***, a solid changes directly to a gas without passing through the liquid state. For example, solid carbon dioxide (dry ice) sublimes to carbon dioxide gas.

change of state

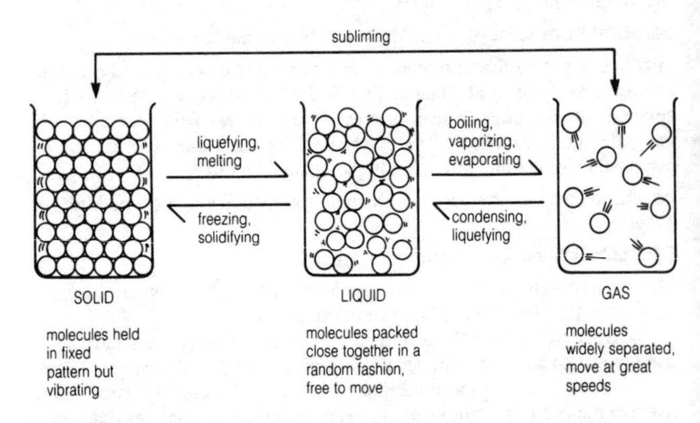

charcoal black, porous form of ◊carbon, produced by heating wood or other organic materials in the absence of air (a process called destructive distillation). It is used as a fuel, for smelting metals such as copper and zinc; in the form of ***activated charcoal***, for purification and filtration of drinking water and other liquids and gases; and by artists for making black line drawings.

charge see ◊electric charge.

chemical element alternative name for ◊element.

chemical equation method of indicating the reactants and products of a chemical reaction by using chemical symbols and formulae. A chemical equation gives two basic pieces of information: (1) the reactants (on the left-hand side) and products (right-hand side); and (2) the reacting proportions (stoichiometry), that is, how many units of each reactant and product are involved. The equation must balance; that is, the total number of atoms of a particular element on the left-hand side must be the same as the number of atoms of that element on the right-hand side.

$$\underbrace{Na_2CO_3 + 2HCl}_{\text{reactants}} \rightarrow \underbrace{2NaCl + CO_2 + H_2O}_{\text{products}}$$

This equation states that one molecule of sodium carbonate combines with two molecules of hydrochloric acid to from two molecules of sodium chloride, one of carbon dioxide, and one of water. ◊State symbols and the energy symbol (ΔH) can be used to provide further information.

$$Na_2CO_{3\,(s)} + 2HCl_{(aq)} \rightarrow 2NaCl_{(aq)} + CO_{2\,(g)} + H_2O_{(l)}\ (-\Delta H)$$

Substituting the relative molecular masses of the substances indicates the proportions of masses involved.

Double arrows indicate that the reaction is reversible—in the formation of ammonia from hydrogen and nitrogen, the direction depends on the temperature and pressure of the reactants.

$$3H_2 + N_2 \leftrightarrow 2NH_3$$

chemical equilibrium condition in which the products of a reversible chemical reaction are formed at the same rate at which they decompose back into the reactants, so that the concentration of each reactant and product remains constant.

chemical family collection of elements that have very similar chemical and physical properties. In the ◊periodic table of the elements such collections are to be found in the vertical columns (groups). The groups that contain the most markedly similar elements are group I, the ◊alkali metals; group II, the ◊alkaline-earth elements; group VII, the ◊halogens; and group 0, the inert or ◊noble gases.

chloride Cl^- negative ion formed when hydrogen chloride dissolves in water, and any salt containing this ion, commonly formed by the action of hydrochloric acid (HCl) on various metals or by direct combination of a metal and chlorine. Sodium chloride (NaCl) is common table salt.

When a silver nitrate solution is added to any chloride solution containing nitric acid, a white precipitate of silver chloride is formed.

$$NaCl_{(aq)} + AgNO_{3\,(aq)} \rightarrow AgCl_{(s)} + NaNO3(aq)$$

chlorine greenish-yellow, gaseous, non-metallic element with a pungent odour, symbol Cl, atomic number 17, relative atomic mass 35.453 (a mixture of two isotopes: 75% Cl-35, 25% Cl-37). It is the second member of the ◊halogen group (group VII of the periodic table). In nature it is widely distributed in combination with the ◊alkali metals, as chlorates or chlorides; in its pure form the gas is a diatomic molecule (Cl_2). It is a very reactive element, and combines with most metals, some non-metals, and a wide variety of compounds.

Industrially, chlorine is prepared by electrolysis of brine (see ◊sodium hydroxide). It is made on a large scale and used in making bleaches, for sterilizing water for drinking and in swimming baths, and in the manufacture of chloro-organic compounds such as chlorinated solvents, CFCs, and PVC.

Some typical reactions are given below.

with metals When dry chlorine is passed over a heated metal, the chloride is formed.

$$Zn + Cl_2 \rightarrow ZnCl_2$$

with non-metals The same reaction occurs with certain non-metals, when the dry gas is passed over the heated element.

$$2P + 5Cl_2 \rightarrow 2PCl_5$$

with compounds With water, chlorine forms a bleaching solution.

$$H_2O + Cl_2 \rightarrow HCl + HOCl$$

$$2OCl^- \rightarrow 2Cl^- + O_2$$

Iron(II) salts are oxidized to iron(III) salts.

$$2FeCl_2 + Cl_2 \rightarrow 2FeCl_3$$

Organic compounds undergo halogenation.

$$C_2H_6 + Cl_2 \rightarrow C_2H_5Cl + HCL$$

Alkalis form chlorides, chlorates, and water.

$$2NaOH + Cl_2 \rightarrow NaCl + NaOCl + H_2O$$

Other halogens are displaced in a redox reaction.

$$2KBr + Cl_2 \rightarrow 2KCl + Br_2$$

chlorofluorocarbon (CFC) non-toxic, odourless chemical, used as a propellant in aerosol cans, as a refrigerant in refrigerators and air conditioners, and in the manufacture of foam boxes for take-away food cartons. CFCs are partly responsible for the destruction of the ozone layer.

When CFCs drift up into the upper atmosphere they break down, under the influence of the Sun's ultraviolet radiation, into chlorine atoms, which destroy the ozone layer and allow harmful radiation from the Sun to reach the Earth's surface. CFCs can remain in the atmosphere for more than 100 years. Production of CFCs is now being phased out in most countries, and replacements developed.

chloroform or ***trichloromethane*** $CHCl_3$ clear, colourless, toxic, carcinogenic liquid with a characteristic, pungent, sickly-sweet smell and taste, formerly used as an anaesthetic (now superseded by less harmful substances). It is used as a solvent and in the synthesis of organic chemical compounds.

chlorophyll pigment present in green plants, responsible for their colour. It is vital because it absorbs sunlight, and allows leaves to make starch from carbon dioxide and water. Without this form of energy-capture, the plant would not be able to perform the complicated chemical reactions of ◊photosynthesis. The pigment absorbs the red and blue-violet parts of sunlight but reflects the green, thus giving plants their most characteristic colour.

Chlorophyll is found within chloroplasts, which are present in large numbers in the ◊palisade cells of leaves. It is similar in structure to ◊haemoglobin, but contains magnesium instead of iron.

chloroplast tiny structures inside plant cells containing the green pigment chlorophyll. Chloroplasts are found in the cells of leaves, and in the surface cells of stems, but, for obvious reasons, not in roots. Within the leaf, they occur mostly in those cells near the top of the leaf, where light intensity is greatest.

cholesterol a waxy, fatty substance abundant in the bodies of animals and found also in some plants. It is a vital constituent of cell membranes and is the starting point in the formation of many hormones, including the sex hormones. Cholesterol is manufactured in the liver and intestine, and is also provided in the human diet by foods such as eggs, meat, and butter. A high level of cholesterol in the blood is thought to contribute to ◊arteriosclerosis (hardening of the arteries).

choroid the black layer found at the rear of the ◊eye beneath the retina. By absorbing light that has already passed through the retina, it stops back-reflection and so aids vision.

chromatography technique for separating a mixture, usually in solution, into its constituent components. This is done by passing the mixture (the 'mobile phase') through another substance (the 'stationary phase'), usually a liquid or solid. The different components of the mixture are absorbed or

chromatography

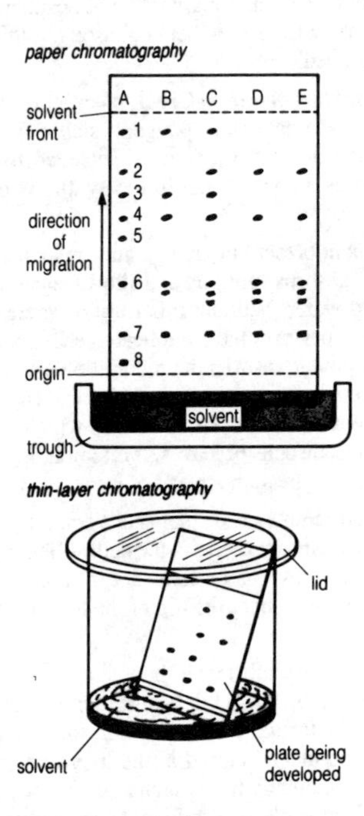

impeded to different extents, and therefore separate. The technique is used for both qualitative and quantitative analyses in biology and chemistry.

In ***paper chromatography***, the mixture separates because the components have differing solubilities in the solvent flowing through the paper and in the chemically bound water of the paper.

In ***thin-layer chromatography***, a wafer-thin layer of absorbent medium on a glass plate replaces the filter paper. The mixture separates because of the differing solubilities of the components in the solvent flowing up the solid layer, and their differing tendencies to stick to the solid (adsorption).

chromium hard, brittle, grey-white, metallic element, symbol Cr, atomic number 24, relative atomic mass 51.996. It takes a high polish, has a high melting point, and is very resistant to corrosion. It is used in chromium electroplating, to make stainless steel and other alloys, and as a catalyst. Its compounds are used for tanning leather and for ◊alums. In human nutrition it is a vital trace element. In nature, it occurs chiefly as a chrome–iron ore ($FeCr_2O_4$).

chromosome a structure in a cell nucleus that carries the ◊genes. Each chromosome consists of one very long strand of DNA, coiled and folded to produce a compact chromosome. The point on a chromosome where a particular gene occurs is known as the gene's locus. Most higher organisms have two copies of each chromosome (they are ◊diploid) but some have only one (they are ◊haploid). See also ◊mitosis and ◊meiosis.

cilia (singular ***cilium***) small threadlike organs on the surface of some cells, composed of contractile fibres that produce rhythmic waving movements. Some single-celled organisms move by means of cilia. In multicellular animals, they keep lubricated surfaces clear of debris. They also move food in the digestive tracts of some invertebrates.

ciliary muscle ring of muscle surrounding and controlling the lens inside the vertebrate eye, used in ◊accommodation (focusing). Suspensory ligaments, resembling the spokes of a wheel, connect the lens to the ciliary muscle and pull the lens into a flatter shape when the muscle relaxes. On contraction, the lens returns to its normal spherical state.

circuit an arrangement of electrical components through which a current can flow. There are two basic types, ◊series circuits and ◊parallel circuits. In a series circuit, the components are connected end-to-end so that the current flows through all components one after the other. In a parallel circuit, components are connected side-by-side so that part of the current passes through each component.

circuit breaker switching device designed to protect an electric circuit from excessive current. It has the same action as a ◊fuse, and many houses now have a circuit breaker between the incoming mains supply and the domestic circuits. Circuit breakers usually work by means of ◊solenoids. Those at electricity-generating stations have to be specially designed to prevent dangerous arcing (the release of luminous discharge) when the high-voltage supply is switched off.

chromosome

the 23 pairs of chromosomes of a normal human male (XY)

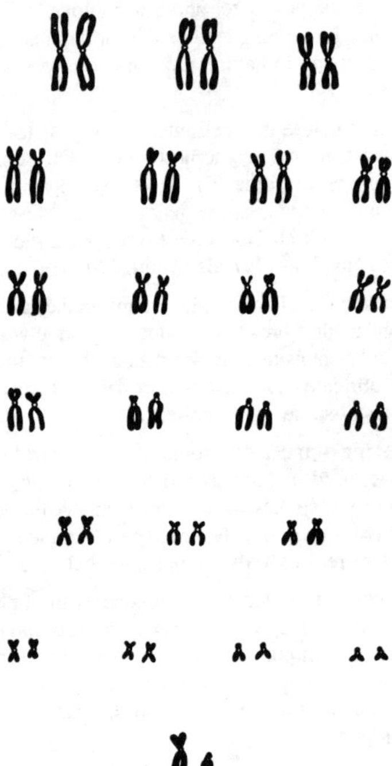

XY

circuit diagram simplified drawing of an electric circuit. The circuit's components are represented by internationally recognized symbols, and the connecting wires by straight lines. A dot indicates where wires join.

circuit symbols internationally agreed symbols for representing the components of an electric circuit.

circulatory system in animals, the system of vessels that transports the blood, and dissolved constituents such as food and oxygen, to and from the different parts of the body. A pump—the heart—keeps the fluid moving within the vessels, although in some animals, for instance the insects, the blood is not necessarily contained within vessels. Circulatory systems evolved as animals got larger. Very small animals can rely on diffusion to bring oxygen and food to the cells, but the process is too slow if cells are deep within the interior of the body.

The main components of the mammalian circulatory system are the ◊blood vessels and the pumping ◊heart. The blood travels in one direction only. Valves in the heart and veins prevent backflow, and the muscular walls of the arteries assist in pushing the blood around the body. In mammals, the blood passes to the lungs and back to the heart before circulating around the remainder of the body (***double circulation***).

citric acid organic acid widely distributed in the plant kingdom, found in high concentrations in citrus fruits, with a sharp, sour taste. At one time it was commercially prepared from concentrated lemon juice, but now the main source is the fermentation of sugar with certain moulds.

climate the average weather conditions of a particular place over a long period of time, usually 30 years. Several different climate zones may be identified. The main factors determining the variations of climate over the surface of the Earth are: (1) the effect of latitude and the tilt of the Earth's axis; (2) the large-scale movement of wind and ocean currents; and (3) the temperature difference between land and sea. Recent research indicates that human activity may influence world climate (see ◊greenhouse effect).

clone group of cells or organisms arising by ◊asexual reproduction from a single 'parent' individual. Clones therefore share exactly the same genetic make-up.

Examples include a group of plants produced from cuttings from a single plant.

coal black or blackish mineral substance of fossil origin, formed over millions of years by the compression of dead plants. It is used as a fuel and in the chemical industry.

coal tar black oily material resulting from the destructive distillation of bituminous coal. Further distillation of coal tar yields a number of fractions: light oil, middle oil, heavy oil, and anthracene oil; the residue is called pitch. On further fractionation a large number of substances are obtained, about 200 of which have been isolated. They are used as dyes and in medicines.

cobalt hard, lustrous, grey, metallic element, symbol Co, atomic number 27, relative atomic mass 58.933. It is found in various ores and occasionally as a free metal, sometimes in metallic meteorite fragments. It is used in the preparation of magnetic, wear-resistant, and high-strength alloys; its compounds are used in inks, paints, and varnishes.

cobalt chloride $CoCl_2$ compound that exists in two forms: the hydrated salt ($CoCl_2.6H_2O$), which is pink, and the anhydrous salt, which is blue. The anhydrous form is used as a chemical test for water, as it forms the hydrated salt and turns pink when water is present. When the hydrated salt is gently heated the anhydrous salt is reformed.

$$CoCl_2 + 6H_2O \leftrightarrow CoCl_26H_2O$$

cochlea the coiled inner section of the ear. It contains thousands of sense cells, detecting sound vibrations and sending electrical signals to the brain via the auditory nerve. The cochlea is therefore responsible for hearing, but can only operate reliably if the outer and middle sections of the ear are in good working order.

coil another name for the contraceptive ◊intrauterine device.

coke clean, light fuel produced by the carbonization of certain types of coal. When this coal is strongly heated in airtight ovens (in order to release all volatile constituents), the brittle, silver-grey remains are coke. Coke comprises 90% carbon together with very small quantities of water, hydrogen, and oxygen, and makes a useful industrial and domestic fuel.

collagen in vertebrates, a strong, rubbery ◊protein important in giving structural support. It is found in skin, and some of the characteristic signs of ageing, for instance wrinkles, are caused by collagen losing some of its elasticity over time. Collagen is an important constituent of bones, with the mineral calcium providing rigidity.

collision theory theory that explains chemical reactions and the way in which the rate of reaction alters when the conditions alter. For a reaction to occur the reactant particles must collide. Only a certain fraction of the total collisions cause chemical change; these are called ***fruitful collisions***. These fruitful collisions have sufficient energy (activation energy) at the moment

collision theory

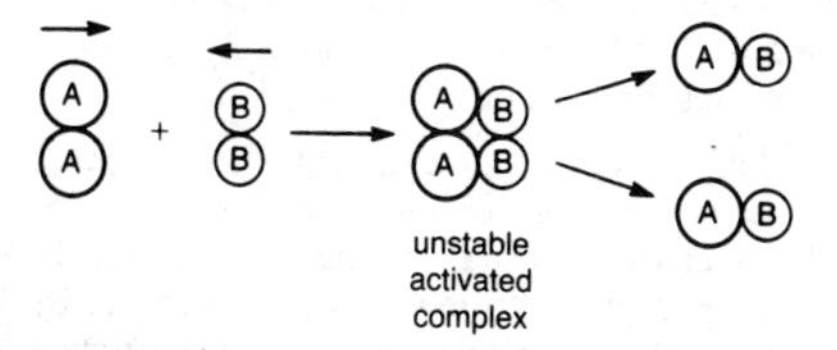

of impact to break the existing bonds and form new bonds, resulting in the products of the reaction. Increasing the concentration of the reactants and raising the temperature bring about more collisions and therefore more fruitful collisions, increasing the rate of reaction.

When a ◊catalyst undergoes collision with the reactant molecules, less energy is required at the moment of impact for the chemical change to occur, and therefore more collisions have sufficient energy for reaction to occur. The reaction rate therefore increases.

colloid substance composed of extremely small particles of one material (the dispersed phase) evenly and stably distributed in another material (the continuous phase). The size of the dispersed particles (1–1,000 nanometres across) is less than that of particles in suspension but greater than that of molecules in true solution.

Colloids involving gases include ***aerosols*** (dispersions of liquid or solid particles in a gas, as in fog or smoke) and ***foams*** (dispersions of gases in liquids). Those involving liquids include ***emulsions*** (in which both the dispersed and the continuous phases are liquids) and ***sols*** (solid particles dispersed in a liquid). Sols in which both phases contribute to a molecular three-dimensional network have a jellylike form and are known as ***gels***; gelatin, starch 'solution', and silica gel are common examples.

Milk is a natural emulsion of liquid fat in a watery liquid; synthetic emulsions such as some paints and cosmetic lotions have chemical emulsifying agents to stabilize the colloid and stop the two phases from separating out.

colon the part of the large intestine between the caecum and rectum, where water and mineral salts are absorbed from digested food, and the residue formed into faeces or faecal pellets for egestion.

colonization the spreading of a species into a new habitat, such as a freshly cleared field, a new motorway verge, or a recently flooded valley. The first species to move in are called ***pioneers***, and may establish conditions that allow other animals and plants to move in (for example, by improving the condition of the soil or by providing shade). Over time a range of species arrives and the habitat matures; early colonizers will probably be replaced, so that the variety of animal and plant life present changes. This is known as ◊succession.

Ecologists can judge the history of a habitat by looking at the organisms present, as certain species of plant and animal are associated with particular stages of colonization and succession.

colour quality or wavelength of light emitted or reflected from an object. Visible white light consists of electromagnetic radiation of various wavelengths, and if a beam is refracted through a prism, it can be spread out into a spectrum, in which the various colours correspond to different wavelengths. From long to short wavelengths (from about 700 to 400 nanometres) the colours are red, orange, yellow, green, blue, indigo, and violet.

When a surface is illuminated, some parts of the white light are absorbed, depending on the molecular structure of the material and the dyes applied to it. A surface that looks red absorbs light from the blue end of the spectrum, but reflects light from the red, long-wave end. Colours vary in brightness, hue, and saturation (the extent to which they are mixed with white).

colourings food ◊additives used to alter or improve the colour of processed foods. They include artificial colours, such as tartrazine and amaranth, which are made from petrochemicals, and 'natural' colours such as chlorophyll, caramel, and carotene. Some of the natural colours are actually synthetic copies of the naturally occurring substances, and some of these, notably the synthetically produced caramels, may be injurious to health.

combustion burning, defined in chemical terms as rapid combination of a substance with oxygen accompanied by the evolution of heat and usually light. A slow-burning candle flame and the explosion of a mixture of petrol vapour and air are examples of combustion.

comet a small icy body, consisting mostly of ice mixed with dust, following a fixed orbit around the ◊Sun. Although they are a common feature of the ◊Solar System, they are rarely seen without a telescope. As they approach the Sun, comets heat up and water vapour streams off to give the characteristic—and remarkable—tail. The most famous of all comets is Halley's Comet; it is extremely bright and reappears without fail every 76 years.

community in ecology, an assemblage of plants, animals, and other organisms living within a circumscribed area. Communities are usually named by reference to a dominant feature such as characteristic plant species (for example, beech wood community), or a prominent physical feature (for example, a freshwater-pond community).

compact disc disc for digital information storage, about 12cm across, mainly used for music, when it can have over an hour's playing time on one side. The compact disc is made of aluminium with a transparent plastic coating; the metal disc underneath is etched by a ◊laser beam with microscopic pits that carry a digital code representing the sounds. During playback, a laser beam reads the code and produces signals that are changed into near-exact replicas of the original sounds.

compass any instrument for finding direction. The most commonly used is the ◊magnetic compass.

competition the struggle by two or more organisms for resources that are in short supply. For instance, plants are commonly said to 'compete' for light, or for minerals from the soil. Similarly, some courtship behaviour can be understood as females and males competing for a mate. However, biologists often argue that direct competition between individuals of different species is rare in nature, because continuous competitive struggle would be extremely expensive in time and energy. Instead, it is believed, evolution has tended to produce adaptations and behaviours (often called 'strategies') which make it less likely that an organism will compete with, or fight, another.

compound chemical substance made up of two or more ◊elements bonded together, so that they cannot be separated by physical means. A compound contains a characteristic, fixed proportion of each element present. Compounds are held together by ◊ionic or ◊covalent bonds.

computer a programmable electronic device that processes data, and performs calculations and other manipulation tasks. Computers are based on ◊integrated circuits.

concave lens lens that possesses at least one surface that curves inwards. It is a ◊diverging lens, spreading out those light rays that have been refracted through it. A concave lens is thinner at its centre than at its edges.

Common forms include the ***biconcave*** lens (with both surfaces curved inwards) and the ***plano-concave*** (with one flat surface and one concave). The whole lens may be further curved overall, making a ***convexo-concave*** or diverging meniscus lens, as in some lenses used for corrective purposes.

concave mirror curved or spherical mirror that reflects light from its inner surface. It may be either circular or parabolic in section. A concave mirror converges light rays to form a reduced, inverted, real image in front, or an enlarged, upright, virtual image seemingly behind it, depending on how close the object is to the mirror.

Only a parabolic concave mirror has a true, single-point ◊principal focus for parallel rays. For this reason, parabolic mirrors are used as reflectors to focus light in telescopes, or to focus microwaves in satellite communication systems. The reflector behind a spot lamp or car headlamp is parabolic.

concave mirror

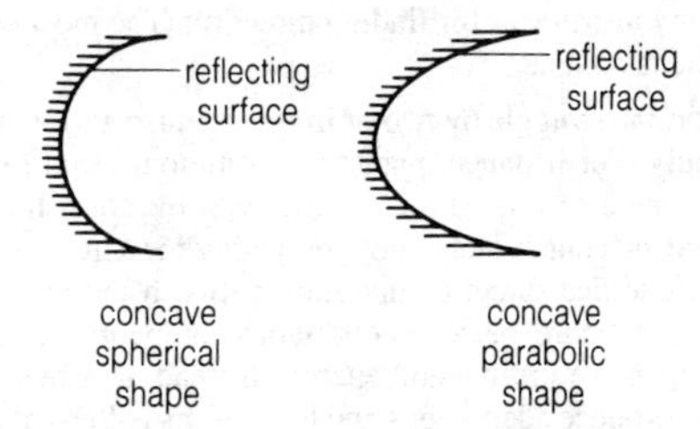

concentration the amount of a substance (◊solute) present in a specified amount of a solution. Either amount may be specified as a mass or a volume (liquids only). Common units used are ◊moles per litre, grams per litre, grams per 100 cubic centimetres, or grams per 100 grams.

concentration gradient change in the concentration of a substance from one area to another. See ◊diffusion.

condensation the conversion of a vapour to a liquid as it loses heat. This is frequently achieved by letting the vapour come into contact with a cold surface. It is an essential step in ◊distillation processes.

condensation polymerization ◊polymerization reaction in which one or more monomers, with more than one reactive functional group, combine to form a polymer with the elimination of water or another small molecule. Polyamides (such as nylon) and polyesters (such as Terylene) are made by condensation polymerization.

condensation reaction or ***addition–elimination reaction*** reaction in which two organic compounds combine to form a larger molecule, accompanied by the removal of a smaller molecule (usually water).

condom or ***sheath*** barrier contraceptive, made of rubber, which fits over an erect penis and holds in the sperm produced by ejaculation. It is an effective means of preventing pregnancy if used carefully, preferably with a ◊spermicide. A condom with spermicide is 97% effective; one without spermicide is 85% effective. Condoms can also give protection against sexually transmitted diseases (STDs).

conduction, electrical the flow of charged particles through a material, forming an electric current. Conduction in metals involves the flow of negatively charged free ◊electrons. Conduction in gases and some liquids involves the flow of ◊ions that carry positive charge in one direction and negative charges in the other. Conduction in a semiconductor such as silicon involves the flow of electrons and positive holes.

conduction, heat flow of heat energy through a material without the movement of any part of the material itself (compare ◊conduction, electric). Heat energy is present in all materials in the form of the energy of their vibrational molecules, and may be conducted from one molecule to the next in the form of this mechanical vibration. In metals, which are particularly good conductors of heat, the free electrons within the material carry heat around very quickly.

conductivity measure of how well a material conducts heat or electricity. A good conductor, such as a metal, has a high conductivity; a poor conductor, called an insulator, has a low conductivity.

conductor material that conducts heat or electricity (as opposed to a nonconductor or insulator). A good conductor has a high electrical or heat conductivity, and is generally a substance rich in mobile electrons such as a metal. A poor conductor (such as the non-metals glass and porcelain) has few mobile electrons. Carbon is exceptional in being non-metallic and yet (in some of its forms) a relatively good conductor of heat and electricity. Substances such as silicon and germanium, with intermediate conductivities that are improved by heat, light, or voltage, are known as ◊semiconductors.

cone cell type of cell found in the ◊retina of the eye that plays a part in colour vision.

conservation of energy principle stating that in a chemical reaction the total amount of energy in the system remains unchanged. For each component there may be changes in energy due to change of physical state, changes in the nature of chemical bonds, and either an input or output of energy. However, there is no net gain or loss of energy.

conservation of mass principle stating that in a chemical reaction the sum of all the masses of the substances involved in the reaction (reactants) is equal to the sum of all of the masses of the substances produced by the reaction (products)—that is, no matter is gained or lost.

consumer any organism that obtains its food by consuming organic material. All animals are consumers, as are some plants, such as the saprophytic fungi, which live off decaying matter.

Herbivores are ***primary consumers***, relying on vegetation. Carnivorous animals are ***secondary consumers***, animals that eat herbivores. ***Tertiary consumers*** eat other carnivores. Secondary and tertiary consumers are less numerous than either the producers or the primary consumers because only about 10% of the energy in any one ◊trophic level can pass into the next level up. See also ◊food chain.

contact process the main industrial method of manufacturing the chemical ◊sulphuric acid. Sulphur dioxide and air are passed over a hot (450°C) catalyst of vanadium(V) oxide. The sulphur trioxide produced is then absorbed in concentrated sulphuric acid to make fuming sulphuric acid (oleum). Unreacted gases are recycled. The oleum is diluted with water to give concentrated sulphuric acid (98%).

$$2SO_2 + O_2 \leftrightarrow 2SO_3$$

$$H_2SO_4 + SO_3 \leftrightarrow H_2S_2O_7 \quad \text{(oleum)}$$

$$H_2S_2O_7 + H_2O \rightarrow 2H_2SO_4$$

contraceptive any drug, device, or technique that prevents pregnancy. The contraceptive ◊Pill contains female hormones that interfere with egg production or the first stage of pregnancy. Barrier contraceptives include ◊condoms (sheaths), which fit over the penis, and ◊diaphragms (caps), which fit over the cervix (neck of the womb); they prevent the sperm from entering the cervix, and are made more effective by the use of spermicidal cream. Intrauterine devices (IUDs or coils) cause a slight inflammation of the lining of the womb, which prevents the fertilized egg from becoming implanted. Other contraceptive methods include sterilization (women) and vasectomy (men); these are usually irreversible.

'Natural' methods include withdrawal of the penis before ejaculation (*coitus interruptus*), and avoidance of intercourse at the time of ovulation (◊rhythm method). These methods are unreliable and normally only used on religious grounds.

convection current movement of heat energy through a liquid or gas that involves the flow of the medium itself (unlike heat conduction; see ◊conduction, heat). Convection is caused by the expansion of the medium as its temperature rises; the expanded material, being less dense, rises above colder and therefore denser material. In some heating systems, convection currents are used to carry hot water upwards in pipes. Such currents arise in the atmosphere above hot land masses or warm seas causing breezes.

conventional current direction in which an electric current is considered to flow in a circuit. By convention, the direction is that in which positive-charge carriers would flow—from the positive terminal of a cell to its negative terminal. In circuit diagrams, the arrows shown on symbols for components such as diodes and transistors point in the direction of conventional current flow.

converging lens ◊convex lens that converges or brings to a focus those light rays that have been refracted by it. Converging lenses are used to form real images in many ◊optical instruments, such as cameras and projectors. A converging lens that forms a virtual, magnified image may be used as a ◊magnifying glass or to correct ◊long-sightedness.

convex lens lens that possesses at least one surface that curves outwards. It is a ◊converging lens, bringing rays of light to a focus. A convex lens is thicker at its centre than at its edges.

convection current

in water

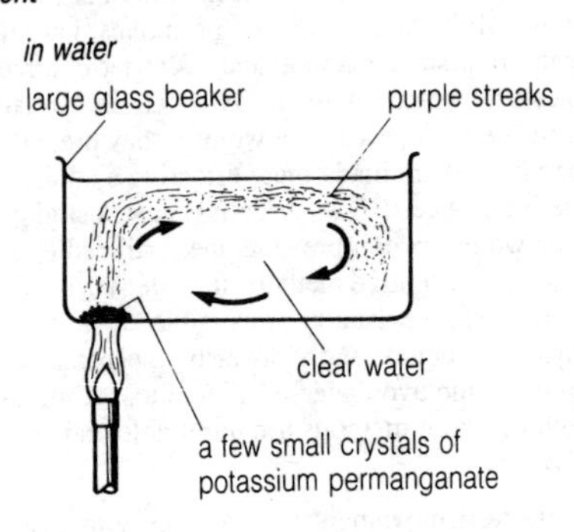

in air

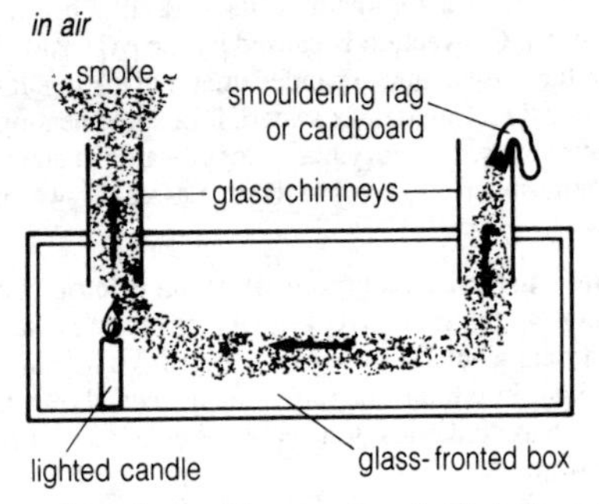

Common forms include the ***biconvex*** lens (with both surfaces curved outwards) and the ***plano-convex*** (with one flat surface and one convex). The whole lens may be further curved overall, making a ***concavo-convex*** or converging meniscus lens, as in some lenses used in corrective eyewear.

convex mirror curved or spherical mirror that reflects light from its outer surface. It diverges reflected light rays to form a reduced, upright, virtual image. Convex mirrors give a wide field of view and are therefore particularly suitable for surveillance purposes in shops.

coordinate bond or ***dative bond*** form of covalent ◊bond in which both electrons are donated by the same atom.

copper orange-pink, very malleable and ductile, metallic element, symbol Cu, atomic number 29, relative atomic mass 63.546. It is used for its toughness, softness, pliability, high thermal and electrical conductivity, and resistance to corrosion.

copper(II) oxide CuO black solid that is readily reduced to copper by carbon, carbon monoxide, or hydrogen if heated with any of these.

$$CuO + C \rightarrow Cu + CO$$

$$CuO + CO \rightarrow Cu + CO_2$$

$$CuO + H_2 \rightarrow Cu + H_2O$$

It is usually made in the laboratory by heating copper(II) carbonate, nitrate, or hydroxide.

$$2Cu(NO_3)_2 \rightarrow 2CuO + 4NO_2 + O_2$$

Copper(II) oxide is a typical basic oxide, dissolving readily in most dilute acids.

copper(II) sulphate $CuSO_4$ substance usually found as a blue, crystalline, hydrated salt $CuSO_4.5H_2O$ (also called blue vitriol). It is made from the action of dilute sulphuric acid on copper(II) oxide, hydroxide, or carbonate.

$$CuO + H_2SO_4 + 4H_2O \rightarrow CuSO_4.5H_2O$$

When the hydrated salt is heated gently it loses its water of crystallization and the blue crystals turn to a white powder. The reverse reaction is used as a chemical test for water.

$$CuSO_4.5H_2O \leftrightarrow CuSO_4 + 5H_2O$$

cornea transparent front section of the vertebrate ◊eye, the part normally visible. It is curved and transparent, focusing light as it passes into the interior of the eye, but able only to give a crude image. Fine focusing (see ◊accommodation), which provides a more useful, sharper image, is controlled by the lens just behind the cornea. This explains why people who do not have a lens (for instance as a result of a cataract operation) still have some vision.

corrosion the slow reaction of metals and alloys with chemicals, and their eventual destruction by them. The ***rusting*** of iron and steel is the commonest form of corrosion. Rusting takes place in moist air: the iron combines with oxygen and water to form a brown–orange deposit of rust (hydrated iron(III) oxide). The rate of corrosion is increased when the atmosphere is polluted with sulphur dioxide, and can be decreased by various methods of ◊rust prevention.

Corrosion is largely an electrochemical process. Acidic and salty conditions favour the establishment of chemical ◊cells on the metal, which cause it to be eaten away.

coulomb SI unit (symbol C) of electrical charge. One coulomb is the quantity of electricity conveyed by a current of one ampere in one second.

covalent bond chemical ◊bond produced when two atoms share one or more pairs of electrons (usually each atom contributes an electron). The bond is often represented by a single line drawn between the two atoms. This type of bonding always produces a ◊molecule. Covalently bonded substances include hydrogen (H_2), water (H_2O), and most organic substances.

Double bonds, seen, for example, in the ◊alkenes, are formed when two atoms share two pairs of electrons (the atoms usually contribute a pair each); triple bonds, seen in the ◊alkynes, are formed when atoms share three pairs of electrons. Such bonds are represented by a double or triple line, respectively, between the atoms concerned.

Covalent compounds have the following general properties: they have low melting and boiling points; never conduct electricity; and are usually insoluble in water and soluble in organic solvents. Compare ◊ionic compound.

cracking reaction where a large ◊alkane molecule is broken down by heat into a smaller alkane and a small ◊alkene molecule. The reaction is carried out at a high temperature (600°C or higher) and often in the presence of a catalyst.

$$C_{12}H_{26} \rightarrow C_8H_{18} + 2C_2H_4$$

Cracking is a commonly used process in the petrochemical industry. It is the main method of preparation of alkenes and is also used to manufacture petrol from the higher-boiling-point fractions obtained from the ◊fractionation of petroleum (crude oil).

cranium the dome-shaped area of the skull, consisting of several fused plates, that protects the brain.

covalent bond

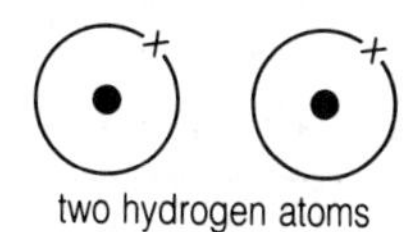
two hydrogen atoms

or H $\overset{\times}{\times}$ H, H–H
a molecule of hydrogen
sharing an electron pair

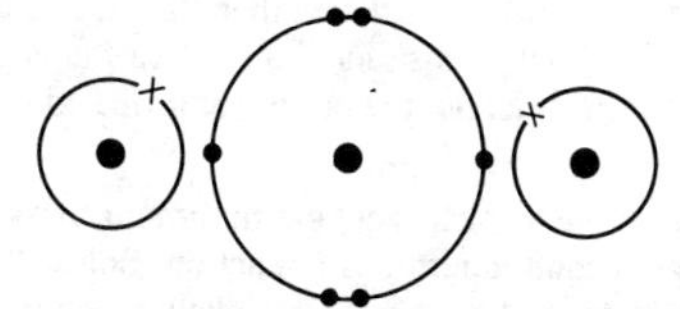
two hydrogen atoms and one
oxygen atom

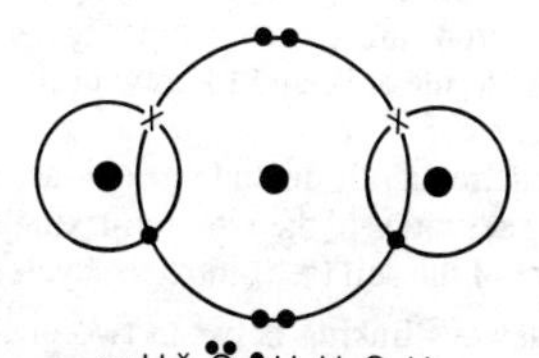
or H $\overset{\times}{\bullet}$ $\ddot{\underset{\bullet\bullet}{O}}$ $\overset{\bullet}{\times}$ H, H–O–H
a molecule of water
showing the two covalent bonds

critical angle

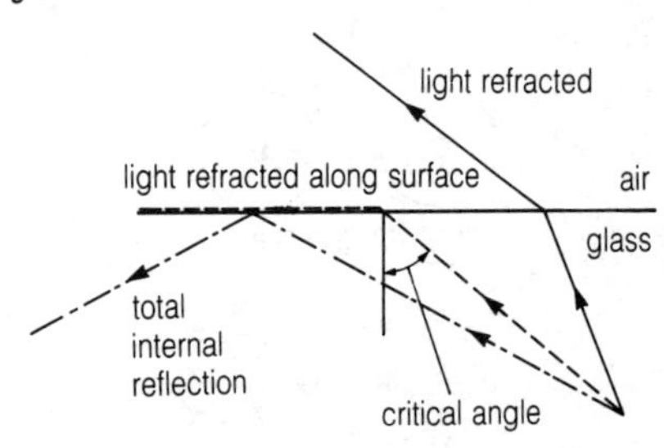

critical angle in optics, for a ray of light passing from a denser to a less dense medium (such as from glass to air), the smallest angle of incidence at which the emergent ray grazes the surface of the denser medium—at an angle of refraction of 90°.

When the angle of incidence is less than the critical angle, the ray does not pass out into the less dense medium; when the angle of incidence is greater than the critical angle, the ray is not reflected back into the denser medium.

critical mass in nuclear physics, the minimum mass of fissile material that can undergo a continuous ◊chain reaction. Below this mass, too many ◊neutrons escape from the surface for a chain reaction to carry on; above the critical mass, the reaction may accelerate into a nuclear explosion.

CRO abbreviation for ◊cathode-ray oscilloscope.

crop rotation the system of regularly changing the types of crop grown on a piece of land. The crops are grown in a particular order so as to use and replace the nutrients in the soil, and to prevent the build-up of insect and fungal pests.

The crops rotated frequently include a legume, such as a bean or pea species, because the activities of the nitrogen-fixing bacteria in its roots add to the nitrate content of the soil (see ◊nitrogen cycle).

cross linking sideways linking between two or more long-chain molecules in a ◊polymer. Cross linking gives the polymer a higher melting point and makes it harder; examples of cross-linked polymers include Bakelite (the first synthetic plastic) and vulcanized rubber.

CRT abbreviation for ◊cathode-ray tube.

crumple zone region at the front and rear of a motor vehicle that is designed to crumple gradually during a collision, so reducing the risk of serious injury to passengers. The progressive crumpling absorbs the kinetic energy of the vehicle more gradually than would a rigid structure, thereby diminishing the forces of deceleration acting on the vehicle and on the people inside.

The crumple zone's effect is based on the principle that the ◊impulse required to stop a vehicle and reduce its momentum to zero is equal to the product of the decelerating force and the time over which that force acts. It follows that if the length of time is increased, the force will be reduced.

crystal substance with an orderly three-dimensional arrangement of its atoms, ions, or molecules, thereby creating an external surface of clearly

crystal

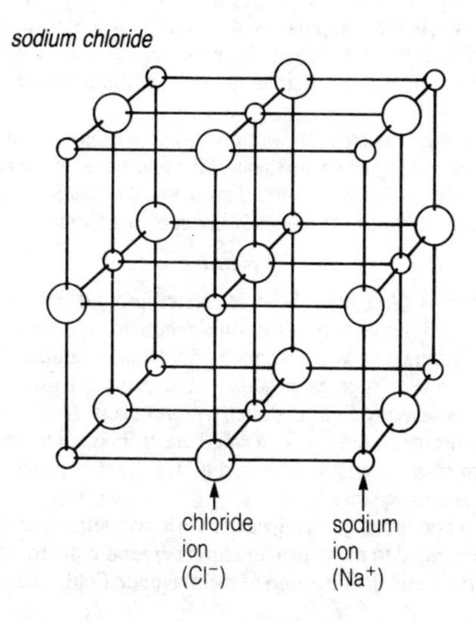

defined smooth faces having characteristic angles between them. Examples are common salt and quartz.

Each geometrical figure or form, many of which may be combined in one crystal, consists of two or more faces—for example, dome, prism, and pyramid.

A mineral can often be identified by the shape of its crystals and the system of crystallization determined. A single crystal can vary in size from a submicroscopic particle to a mass some 30m/100ft in length.

current, electric the flow of electrically charged particles through a conducting circuit due to the presence of a ◊potential difference. The current at any point in a circuit is the amount of charge flowing per second; it is measured in amperes (coulombs per second). If the amount of charge is Q coulombs and the time over which charge flow is measured is t seconds then the current I (in amperes) is given by the formula:

$$I = Q/t$$

Current carries electrical energy from a power supply, such as a battery of electrical cells, to the components of the circuit where it is converted into other forms of energy, such as heat, light, or motion. It may be either direct (DC, see ◊direct current) or alternating (AC, see ◊alternating current). See also ◊Ohm's law.

heating effect When current flows in a component possessing resistance, electrical energy is converted into heat. If the resistance of the component is R ohms and the current through it is I amperes, then the heat energy W (in joules) generated in a time t seconds is given by the formula:

$$W = I^2Rt$$

magnetic effect A ◊magnetic field is created around all conductors that carry a current. When a current-bearing conductor is made into a coil it forms an ◊electromagnet with a magnetic field that is similar to that of a bar magnet, but which disappears as soon as the current is switched off. The strength of the magnetic field is directly proportional to the current in the conductor—a property that allows a small electromagnet to be used to produce a pattern of magnetism on recording tape that accurately represents the sound or data to be stored.

motor effect A conductor carrying current in a magnetic field experiences a force, and is impelled to move in a direction perpendicular to both the direction of the current and the direction of the magnetic field. The magnitude of the force experienced depends on the length of the conductor and on the strengths of the current and the magnetic field, and is greatest when the conductor is at right angles to the field. A conductor wound into a coil that can rotate between the poles of a magnet forms the basis of an ◊electric motor.

cutting technique used in the propagation of plants, in which a piece of stem, usually with leaves attached, is removed from a plant and grown separately.

cyclic compound any of a group of organic chemicals that have rings of atoms in their molecules, giving them a closed-chain structure. They may be alicyclic (for example, cyclopentane), aromatic (for example, benzene), or heterocylic (for example, pyridine).

cytoplasm the part of the cell outside the ◊nucleus but inside the cell membrane. It therefore includes the main structures of the cell; under high magnification it has a granular, complicated appearance

D

data facts, figures, and symbols, especially those stored in computers. The term is often used to mean raw, unprocessed facts, as distinct from information to which a meaning or interpretation has been applied.

DC abbreviation for ◊direct current.

death the ending of all life functions, so that the molecules and structures associated with living things become disorganized and indistinguishable from similar molecules found in non-living things. Living organisms expend large amounts of energy preventing their complex molecules from breaking up; cellular repair and replacement are vital processes in multicellular organisms. At death this energy is no longer available, and the processes of disorganization become inevitable.

decay, radioactive see ◊radioactive decay.

deciduous describing trees and shrubs that shed their leaves before the onset of winter or a dry season. In temperate regions there is little water available during winter, and leaf fall is an adaptation to reduce ◊transpiration. Examples of deciduous trees are oak and beech.

decomposer any organism that breaks down dead matter. Decomposers play a vital role in the ◊ecosystem by freeing important chemical substances, such as nitrogen compounds, locked up in dead organisms or excrement. They feed on some of the released organic matter, but leave the rest to filter back into the soil or pass in gas form into the atmosphere. The principal decomposers are bacteria and fungi, but earthworms and many other invertebrates are often included in this group. The ◊nitrogen cycle relies on the actions of decomposers.

deep freezing method of preserving food by rapid freezing and storage at –18°C. See ◊food technology.

deforestation the destruction of forest for timber or for agriculture, without planting new trees to replace those lost (reafforestation). Deforestation allows fertile soil to be blown away or washed into rivers, because tree roots, which normally bind soil particles together, decay once the tree has been cut. This leads to ◊soil erosion, drought, and flooding. The most spectacular example of deforestation is the burning of the Amazon rainforests. Deforestation in the Himalayas is held responsible for several recent floods in Bangladesh.

dehydration removal of water from a substance to give a product with a new chemical formula; it is not the same as ◊drying.

There are two types of dehydration. For substances such as hydrated copper sulphate ($CuSO_4.5H_2O$) that contain ◊water of crystallization, dehydration means removing this water to leave the anhydrous substance. This may be achieved by heating, and is reversible.

Some substances, such as ethanol, contain the elements of water (hydrogen and oxygen) joined in a different form. ***Dehydrating agents*** such as concentrated sulphuric acid will remove these elements in the ratio 2:1.

$$C_2H_5OH \rightarrow CH_2 = CH_2 + H_2O$$

deliquescence the phenomenon of a substance absorbing so much moisture from the air that it ultimately dissolves in it to form a solution. Deliquescent substances make very good drying agents in the bottoms of ◊desiccators. Calcium chloride ($CaCl_2$) is one of the commonest.

denaturation irreversible changes occurring in the structure of proteins such as enzymes, usually caused by changes in pH or temperature. An example is the heating of egg albumen resulting in solid egg white.

The enzymes associated with digestion and metabolism become inactive under abnormal conditions. Heat damages their complex structure so that the usual interactions between enzyme and substrate can no longer occur.

denitrification a process occurring naturally in soil by which bacteria break down ◊nitrates to give nitrogen gas, which returns to the atmosphere.

density measure of the compactness of a substance; it is equal to its mass per unit volume, and is measured in kilograms per cubic metre. The density D of a mass m kg occupying a volume $V\text{m}^3$ is given by the formula:

$$D = m/V$$

◊Relative density is the ratio of the density of a substance to that of water at 4°C.

dental formula a way of describing the pattern of an animal's teeth. The dental formula consists of eight numbers separated by a line into two rows.

The four above the line represent the teeth in one side of the upper jaw, starting at the front. If this reads 2 1 2 3 (as for humans) it means two incisors, one canine, two premolars, and three molars (see ◊tooth). The numbers below the line represent the lower jaw. The total number of teeth can be calculated by adding up all the numbers and multiplying by two.

dentition the type and number of teeth in a species. Different kinds of teeth have different functions, and a grass-eating animal will have well-developed molars for grinding its food, whereas a meat-eater will need large canines for catching and killing its prey. Less useful teeth may be

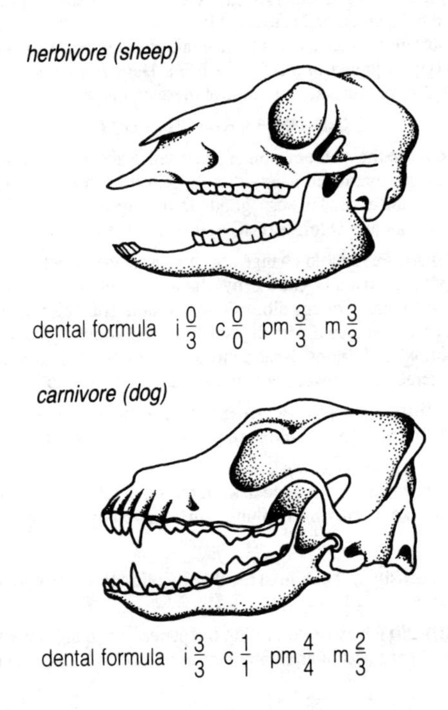

reduced in size or missing altogether. An animal's dentition is represented diagrammatically by a ◊dental formula.

deoxyribonucleic acid the full name of ◊DNA.

depression or *cyclone* region of low atmospheric pressure. Depressions form as warm air from the tropics spirals around cold polar air, producing cold and warm fronts. The warm air rises above the cold air to produce the area of low pressure. Depressions bring unstable weather with cloud and rain.

A deep depression is one in which the pressure at the centre is very much lower than that round about; it produces very strong winds, as opposed to a shallow depression in which the winds are comparatively light. Depressions tend to travel eastwards and can remain active for several days.

desiccator airtight vessel, traditionally made of glass, in which materials may be stored either to dry them or to prevent them, once dried, from reabsorbing moisture.

The base of the desiccator is a chamber in which is placed a substance with a strong affinity for water (such as calcium chloride or silica gel) that removes water vapour from the desiccator atmosphere and from substances placed in it.

destarching in photosynthesis experiments, the process by which all starch is removed from a plant so that the effects of increased light or carbon dioxide concentration can be studied. It is achieved by placing the plant in the dark for 24 hours. During this time the plant cannot photosynthesize and therefore uses up its reserves of food.

detergent surface-active cleansing agent. The common detergents are made from fats or certain hydrocarbons and sulphuric acid, and their long-chain molecules have a structure similar to that of soap molecules: a salt group at one end attached to a long hydrocarbon 'tail'. They have the advantage over soap in that they do not produce scum by forming insoluble salts with the calcium and magnesium ions present in hard water.

To remove dirt, which is generally attached to materials by oil or grease, the hydrocarbon 'tails' (soluble in oil or grease) penetrate the oil or grease drops, while the 'heads' (soluble in water but insoluble in grease) remain in the water and, being salts, become ionized. Consequently the oil drops become negatively charged and tend to repel one another; thus they remain in suspension and are washed away with the dirt.

When detergents escape the normal processing of sewage, they cause troublesome foam in rivers; phosphates in some detergents can also enrich the vegetation in rivers and lakes, causing ◊eutrophication.

deuterium naturally occurring heavy isotope of hydrogen, mass number 2 (one proton and one neutron), discovered by Harold Urey in 1932. In nature, about one in every 6,500 hydrogen atoms is deuterium. The symbol D is sometimes used for it. Combined with oxygen, it produces 'heavy water' (D_2O), used in the nuclear industry.

development the process whereby a living thing transforms itself from a single cell into a vastly complicated multicellular organism. Cells, tissues and organs must be formed in the correct place and at the correct time, with a particular sequence of events being followed. The control of this highly complicated process rests with the genetic code, but is only poorly understood.

diabetes the disease ***diabetes mellitus*** in which a disorder of the islets of Langerhans in the ◊pancreas prevents the body producing the hormone ◊insulin, so that sugars cannot be used properly. Treatment is by strict dietary control and oral or injected insulin. See ◊blood-glucose regulation.

diamond generally colourless, transparent mineral, the hard crystalline form of ◊carbon. It has a ◊giant molecular structure, with its carbon atoms linked by covalent bonds to form tetrahedra. It is regarded as a precious gemstone, and is the hardest natural substance known. Industrial diamonds are used for cutting, grinding, and polishing.

diaphragm or ***cap*** or ***Dutch cap*** barrier ◊contraceptive that is pushed into the vagina and fits over the cervix (neck of the uterus), preventing sperm from entering the uterus. For a cap to be effective, a ◊spermicide must be used and the diaphragm left in place for 6-8 hours after intercourse. This method is 97% effective if practised correctly.

diaphragm in mammals, a muscular sheet separating the thorax from the abdomen. Its rhythmical movements affect the size of the thorax and cause the pressure changes within the lungs that result in inhalation or exhalation.

diatomic molecule molecule composed of two identical atoms joined together, such as oxygen (O_2).

dibasic acid acid containing two replaceable hydrogen atoms, such as sulphuric acid (H_2SO_4). The acid can form two series of salts, the normal salt (sulphate, SO_4^{2-}) and the acid salt (hydrogensulphate HSO_4^-).

diesel oil the fuel oil used in diesel engines. Like petrol, it is a petroleum product. When used in vehicle engines, it is also known as ***derv*** (*d*iesel-*e*ngine *r*oad *v*ehicle).

diet the range of foods eaten by an animal. The basic components of a diet are a group of chemicals: proteins, carbohydrates, fats, vitamins, minerals, and water. Different animals require these substances in different proportions, but the necessity of finding and processing an appropriate diet is a very basic drive in animal evolution. For instance, all guts are adapted for digesting and absorbing food, but different guts have adapted to cope with particular diets.

The diet an animal needs may vary over its lifespan, according to whether it is growing, reproducing, highly active, or approaching death. An animal may need increased carbohydrate for additional energy, or increased minerals during periods of growth.

In humans, ◊roughage (dietary fibre) is an important part of the diet, even though it has no nutritional value; it works to keep the gut healthy.

diffraction the slight spreading of a light beam into a pattern of light and dark bands when it passes through a narrow slit or past the edge of an obstruction. The resulting patterns are known as interference phenomena. A ***diffraction grating*** is a device for separating a wave train such as a beam of incident light into its component frequencies (white light results in a spectrum).

diffusion the random movement of molecules from a region in which they are at a high concentration to a region in which they are at a low concentration until an equal concentration is achieved throughout. The change in concentration that gives rise to diffusion is called a ***concentration gradient***. For instance, if sugar is added to water the sugar molecules will diffuse along the concentration gradient until they become evenly distributed throughout.

In biological systems, diffusion plays an essential role in the transport, over short distances, of molecules such as nutrients and respiratory gases (carbon dioxide and oxygen). Organs such as the gut and the lungs have a large surface area and are constructed so as to make diffusion effective. See also ◊gas exchange.

digestion the process whereby food eaten by an animal is altered physically, and chemically by ◊enzymes. In animals with an alimentary canal, the effect of digestion is to produce molecules small enough to pass through

diffusion

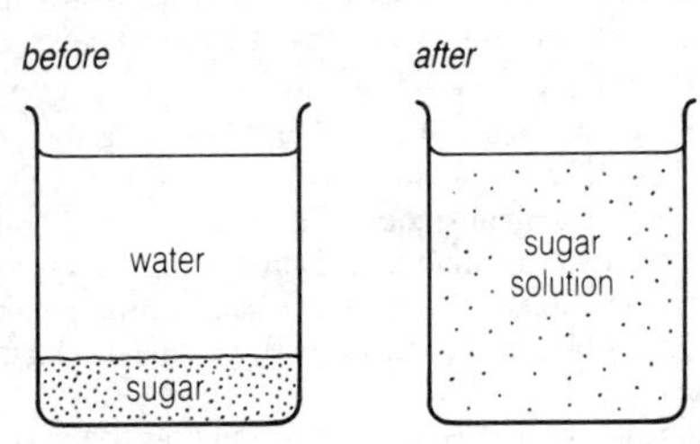

sugar and water molecules become evenly mixed

gas exchange in Amoeba

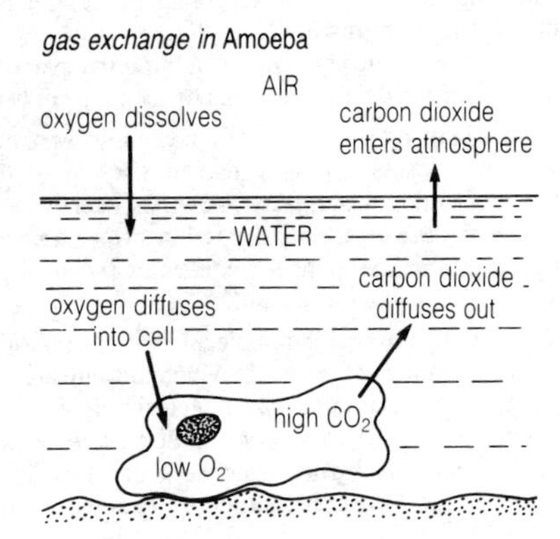

the gut wall and into the bloodstream. Starch is broken down into sugar molecules while protein is broken down into amino acids. Fats are digested into fatty acids. See ◊gut. In some single-celled organisms, such as *Amoeba*, a food particle is engulfed, and digested in a ◊vacuole within the cell.

digestive system the mouth, stomach, small and large intestines, and associated glands of animals, which are responsible for digesting food. The food is broken down by physical and chemical means in the stomach; digestion is completed, and most nutrients are absorbed in the small intestine;

what remains is stored and concentrated into faeces in the large intestine. See ◊gut.

digital (of a a quantity or device) changing in a series of distinct steps; by contrast, an ◊analogue quantity or device varies continuously. For example, a digital clock measures time with a numerical display that changes in a series of discrete steps, whereas an analogue clock measures time by means of a continuous movement of hands around a dial.

Computers are digital devices because their electronic circuits can distinguish between just two values, 0 and 1 (representing two states: on and off, or high-voltage and low-voltage pulses). All the data that a computer stores, processes, and transmits must therefore be encoded digitally, as a series of 0s and 1s, in binary number code.

dihydroxyethane or ***ethylene glycol*** or ***ethane-1,2-diol*** CH_2OHCH_2OH thick, colourless, odourless, sweetish liquid used in antifreeze solutions. It is also used in the preparation of ethers and esters (used for explosives), as a solvent, and as a substitute for glycerine.

dilution the process of reducing the ◊concentration of a solution by the addition of ◊solvent.

The extent of a dilution normally indicates the final volume of solution required. A fivefold dilution would mean the addition of sufficient solvent to make the final volume five times the original.

diode a cold anode and a heated cathode (or the semiconductor equivalent, which incorporates a p–n junction). Either device allows the passage of direct current in one direction only, and so is commonly used in a rectifier to convert alternating current (AC) to direct current (DC). See ◊rectification.

dioxyethane or ***diethyl ether*** $C_2H_5OC_2H_5$ colourless, volatile, inflammable liquid, slightly soluble in water and miscible with ethanol. It is prepared by treatment of ethanol with excess concentrated sulphuric acid at 140°C. Dioxyethane is used as an anaesthetic by vapour inhalation ('ether') and as an external cleansing agent before surgical operations. It is also used as a solvent, and in the extraction of oils, fats, waxes, resins, and alkaloids.

diploid having two sets of ◊chromosomes in each cell. In sexually reproducing species, one set is derived from each parent.

direct combination method of making a simple salt by heating its two constituent elements together. For example, iron(II) sulphide can be made by heating iron and sulphur, and aluminium chloride by passing chlorine over hot aluminium.

$$Fe + S \rightarrow FeS$$

direct current an electric current that flows in one direction, and does not reverse its flow as ◊alternating current does. The electricity produced by a battery is direct current.

disaccharide a ◊sugar made up of two monosaccharide units. Sucrose ($C_{12}H_{22}O_{11}$), or table sugar, is a disaccharide.

disinfectant agent that kills, or prevents the growth of, bacteria and other microorganisms. Chemical disinfectants include carbolic acid (phenol, used in surgery in the 1870s), ethanal, methanal, chlorine, and iodine.

dislocation an injury to a ◊joint, in which the two moving surfaces jump out of position and can no longer move normally. Most dislocations are easily corrected, but require time to heal as ligaments, tendons and muscles may be stretched or damaged.

dispersal the phase of reproduction during which gametes, eggs, seeds, or offspring move away from the parents into other areas. The result is that overcrowding may be avoided and parents will not find themselves in competition with their own offspring. The mechanisms are various, including passive dispersal by means of wind or water currents or animal carriers (vectors) and active dispersal by locomotion. The ability of a species to spread widely through an area and to colonize new habitats has survival value in evolution.

dispersion in optics, the splitting of white light into a spectrum; for example, when it passes through a prism or a diffraction grating. It occurs because the prism (or grating) bends each component wavelength to a slightly different extent. The natural dispersion of light through raindrops creates a rainbow.

displacement reaction reaction in which a less reactive element is replaced in a compound by a more reactive one. For example, the addition of powdered zinc to a solution of copper(II) sulphate displaces copper metal, which can be detected by its characteristic colour (see ◊reactivity series).

$$Zn_{(s)} + CuSO_{4(aq)} \rightarrow ZnSO_{4\,(aq)} + Cu_{(s)}$$

dissociation process whereby a single compound splits into two or more smaller products that can easily recombine to form the reactant. This can be achieved in two ways.

thermal dissociation Some compounds dissociate on heating.

$$NH_4Cl_{(g)} \leftrightarrow NH_{3(g)} + HCl_{(g)}$$

ionization Some compounds dissociate when dissolved in water to form ions.

$$CH_3COOH + aq \leftrightarrow H^+_{(aq)} + CH_3COO^-_{(aq)}$$

In the dissociation process, a covalent bond is broken. In some instances the two portions retain their bonding electron, so no ions are formed. This usually occurs when heating is used. In a solvent such as water, the covalent bond breaks but one product retains both electrons of the bond, so forming a negative ion. The other product is therefore a positive ion.

$$A–B \rightarrow A + B$$

$$A–B \rightarrow A+ + B^-$$

Where dissociation is incomplete, a ⇨chemical equilibrium exists between the chemical compound and its dissociation products. The extent of incomplete dissociation is defined by a numerical value (dissociation constant).

distance the extent of a journey or space between two points. It is a scalar quantity, since it possesses magnitude but not direction. The SI unit of distance is the metre.

distance–time graph

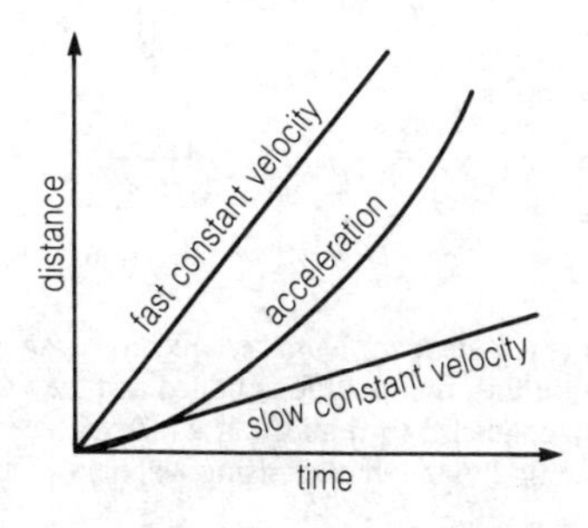

distance–time graph graph used to describe the motion of a body by illustrating the relation between the distance that it travels and the time taken. Plotting distance (on the vertical axis) against time (on the horizontal axis) produces a graph the gradient of which is the body's speed. If the gradient is constant (the graph is a straight line), the body has uniform or constant speed; if the gradient varies (the graph is curved), then so does the speed and the body may be said to be accelerating.

distillation technique used to separate and purify substances that are either in solution or in a mixture of liquids that have different boiling points. The liquid is boiled; the vapours are cooled and condensed in a separate piece of apparatus (a condenser); and the liquid produced (the distillate) is collected. This form of simple distillation is used in the recovery of solvents.

distillation

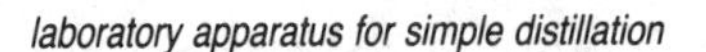

laboratory apparatus for simple distillation

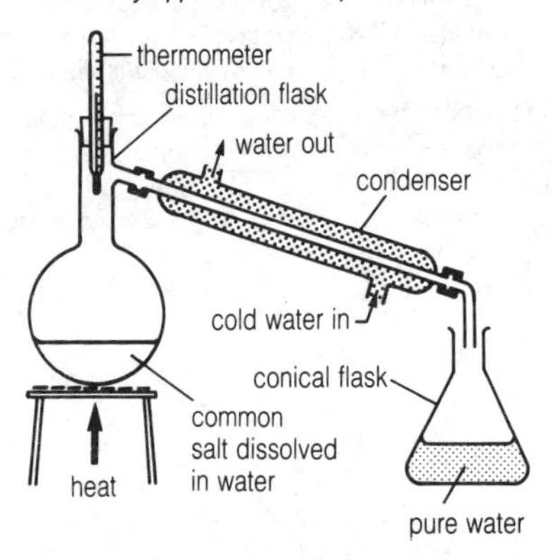

Mixtures of liquids (such as ◊petroleum or aqueous ethanol) require a ◊fractionating column. In this, the mixture is boiled and the vapours enter the column. Here they condense to liquid, but as they descend they are reheated to boiling point by the hotter rising vapours. This boiling–

condensing process occurs repeatedly inside the column. As the column is ascended, progressive enrichment by the lower-boiling-point components occurs; there is thus a temperature gradient inside the column. In crude-oil fractional distillation, groups of compounds of similar relative molecular masses and boiling points (the fractions) are tapped off from the column.

diverging lens lens that diverges or spreads out those light rays that have been refracted by it. It is a ◊concave lens, with one or both of its surfaces curving inwards. Such a lens is thinner at the centre than at the edge. Diverging lenses are used to correct ◊short-sightedness.

DNA (abbreviation for ***deoxyribonucleic acid***) a complex two-stranded molecule that contains, in chemically coded form, all the information needed to build, control, and maintain a living organism. DNA is a ladder-like double-stranded ◊nucleic acid that forms the basis of genetic inheritance in all organisms, except for a few viruses. It is organized into ◊chromosomes and is contained in the cell nucleus.

dominant in genetics (of an ◊allele), masking another. For example, if a ◊heterozygous person has one allele for blue eyes and one for brown eyes, their eye colour will be brown. The allele for blue eyes is described as ◊recessive and the allele for brown eyes as dominant.

Doppler effect change in observed frequency (or wavelength) of waves due to relative motion between wave source and observer. It is responsible for the perceived change in pitch of a siren as it approaches and then recedes, and for the ◊red shift of light from distant stars.

dormancy in plants, a phase of reduced activity shown by certain buds, seeds, and spores. Dormancy can help a plant to survive unfavourable conditions. Annual plants survive the cold winter season as dormant seeds.

double bond two covalent bonds between adjacent atoms, as in the ◊alkenes (–C=C–) and ◊ketones (–C=O–).

double decomposition reaction between two chemical substances (usually ◊salts in solution) that results in the exchange of a constituent from each compound to create two different compounds.

For example, if silver nitrate solution is added to a solution of sodium chloride, there is an exchange of ions yielding sodium nitrate and silver chloride.

$$AgNO_{3(aq)} + NaCl_{(aq)} \rightarrow NaNO_{3(aq)} + AgCl_{(s)}$$

downward displacement method of gas collection where the gas is less dense than air, so the air in an inverted gas jar is displaced downwards by the less dense gas.

drug any of a range of chemicals voluntarily or involuntarily introduced into the bodies of humans and animals. Most drugs in use are medicines and are seen as beneficial, at least if administered in the correct amount. Such drugs include antibiotics, stimulants, sedatives, and pain-relievers (analgesics). The most widely used drugs, tobacco and alcohol, operate on the nervous system and can be considered dangerous.

dry ice solid carbon dioxide (CO_2), used as a refrigerant. At temperatures above −79°C, it sublimes to gaseous carbon dioxide. Water vapour in the cooled air condenses to a dense mist, and the effect is used to generate mist in the theatre.

drying removal of liquid water from a substance without altering its chemical composition (unlike ◊dehydration); this is generally done by heating in an oven. Drying agents include deliquescent substances (see ◊deliquescence) such as calcium chloride, concentrated acids such as sulphuric and nitric acid, and silica gel.

ductile material material that can sustain large deformations beyond its elastic limit (see ◊elasticity) without fracture. Metals are very ductile, and may be pulled out into wires, or hammered or rolled into thin sheets

duodenum short length of gut (alimentary canal) found between the stomach and the ileum. Its role is in the digestion of carbohydrates, fats, and proteins. The smaller molecules formed are then absorbed, either in the duodenum itself or in the ileum.

Entry to the duodenum is controlled by the ***pyloric sphincter***, a muscular ring at the base of the stomach. Once food has passed into the duodenum it is mixed with bile from the liver and with a range of enzymes secreted from the pancreas, a digestive gland near the top of the intestine. The bile neutralizes the acidity of the gastric juices passing out of the stomach and aids fat digestion.

dynamo a simple generator, or machine for transforming mechanical energy into electrical energy. A dynamo in basic form consists of a powerful field magnet, between the poles of which a suitable conductor, usually in the form of a coil (armature), is rotated. The mechanical energy of rotation is thus converted into an electric current in the armature.

E

ear the organ used for hearing. It collects sound vibrations and transforms them into electrical signals, which are passed to the brain. Here the information is processed, and the animal becomes aware of a sound. The whole process takes place in microseconds.

A mammal's ear consists of three parts: outer ear, middle ear, and inner ear. The ***outer ear*** is a funnel that collects sound, directing it down a tube to the ***ear drum*** (tympanic membrane), which separates the outer and ***middle ear***. Sounds vibrate this membrane, and this movement is transferred to a smaller membrane leading to the ***inner ear*** by three tiny bones, the ossicles. Vibrations of the inner ear membrane move fluid contained in the snail-shaped cochlea, which vibrates hair cells that stimulate the auditory nerve. Electrical signals then pass along this nerve to the brain.

ear

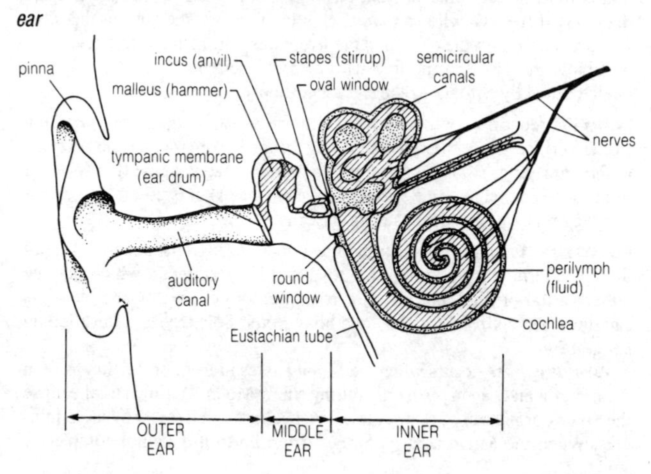

Earth third planet from the Sun. It is almost spherical, flattened slightly at the poles. 70% of the surface (including the north and south polar icecaps) is covered with water. The Earth is surrounded by a life-supporting atmosphere and is the only planet on which life is known to exist.
structure the Earth's interior is thought to be composed of a number of concentric layers: an inner core of solid iron and nickel; an outer core of molten iron and nickel; a mantle of mostly solid rock; and a solid crust. Evidence for the layered structure has been gathered by scientists surveying the paths taken by seismic waves (earthquake waves). The crust and the topmost layer of the mantle form about 12 major moving plates, some of which carry the continents. According to the theory of ◊plate tectonics, the plates are in constant, slow motion, called tectonic drift.

earth wire safety wire that connects the metal components of an electrical appliance to the earth or ground. It forms the third wire in a mains cable (the other two being the live and the neutral wires) and its insulator is usually coloured green with a yellow stripe.

If an appliance develops a fault so that current flows from the live wire to components that may be touched by the user, the earth wire will conduct that current to earth and prevent electric shock. It has a resistance lower than that of the live wire so that when a fault occurs, the current that now flows through both the earth and the live wires will increase to a level that will blow the ◊fuse in the live wire or trip a ◊circuit breaker (switching device), thereby cutting off the electrical supply.

echo the repetition of a sound wave, or of a ◊radar signal, by reflection from a surface. By accurately measuring the time taken for an echo to return to the transmitter, and by knowing the speed of a radar signal (the speed of light) or a sonar signal (the speed of sound in water), it is possible to calculate the range of the object causing the echo.

eclipse the passage of an astronomical body through the shadow of another. The term is usually used for solar and lunar eclipses, which may be either partial or total, but also, for example, for eclipses by Jupiter of its satellites. An eclipse of a star by a body in the Solar System is called an ***occultation***.

A ***solar eclipse*** occurs when the Moon passes in front of the Sun as seen from Earth, and can happen only during a new Moon. During a total eclipse the Sun's corona can be seen. A total solar eclipse can last just over 7.5 minutes. When the Moon is at its farthest from Earth it does not completely cover the face of the Sun, leaving a ring of sunlight visible. This is an ***annular eclipse*** (from the Latin word *annulus* 'ring'). Between two and five solar eclipses occur each year.

A ***lunar eclipse*** occurs when the Moon passes into the shadow of the Earth, becoming dim until emerging from the shadow. Lunar eclipses may be partial or total, and they can happen only at full Moon. Total lunar eclipses last for up to 100 minutes; the maximum number each year is three.

ecology the study of the relationship between an organism and the environment in which it lives, including other living organisms and the non-living surroundings.

Ecology may be concerned with individual organisms, with populations or species, or with entire ◊communities (for example, competition between species for access to resources in an ecosystem, or predator–prey relationships). Applied ecology is concerned with the management and conservation of habitats and the consequences and control of pollution.

ecosystem in ecology, a unit made up of a group (community) of living things interacting with the physical environment. It is therefore made up of two components: the ***biotic***, or living, and the ***abiotic***, or non-living. The relationship between all these components is finely balanced, and the continuation of that balance is essential for the maintenance of the ecosystem. The alteration of any one component can have a disastrous effect – for instance, the removal of a major predator might allow the number of herbivores to rise to such a level that the vegetation is severely damaged.

Energy and nutrients from the abiotic component of the ecosystem pass through the organisms of the biotic component in a particular sequence (see ◊food chain). The Sun's energy is captured through ◊photosynthesis, and nutrients are taken up from the soil or water by green plants (primary producers); both are passed to the herbivores that eat the plants and then to carnivores that feed on herbivores. These nutrients are returned to the soil through the decomposition of excrement and dead organisms, thus completing a cycle that is crucial to the ecosystem's stability and survival (see ◊decomposer).

Ecosystems may be aquatic (as in lakes and rockpools) or terrestrial (as in forests and grasslands).

ectoparasite a ◊parasite that lives on the outer surface of its host.

ectotherm a 'cold-blooded' animal, such as a lizard, that relies on external warmth (ultimately from the Sun) to raise its body temperature so that it can become active. To cool the body, ectotherms seek out a cooler environment.

efficiency in a machine, the useful work output (work done by the machine) divided by the work input (work put into the machine), usually expressed as a percentage. In formula terms:

$$\text{efficiency} = \frac{\text{useful work output}}{\text{work input}} \times 100\%$$

or, because power is the rate at which work is done:

$$\text{efficiency} = \frac{\text{useful power output}}{\text{power input}} \times 100\%$$

Losses of energy caused by friction mean that efficiency is always less than 100%, although it can approach this for electrical machines with no moving parts (such as a transformer).

Because work output may be defined as the product of the machine's load and the distance moved by that load, and work input as the product of the effort and the distance moved by the effort, efficiency may be also be expressed as:

$$\text{efficiency} = \frac{\text{load} \times \text{distance load moved}}{\text{effort} \times \text{distance effort moved}} \times 100\%$$

or, because the ◊mechanical advantage (MA) of a machine is the ratio of the load to the effort, and its ◊velocity ratio (VR) is the distance moved by the effort divided by the distance moved by the load, as:

$$\text{efficiency} = \frac{\text{mechanical advantage}}{\text{velocity ratio}} \times 100\%$$

efflorescence the loss of water of crystallization from crystals on standing in air, resulting in a dry powdery surface.

egg in animals, another word for the ◊ovum, or female gamete (sex cell), found in both plants and animals. The term is also applied to a fertilized ovum that develops outside the mother's body—for instance, the shell-covered eggs of birds and reptiles.

egg cell in plants, another word for the ◊ovum or oosphere, the female gamete.

elasticity the ability of a solid to recover its shape once deforming forces (stresses modifying its dimensions or shape) are removed. An elastic material obeys ◊Hooke's law: that is, its deformation is proportional to the applied stress up to a certain point, called the ***elastic limit***, beyond which additional stress will deform it permanently. Elastic materials include metals and rubber; however, all materials have some degree of elasticity.

electrical energy form of potential energy carried by an electric current. It may be converted into other forms of energy such as heat, light and motion. The electrical energy W watts converted in a circuit component through which a charge Q coulombs passes and across which there is a potential difference of V volts is given by the formula:

$$W = QV$$

electrical safety measures taken to protect human beings from electric shock or from fires caused by electrical faults. They are of paramount importance in the design of electrical equipment. Safety measures include the fitting of earth wires, and fuses or circuit breakers; the insulation of wires; the double insulation of portable equipment; and the use of residual-current devices (RCDs), which will break a circuit and cut off all currents if there is any imbalance between the currents in the live and neutral wires connected to an appliance (caused, for example, if some current is being conducted through a person).

The effects of electric shock vary from a tingling sensation to temporary paralysis and even death, and depend upon the amount of current passing through the body, and upon whether it passes through the central nervous system thereby affecting brain and heart function. Fires are usually caused by overheated cables or loose connections.

electric charge property of some bodies that causes them to exert forces on each other. Two bodies both with positive or both with negative charges repel each other, whereas bodies with opposite or 'unlike' charges attract each other, since each is in the ◊electric field of the other. In atoms, ◊electrons possess a negative charge, and ◊protons an equal positive charge. The unit of electric charge is the coulomb (symbol C).

Atoms have no charge but can sometimes gain electrons to become negative ***ions*** or lose them to become positive ions. So-called ◊***static electricity***, seen in such phenomena as the charging of nylon shirts when they are pulled on or off, or in brushing hair, is in fact the gain or loss of electrons from the surface atoms. A flow of charge (such as electrons through a copper wire) constitutes an ***electric current***; the flow of current is measured in ***amperes*** (symbol A).

electric field region in which a particle possessing electric charge experiences a force owing to the presence of another electric charge. It is a type of electromagnetic field.

electricity all phenomena caused by ◊electric charge, whether static or in motion. Electric charge is caused by an excess or deficit of electrons in the charged substance, and an electric current by the movement of electrons around a circuit. Substances may be electrical conductors, such as metals, which allow the passage of electricity through them, or insulators, such as rubber, which are extremely poor conductors. Substances with relatively poor conductivities that can be improved by exposure to heat or light are known as ◊semiconductors.

Electricity is the most useful and most convenient form of energy, readily convertible into heat and light and used to power machines. It can be generated in one place and distributed anywhere because it readily flows through wires. It is generated at power stations where a suitable energy source is harnessed to drive ◊turbines that spin electricity generators. Current energy sources are coal, oil, water power (hydroelectricity), natural gas, and ◊nuclear energy. Research is under way to increase the contribution of wind, tidal, and geothermal power. Nuclear fuel has proved a more expensive source of electricity than initially anticipated and worldwide concern over radioactivity may limit its future development.

Electricity is generated at power stations at a voltage of about 25 kilovolts, which is not a suitable voltage for long-distance transmission. For minimal power loss, transmission must take place at very high voltage (400 kilovolts or more). The generated voltage is therefore increased ('stepped up') by a ◊transformer. The resulting high-voltage electricity is then fed into the main arteries of the ◊national grid system, an interconnected network of power stations and distribution centres covering a large area. After transmission to a local substation, the line voltage is reduced by a step-down transformer and distributed to consumers.

electric motor a machine that converts electrical energy into mechanical energy. There are various types, including direct-current and induction motors, most of which produce rotary motion. A linear induction motor produces linear (sideways) rather than rotary motion.

electric motor

simple direct-current motor

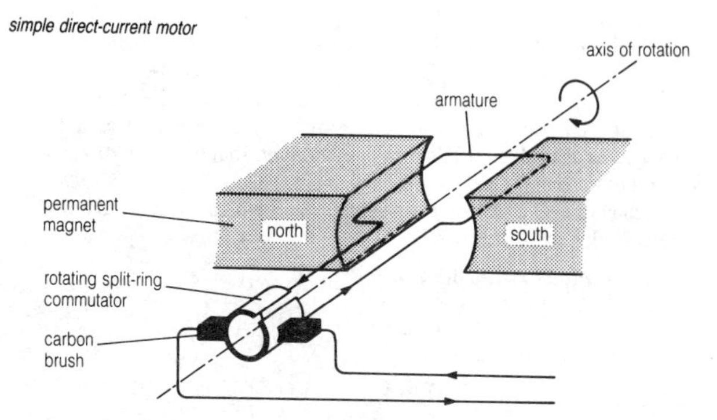

A simple ***direct-current motor*** consists of a horseshoe-shaped permanent ◊magnet with a wire-wound coil mounted so that it can rotate between the poles of the magnet. A commutator is a device added to convert the AC current produced into a DC current.

An ***induction motor*** uses ◊alternating current. It comprises a stationary current-carrying coil (stator) surrounding another coil (rotor), which rotates because of the current induced in it by the magnetic field created by the stator; it requires no commutator.

electrochemistry the branch of science that studies chemical reactions involving electricity. ◊Electrolysis, for example, is used in many industrial processes, such as the manufacture of chlorine and the extraction of aluminium. The use of chemical reactions to produce electricity is the basis of batteries; see ◊cell, chemical.

electrode in electrolysis, any terminal by which electric current passes into or out of the conducting substance (electrolyte). A positively charged electrode is called an ***anode***, because negative ions (anions or electrons) are attracted towards it; a negatively charged electrode is a ***cathode***, and cations (positive ions) are attracted towards it.

The terminals that emit and collect the flow of electrons in electronic devices such as cathode-ray tubes and diodes are also called electrodes.

electrolysis the production of chemical changes by passing an electric current through a solution (the electrolyte), resulting in the migration of the ions to the electrodes: positive ions (cations) to the negative electrode (cathode) and negative ions (anions) to the positive electrode (anode).

During electrolysis, the ions react at the electrode, either receiving or giving up electrons. The resultant atoms may be liberated as a gas, or deposited as a solid on the electrode, in amounts that are proportional to the amount of current passed.

When acidified water is electrolyzed, the chemical changes that occur at the electrodes are as follows.

negative electrode (reduction):

$$4H^+ + 4e^- \rightarrow 2H_2$$

positive electrode (oxidation):

$$4OH^- - 4e^- \rightarrow 2H_2O + O_2$$

electrolysis

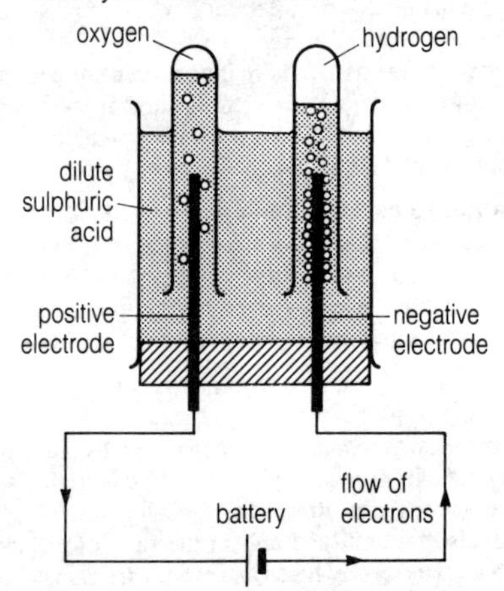

One application of electrolysis is ***electroplating***, in which a solution of a salt, such as silver nitrate ($AgNO_3$), is used and the object to be plated acts as the negative electrode, thus attracting silver ions (Ag^+). Electrolysis is used in many industrial processes, such as coating metals for vehicles and ships, and refining bauxite into aluminium.

electrolyte molten substance or solution in which an electric current is made to flow by the movement and discharge of ions.

electromagnet an iron bar with coils or wire around it, which acts as a magnet when an electric current flows through the wire. Electromagnets have many uses: in switches, electric bells, solenoids, and metal-lifting cranes.

electromagnetic field the agency by which a particle with an ◊electric charge experiences a force in a particular region of space. If it does so only when moving, it is in a pure ***magnetic field***; if it does so when stationary, it is in an ***electric field***. Both can be present simultaneously.

electromagnetic induction the production of an electric current in a circuit by a change of magnetic flux through the circuit or by relative motion of the circuit and the magnetic flux. In a closed circuit an ◊induced current will be produced.

electromagnetic radiation transfer of energy in the form of ◊electromagnetic waves.

electromagnetic spectrum the complete range, over all wavelengths from the lowest to the highest, of ◊electromagnetic waves.

electromagnetic waves oscillating electric and magnetic fields travelling together through space at a speed of nearly 300,000 kilometres per second. The (limitless) range of possible wavelengths or ◊frequencies of electromagnetic waves, which can be thought of as making up the ***electromagnetic spectrum***, includes radio waves, infrared radiation, visible light, ultraviolet radiation, X-rays, and gamma radiation.

electromotive force (emf) the energy supplied by a source of electric power in driving a unit charge around an electrical circuit. The unit is the ◊volt.

When the source is connected in circuit some of the energy it supplies will be lost in driving current across its own ◊internal resistance, and so its ◊terminal voltage (the potential difference across its terminals) will be less than its emf. If a source's terminal voltage is *V* volts, the current it supplies to a circuit is *I* amperes, and its internal resistance is *r* ohms, then its emf *E* can be expressed as:

$$E = V + Ir$$

or, where *R* is the total circuit resistance, as:

$$E = I(R + r)$$

electron stable, negatively charged particle, a constituent of all ◊atoms and the basic particle of electricity. Its mass is about $^1/_{1,840}$ that of a proton.

A beam of electrons will undergo ◊diffraction (scattering), and produce interference patterns, in the same way as ◊electromagnetic waves such as light; hence they may also be regarded as waves.

electronics the branch of science that deals with the emission of ◊electrons from conductors and ◊semiconductors, with the subsequent manipulation of these electrons, and with the construction of electronic devices. The first electronic device was the thermionic valve, or vacuum tube, in which electrons moved in a vacuum, and led to such inventions as ◊radio, ◊television, ◊radar, and the digital ◊computer. Replacement of valves with the comparatively tiny and reliable transistor in 1948 revolutionized electronic development. Modern electronic devices are based on minute ◊integrated circuits (silicon chips), wafer-thin crystal slices holding tens of thousands of electronic components.

By using solid-state devices such as integrated circuits, extremely complex electronic circuits can be constructed, leading to digital watches, pocket calculators, powerful microcomputers, and word processors.

electroplating deposition of metals upon metallic surfaces by electrolysis for decorative and/or protective purposes. It is used in the preparation of printers' blocks, 'master' audio discs , and in many other processes.

A current is passed through a bath containing a solution of a salt of the plating metal, the object to be plated being the cathode (negative electrode); the anode (positive electrode) is either an inert substance or the plating metal. Among the metals most commonly used for plating are zinc, nickel, chromium, cadmium, copper, silver, and gold.

electroscope an apparatus for detecting ◊electric charge. The simple gold-leaf electroscope consists of a vertical conducting (metal) rod ending in a pair of rectangular pieces of gold foil, mounted inside and insulated

electroscope

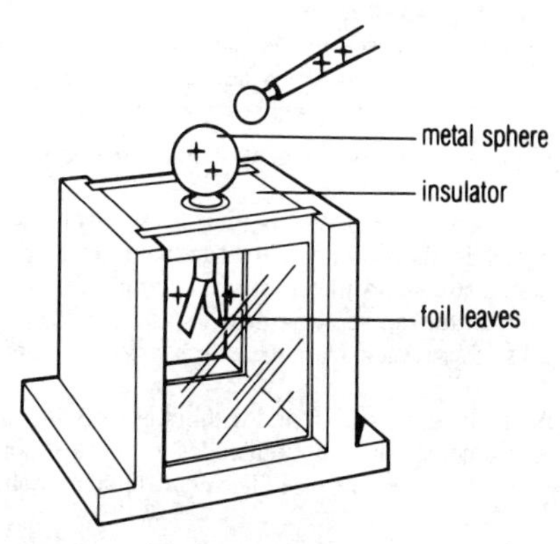

from an earthed metal case. An electric charge applied to the end of the metal rod makes the gold leaves diverge, because they each receive a similar charge (positive or negative) and so repel each other.

The polarity of the charge can be found by bringing up another charge of known polarity and applying it to the metal rod. A like charge has no effect on the gold leaves, whereas an opposite charge neutralizes the charge on the leaves and causes them to collapse.

electrostatics the study of electric charges from stationary sources (not currents).

element substance that cannot be split chemically into simpler substances. The atoms of a particular element all have the same number of protons in their nuclei (their atomic number). Of the 109 known elements, 95 are known to occur in nature (those with atomic numbers 1–95). Eighty-one of the elements are stable; all the others, which include atomic numbers 43, 61, and from 84 up, are radioactive. Those from 96 to 109 do not occur in nature and can only be synthesized in particle accelerators.

Symbols are used to denote the elements; the symbol is usually the first letter or letters of the English or Latin name (for example, C for carbon, Ca for calcium, Fe for iron, *ferrum*). The symbol represents one atom of the element.

elevation of boiling point raising of the boiling point of a liquid above that of the pure solvent, caused by a substance being dissolved in it. The phenomenon is observed when salt is added to boiling water; the water ceases to boil because its boiling point has been elevated.

embryo stage in the early development of an animal or a plant following fertilization of an egg.

In animals the embryo exists either within an egg (where it is nourished by food contained in the yolk), or, in mammals, in the ◊uterus of the mother. In mammals (except marsupials) the embryo is fed through the ◊placenta. In humans the term embryo describes the fertilized egg during its first seven weeks of existence; from the eighth week onwards it is referred to as a fetus.

In higher plants the embryo is found within the seed. It sometimes consists of only a few cells, but usually includes a root, a shoot (or primary bud), and one or two cotyledons (seed leaves), which nourish the growing seedling.

emf abbreviation for ◊***electromotive force***.

emulsifaction the process in the small intestine by which ◊bile, secreted from the liver, breaks down large globules of fat into microscopically small particles. These tiny droplets, less than 0.5 micrometres in diameter, can then be attacked by the digestive enzyme lipase.

emulsifier a food ◊additive used to keep oils dispersed and in suspension, in products such as mayonnaise and peanut butter. Egg yolk is a naturally occurring emulsifier.

emulsion type of ◊colloid, consisting of a stable dispersion of a liquid in another liquid—for example, oil and water in some cosmetic lotions.

endangered species plant or animal species whose numbers are so few that it is at risk of becoming extinct.

An example of an endangered species is the Javan rhinoceros. There are only about 50 alive today and, unless active steps are taken to promote this species' survival, it will probably be extinct within a few decades.

endocrine gland gland that secretes ◊hormones into the bloodstream in order to regulate body processes. In humans the main endocrine glands are the pituitary, thyroid, adrenal, pancreas, ovary, and testis.

endoparasite ◊parasite that lives inside the body of its host.

endocrine gland

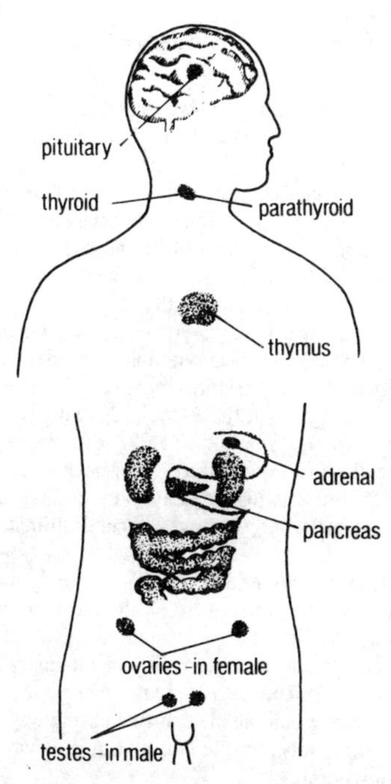

endoskeleton the internal supporting structure of vertebrates, made up of cartilage or bone. It provides support, and acts as a system of levers to which muscles are attached to provide movement. Certain parts of the skeleton (the skull and ribs) give protection to vital body organs.

endotherm warm-blooded animal; for example, a mammal. Endotherms have internal mechanisms for regulating their body temperatures to levels different from the environmental temperature.

endothermic reaction physical or chemical change where energy is absorbed by the reactants from the surroundings. The energy absorbed is represented by the symbol $+\Delta H$. The dissolving of sodium chloride in water and the process of photosynthesis are both endothermic changes. See ◊energy of reaction.

energy the capacity for doing ◊work. ***Potential energy*** (PE) is energy deriving from position; thus a stretched spring has elastic PE; an object raised to a height above the Earth's surface, or the water in an elevated reservoir, has gravitational PE; a lump of coal and a tank of petrol, together with the oxygen needed for their combustion, have chemical PE (due to relative positions of atoms). Other sorts of PE include electrical and nuclear. Moving bodies possess ***kinetic energy*** (KE). Energy can be converted from one form to another, but the total quantity stays the same (in accordance with the conservation laws that govern many natural phenomena). For example, as an apple falls, it loses gravitational PE but gains KE.

So-called energy resources are stores of convertible energy. Non-renewable resources include the fossil fuels (coal, oil, and gas) and ◊nuclear fission 'fuels'—for example, uranium-235. Renewable resources, such as wind, tidal, and geothermal power, have so far been less exploited. Hydroelectric projects are well established, and wind turbines and tidal systems are being developed. All energy sources depend ultimately on the Sun's energy.

Burning fossil fuels causes acid rain and is gradually increasing the carbon dioxide content in the atmosphere, with unknown consequences for future generations. Coal-fired power stations also release significant amounts of radioactive material, and the potential dangers of nuclear power stations are greater still. The ultimate non-renewable but almost inexhaustible energy source would be nuclear fusion (the way in which energy is generated in the Sun), but controlled fusion is a long way off. (The hydrogen bomb is a fusion bomb.)

Harnessing resources generally implies converting their energy into electrical form, because electrical energy is easy to convert to other forms and to transmit from place to place, though not to store.

energy of reaction energy released or absorbed during a chemical reaction, also called ***enthalpy of reaction*** or ***heat of reaction***; it has the symbol ΔH.

In a chemical reaction, the energy stored in the reacting molecules is rarely the same as that stored in the product molecules. Depending on which is the greater, energy is either released (an exothermic reaction) or absorbed (an endothermic reaction) from the surroundings (see ◊conservation of energy). The amount of energy released or absorbed by the quantities of substances represented by the chemical equation is the energy of reaction.

engine a device for converting stored energy into useful work or movement. Most engines use a fuel as their energy store. The fuel is burnt to produce heat energy, which is then converted into movement. Heat engines can be classified according to the fuel they use (petrol engine or diesel engine), or according to whether the fuel is burnt inside (◊internal-combustion engine) or outside (steam engine) the engine.

enthalpy alternative term for ◊energy of reaction, the heat energy associated with a chemical change.

environment in ecology, the sum of conditions affecting a particular organism or ◊ecosystem, including physical surroundings, climate, and influences of other living organisms. See also ◊biosphere and ◊habitat.

enzyme a biological ◊catalyst produced in cells, capable of speeding up the chemical reactions necessary for life. Enzymes take part in reactions, but are not themselves changed or destroyed during those reactions. They are therefore extremely efficient. Enzymes are usually described as 'reaction-specific' because each reaction in the cell requires its own particular, customized, enzyme. Within a cell therefore thousands of different enzymes can be found. They are large proteins, and each has a characteristic shape. This complex protein structure explains many of the characteristics of enzymes—for instance, their specificity, their destruction by temperatures above 60°C, and their vulnerability to changes in acidity or alkalinity.

The lock-and-key model is an attempt to explain the activity of enzymes. A large molecule about to be digested is seen as a padlock; the enzyme is the key. Only one type of key has the right shape to fit the lock. Once the padlock has been split, the key can be removed unchanged and can repeat the process.

Digestive enzymes include ◊amylases (which digest starch), ◊lipases (which digest fats), and ◊proteases (which digest protein). Other enzymes play a part in the conversion of food energy into ◊ATP; the manufacture of all the molecular components of the body; the replication of ◊DNA when a cell divides; the production of hormones; and the control of movement of substances into and out of cells.

enzyme

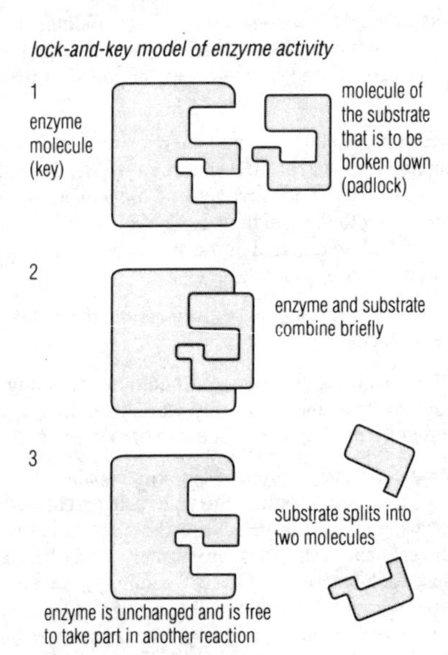

Enzymes have many medical and industrial uses, from washing powders to drug production, and as research tools in molecular biology. They can be extracted from bacteria and moulds, and ◊genetic engineering now makes it possible to tailor an enzyme for a specific purpose.

epiglottis a small, cartilaginous flap in the ◊pharynx; it moves during swallowing to prevent food from passing into the trachea (windpipe) and causing choking.

The action of the epiglottis is a highly complex reflex process involving two phases. During the first stage a mouthful of chewed food is lifted by the tongue towards the top and back of the mouth. This is accompanied by the stopping of breathing and by the blocking of the nasal areas from the mouth. The second phase involves the epiglottis moving over the larynx while the food passes down into the oesophagus.

epithelium in animals, closely packed cells that form a surface. It may be protective, as in skin, or secretory, as in the wall of the gut.

The disease scurvy, caused by a lack of vitamin C, affects the health of epithelial surfaces including the gums.

equation the representation of a reaction by symbols and numbers; see ◊chemical equation.

equilibrium the state of equilibrium or balance achieved by a body when the forces acting on it cancel each other (there is no resultant force), and the moments of those forces are in equilibrium. In accordance with Newton's first law of motion, a body in equilibrium remains at rest or moves with constant velocity; it does not accelerate. See also ◊stability.

erosion the wearing away of the Earth's surface, caused by the breakdown and transportation of particles of rock or soil (by contrast, ◊weathering does not involve transportation). Agents of erosion include the sea, rivers, glaciers, and wind.

erythrocyte another name for ◊red blood cell.

escape velocity minimum velocity with which an object must be projected for it to escape from the gravitational pull of a planetary body.

ester organic compound formed by the reaction between an alcohol and an acid, with the elimination of water. Unlike ◊salts, esters are covalent compounds. Naturally occurring esters are the basis of ◊oils and ◊fats.

ethane CH_3CH_3 colourless, odourless gas, the second member of the ◊alkane series of hydrocarbons (paraffins).

ethanoate or ***acetate*** CH_3COO^- salt of ethanoic (acetic) acid. In textiles, acetate rayon is a synthetic fabric made from modified cellulose (wood pulp) treated with ethanoic acid; in photography, acetate film is a non-flammable film made of cellulose ethanoate.

ethanoic acid or ***acetic acid*** CH_3CO_2H one of the simplest ◊carboxylic acids. In the pure state it is a colourless liquid with an unpleasant pungent odour; it solidifies to an icelike mass of crystals at 16.7°C, and hence is often called ***glacial ethanoic acid***. In aqueous solution it behaves as a weak, monobasic acid. Vinegar is 3–6% ethanoic acid.

ethanol or ***ethyl alcohol*** C_2H_5OH alcohol found in beer, wine, cider, spirits, and other alcoholic drinks. When pure, it is a colourless liquid with a pleasant odour, miscible with water or ether, and which burns in air with a pale blue flame. The vapour forms an explosive mixture with air and may be used in high-compression internal combustion engines. It is produced naturally by the fermentation of carbohydrates by yeast cells. Industrially, it can be made by absorption of ethene in concentrated sulphuric acid and subsequent reaction with water, but most is made from the catalysed hydration of ethene by steam at 600°C. It is widely used as a solvent.

ethanolic solution solution produced when a solute is dissolved in ethanol; for example, ethanolic potassium hydroxide is a solution of KOH in ethanol.

ethene or ***ethylene*** C_2H_4 colourless, flammable gas, the first member of the ◊alkene series of hydrocarbons. It is the most widely used synthetic organic chemical and is used to produce polyethene (Polythene), dichloroethane, and polyvinyl chloride (PVC). It is obtained by ◊cracking of oil fractions, from natural gas or coal gas, or by the dehydration of ethanol.

ether any of a series of organic compounds having an oxygen atom linking the carbon atoms of two hydrocarbon radical groups (general formula R-O-R'), for example ◊dioxyethane.

ethyl alcohol alternative name for ◊ethanol.

ethylene alternative name for ◊ethene.

ethylene glycol alternative name for ◊dihydroxyethane.

ethyne or ***acetylene*** C_2H_2 colourless, inflammable gas produced by mixing calcium carbide and water. The simplest member of the ◊alkyne series of hydrocarbons, it is used in the manufacture of the synthetic rubber neoprene and in oxyacetylene welding and cutting.

Its combustion provides more heat, relatively, than almost any other fuel known (its calorific value is five times that of hydrogen). This means that the gas gives an intensely hot flame.

Eustachian tube small air-filled canal connecting the middle ◊ear with the back of the throat. It is found in all land vertebrates, and serves to equalize the pressure on both sides of the tympanic membrane (ear drum).

eutrophication the excessive enrichment of lake waters, primarily by artificial, nitrate fertilizers, washed from the soil by rain, and by phosphates from detergents. These encourage the growth of algae and bacteria, which use up the oxygen in the water—making it uninhabitable for fishes and other animal life.

evaporation process in which a liquid turns to a vapour without its temperature reaching boiling point. A liquid left to stand in a saucer eventually evaporates because, at any time, a proportion of its molecules will be fast enough (have enough kinetic energy) to escape through the attractive intermolecular forces at the liquid surface and into the atmosphere. The rate of evaporation rises with increased temperature because as the mean kinetic energy of the liquid's molecules rises so will the number possessing enough energy to escape.

A fall in the temperature of the liquid, known as the ***cooling effect***, accompanies evaporation because as the faster proportion of the molecules escapes through the surface the mean energy of the remaining molecules falls. The effect may be noticed when wet clothes are worn, or as perspiration evaporates. ◊Refrigeration makes use of the cooling effect to extract heat from foodstuffs.

evergreen plant, such as pine, spruce, or holly, that bears its leaves all year round. Most conifers are evergreen. Plants that shed their leaves in autumn or a dry season are described as ◊deciduous.

evolution a slow process of change from one form to another, as in the evolution of the universe from its formation in the Big Bang to its present state, or in the evolution of life on Earth.

With respect to the living world, the idea of continuous evolution can be traced as far back as the 1st century BC, but it did not gain wide acceptance until the 19th century following the work of Charles Darwin. Darwin assigned the chief role in evolutionary change to ◊natural selection acting on randomly occurring variations (now known to be produced by spontaneous changes or ◊mutations in the genetic material of organisms). Natural selection occurs because those individuals better adapted to their particular environments reproduce more effectively, thus contributing their characteristics (in the form of genes) to future generations.

excretion the removal of waste products from the cells of living organisms. In plants and simple animals, waste products are removed by diffusion, but in higher animals specialized organs are required. In mammals, for example, carbon dioxide is removed via the lungs, and excess water and nitrogenous compounds such as urea are removed as urine via the kidneys.

exoskeleton the hardened external skeleton of insects, spiders, crabs, and other arthropods. It provides attachment for muscles and protection for the internal organs, as well as support. To permit growth it is periodically shed in a process called ecdysis.

exothermic reaction reaction during which heat is given out (see ◊energy of reaction).

expansion the increase in size of a constant mass of substance (a body) caused by, for example, increasing its temperature or its internal pressure. The ***expansivity***, or coefficient of thermal expansion, of a material is its expansion (per unit volume, area, or length) per degree rise in temperature.

extensor a muscle that straightens a limb.

extinction the complete disappearance of a species.

In the past, extinctions are believed to have occurred because species were unable to adapt quickly enough to a naturally changing environment. Today, most extinctions are due to human activity. Some species, such as the dodo of Mauritius, the moas of New Zealand, and the passenger pigeon of North America, were exterminated by hunting. Others become extinct when their habitat is destroyed. See also ◊endangered species.

Mass extinctions are episodes during which whole groups of species become extinct, the best known being that of the dinosaurs, other large reptiles, and various marine invertebrates about 65 million years ago.

eye in animals, the organ of vision. The human eye is a roughly spherical structure contained in a bony socket. Light enters it through the ***cornea***, and passes through the circular opening (***pupil***) in the iris (the coloured part of the eye). The light is focused by the combined action of the curved cornea, the internal fluids (aqueous and vitreous humours), and the ***lens*** (the rounded transparent structure behind the iris). The ciliary muscles act on the lens to change its shape, so that images of objects at different distances can be focused on the ***retina***. This is at the back of the eye, and is packed with light-sensitive cells (rods and cones), connected to the brain by the optic nerve.

eye, defects of the abnormalities of the eye that impair vision. Glass or plastic lenses, in the form of spectacles or contact lenses, are the usual means of correction. Common optical defects are ◊short-sightedness or myopia; ◊long-sightedness or hypermetropia; lack of ◊accommodation or presbyopia; and astigmatism.

eye

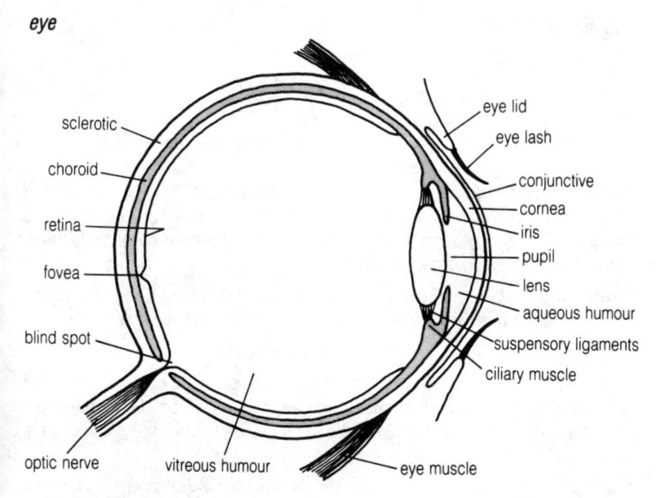

In ◊***short-sightedness*** or myopia, a person can see clearly only those objects at distances of a few metres or less. It can be corrected with a ◊diverging lens.

In ◊***long-sightedness*** or hypermetropia, a person can see clearly only distant objects. This is corrected with a converging lens.

In ***lack of accommodation*** or presbyopia, the eye's lens is unable to adjust adequately in order to focus objects at different distances. This condition develops in almost all eyes from the age of 40 onwards, and can be corrected only by using different lenses for seeing short and long distances.

In ***astigmatism***, the curvature of the ◊cornea is uneven. It is corrected by means of a cylindrical lens.

faeces remains of food and other debris passed out of the digestive tract of animals. Faeces consist of quantities of cellulose material, bacteria and other microorganisms, rubbed-off lining of the digestive tract, bile fluids, undigested food, minerals, and water.

Fahrenheit scale temperature scale invented 1714 by Gabriel Fahrenheit, no longer in scientific use. Intervals are measured in degrees (°F);

$$°F = (°C \times {}^{9}/_{5}) + 32$$

Fahrenheit took as the zero point the lowest temperature he could achieve anywhere in the laboratory, and, as the other fixed point, body temperature, which he set at 96°F. On this scale, water freezes at 32°F and boils at 212°F.

Fallopian tube or ***oviduct*** in mammals, one of two tubes that carry ova (eggs) from the ovary to the uterus. An ovum is fertilized by sperm in the Fallopian tubes, which are lined with cells whose ◊cilia move the ovum towards the uterus.

farad SI unit (symbol F) of electrical capacitance (how much electricity a ◊capacitor can store for a given voltage). One farad is a capacitance of one coulomb per volt. For practical purposes the microfarad (one millionth of a farad) is more commonly used.

fat in the broadest sense, a mixture of lipids—chiefly triglycerides (these consist of three ◊fatty acid molecules linked to a molecule of glycerol). More specifically, the term refers to a lipid mixture that is solid at room temperature (20°C); lipid mixtures that are liquid at room temperature are called ***oils***. The higher the proportion of saturated fatty acids in a mixture, the harder the fat.

Fats are essential sources of energy for many animals; they have a ◊calorific value twice that of carbohydrates, but are more difficult to digest and respire. In many animals and plants, excess carbohydrates and proteins are converted into fats for storage. Vertebrates, particularly mammals, store fats in specialized connective tissues (called adipose tissues), which not only act as energy reserves but also insulate the body and cushion body organs.

fatty acid organic acid consisting of a hydrocarbon chain, up to 24 carbon atoms long, with a carboxyl group (–COOH) at one end.

The bonds may be single or double; where a double bond occurs the carbon atoms concerned carry one instead of two hydrogen atoms. Chains with only single bonds have all the hydrogen they can carry, so they are said to be ***saturated*** with hydrogen. Chains with one or more double bonds are said to be ***unsaturated*** (see ◊polyunsaturates). Fatty acids are generally found combined with glycerol in triglycerides (see ◊fat).

feedback modification or control of a biological or chemical system by its results or products.

femur or ***thigh bone*** the upper bone in the hind limb of a four-limbed vertebrate.

fermentation the breakdown of sugars by microbes. Fermentation processes have long been used in baking bread, making beer and wine, and producing cheese, yoghurt, soy sauce, and many other foodstuffs. The chemical reactions are those of ◊anaerobic respiration; when yeast is the microbe used then the main products are alcohol and carbon dioxide. The importance of yeast, and of fermentation, can be seen in the brewing industry where alcohol is usually the product of interest, although sparkling drinks such as champagne often use the carbon dioxide as well. Bakers use yeast because the carbon dioxide liberated by fermentation causes the bread to rise.

Anaerobic or partially anaerobic respiration by bacteria is another example of fermentation, and is becoming increasingly industrialized. The controlled fermentation by bacteria can be used for instance to produce medicines, dyes and even foods.

ferric ion traditional name for the trivalent condition of iron, Fe^{3+}; the modern name is iron(III). Ferric salts are usually reddish or yellow in colour and form reddish-yellow solutions. $Fe_2(SO_4)_3$ is iron(III) sulphate (ferric sulphate).

ferrous ion traditional name for the divalent condition of iron, Fe^{2+}; the modern name is iron(II). Ferrous salts are usually green, and form very pale green solutions. $FeSO_4$ is iron(II) sulphate (ferrous sulphate).

ferrous metal metal affected by magnetism. Iron, cobalt, and nickel are the three ferrous metals.

fertilization in sexual reproduction, the union of two ◊gametes (sex cells, often called ovum and sperm) to produce a zygote, which combines the genetic material contributed by each parent. In ***self-fertilization*** the male and female gametes come from the same plant; in ***cross-fertilization*** they come from different plants. Self-fertilization rarely occurs in animals; usually even ◊hermaphrodite animals cross-fertilize each other.

fertilization

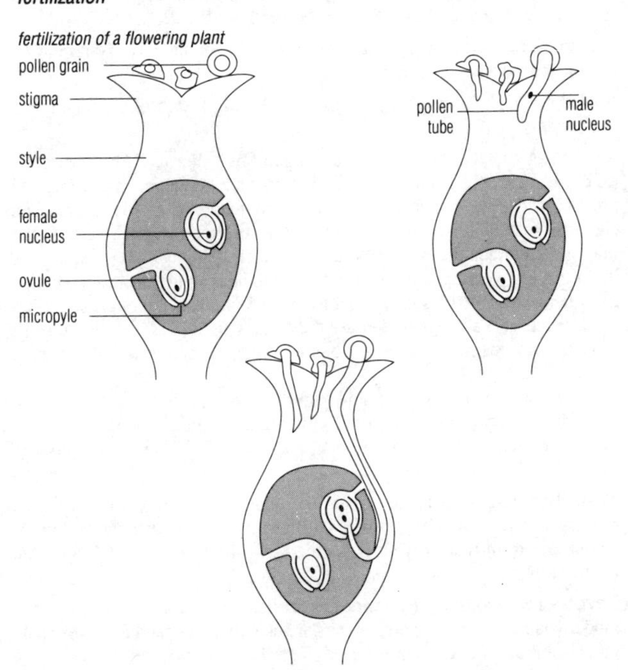

In insects, mammals, reptiles, and birds, fertilization occurs within the female's body; in most fish and amphibians, and most aquatic invertebrates, it occurs externally, when both sexes release their gametes freely into the water. In most fungi, gametes are not released, but the hyphae of the two parents grow towards each other and fuse to achieve fertilization. In higher plants, ◊pollination precedes fertilization.

fertilizer substance containing a range of about 20 chemical elements necessary for healthy plant growth, used in agriculture and horticulture to compensate for the deficiencies of poor or depleted soil and to improve plant growth. Fertilizers may be ***organic*** (derived from living things), for example, manure, compost, or ashes; or ***inorganic*** (artificial), mainly in the form of compounds of nitrogen, potassium, and phosphorus, which have been used on a greatly increased scale since 1945.

Over-use of artificial fertilizers can lead to contamination of local lakes and rivers; see ◊eutrophication.

fetus or ***foetus*** a stage in mammalian ◊embryo development. The human embryo is usually called a fetus after the eighth week of development, when the limbs and external features of the head are recognizable.

fibre optics the branch of physics dealing with the transmission of light and images through glass or plastic fibres known as ◊optical fibres.

fibrin an insoluble blood protein used by the body to stop bleeding. When an injury occurs a mesh of fibrin is deposited around the cut blood vessels so that bleeding stops. See ◊blood clotting.

field the region of space in which an object exerts a force on another separate object because of certain properties they both possess. For example, there is a force of gravitational attraction between any two objects that have mass when one is in the gravitational field of the other.

Other fields of force include ◊electric fields (caused by electric charges) and ◊magnetic fields (caused by magnetic poles), either of which can involve attractive or repulsive forces.

field of view angle over which an image may be seen in a mirror or an optical instrument such as a telescope. A wide field of view allows a greater area to be surveyed without moving the instrument, but has the disadvantage that each of the objects seen is smaller. A ◊convex mirror gives a larger field of view than a plane or flat mirror. The field of view of an eye is called its ***field of vision*** or visual field.

field of view

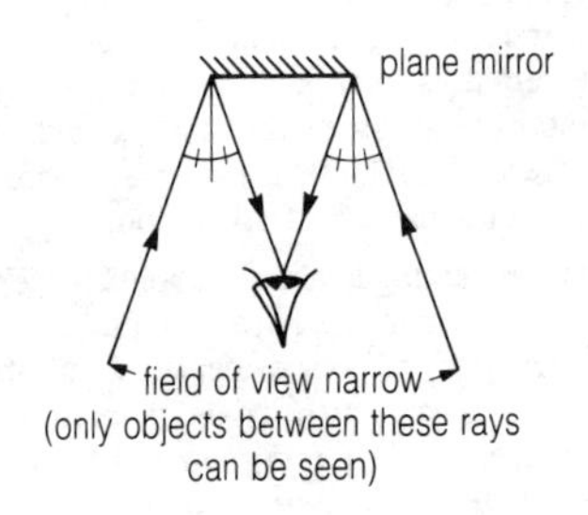

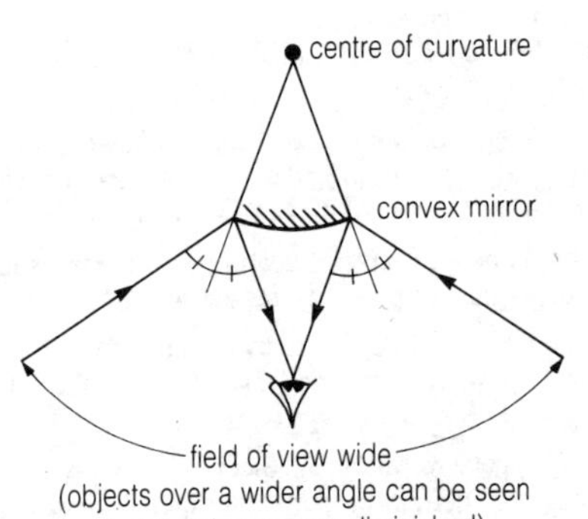

filtrate liquid or solution that has passed through the filter paper or cloth in the filtration process.

filtration technique where suspended solid particles in a fluid are removed by passing the mixture through a porous barrier, usually paper or cloth. The particles are retained by the paper or cloth to form a residue and the fluid passes through to make up the filtrate. Soot may be filtered from air, and suspended solids may be filtered from water.

firedamp gas that occurs in coal mines and is explosive when mixed with air in certain proportions. It consists chiefly of methane (CH_4, natural gas or

filtration

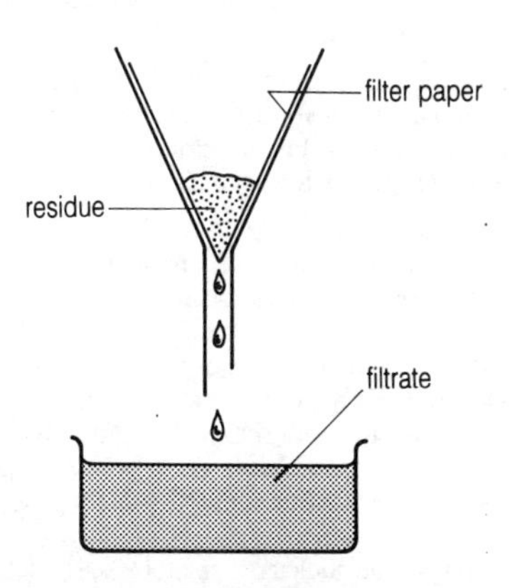

marsh gas) but always contains small quantities of other gases, such as nitrogen, carbon dioxide, and hydrogen, and sometimes ethane and carbon monoxide.

fire extinguisher device for putting out a fire. Fire extinguishers work by removing one of the three conditions necessary for fire to continue (heat, oxygen, and fuel), either by cooling the fire or by excluding oxygen.

The simplest fire extinguishers contain water, which when propelled onto the fire cools it down. Water extinguishers cannot be used on electrical fires, as there is a danger of electrocution, or on burning oil, as the oil will float on the water and spread the blaze.

Many domestic extinguishers contain liquid carbon dioxide under pressure. When the handle is pressed, carbon dioxide is released as a gas that blankets the burning material and prevents oxygen reaching it. Dry extinguishers spray powder, which then releases carbon dioxide gas. Wet extinguishers are often of the soda-acid type; when activated, sulphuric acid mixes with sodium bicarbonate, producing carbon dioxide. The gas pressure forces the solution out of a nozzle, and a foaming agent may be added to produce foam.

Some extinguishers contain halons (hydrocarbons with one or more hydrogens substituted by a halogen such as chlorine, bromine or fluorine).

These are very effective at smothering fires, but cause damage to the ozone layer in the atmosphere (see ◊ozone).

fire triangle the three essential ingredients needed to cause a fire: heat, air, and fuel. Fire-prevention strategies attempt to ensure these three conditions do not occur together. Fire control concentrates on one or more of these ingredients: water cools the temperature while foam excludes the air.

fish aquatic vertebrate that uses gills for obtaining oxygen from water. There are three main groups, not closely related: the bony fishes (goldfish, cod, tuna), the cartilaginous fishes (sharks, rays), and the jawless fishes (hagfishes, lampreys).

The bony fishes constitute the majority of living fishes (about 20,000 species). The skeleton is bone, movement is controlled by mobile fins, and the body is usually covered with scales. The gills are covered by a single flap. Many have a swim bladder with which the fish adjusts its buoyancy. Most lay eggs, sometimes in vast numbers; some cod can produce as many as 28 million.

fission the splitting of the nucleus of an atom; see ◊nuclear fission.

flaccidity the loss of rigidity (◊turgor) in plant cells, caused by loss of water from the central vacuole so that the cytoplasm no longer pushes against the cellulose cell wall. If this condition occurs throughout the plant then wilting is seen.

Flaccidity can be brought about in the laboratory by immersing the plant cell in a strong salt or sugar solution. Water leaves the cell by ◊osmosis, causing the vacuole to shrink. In extreme cases the actual cytoplasm pulls away from the cell wall, a phenomenon known as ***plasmolysis***.

flame test the use of a flame to identify metal ions present in a solid. A nichrome or platinum wire is moistened with acid, dipped in the test substance, and then held in a hot flame. The colour produced in the flame is characteristic of metals present; for example, sodium burns with a yellow flame, and potassium with a lilac one.

flash point the lowest temperature at which a liquid or volatile solid heated under standard conditions gives off sufficient vapour to ignite on the application of a small flame.

The ***fire point*** of a material is the temperature at which full combustion occurs. For safe storage of materials such as fuel or oil, conditions must be well below the flash and fire points to reduce fire risks to a minimum.

flexor any muscle that bends a limb. Flexors usually work in opposition to other muscles, the extensors, an arrangement known as antagonistic pairing.

flip-flop in electronics, another term for a ◊bistable circuit.

floating state of equilibrium in which a body rests on or is suspended in the surface of a fluid (liquid or gas). According to Archimedes' principle, a body wholly or partly immersed in a fluid will be subjected to an upward force, or upthrust, equal in magnitude to the weight of the fluid it has displaced. If the ◊density of the body is greater than that of the fluid, then its weight will be greater than the upthrust and it will sink. However, if the body's density is less than that of the fluid, the upthrust will be the greater and the body will be pushed upwards towards the surface. As the body rises above the surface the amount of fluid that it displaces (and therefore the magnitude of the upthrust) decreases. Eventually the upthrust acting on the

floating

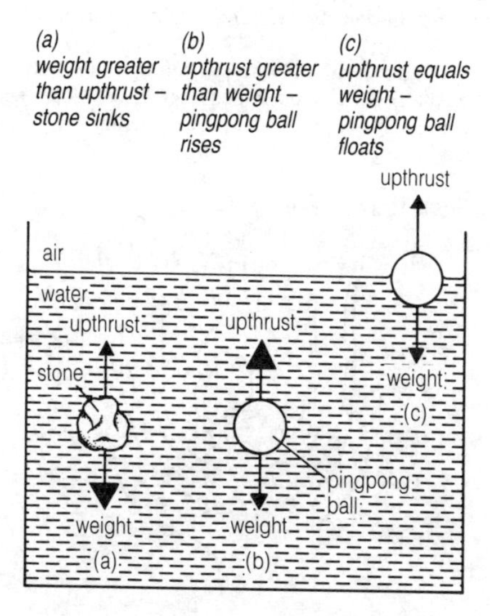

submerged part of the body will equal the body's weight, equilibrium will be reached, and the body will float.

flotation, law of law stating that a floating object displaces its own weight of the fluid in which it floats. It provides an explanation of how an object as large and as heavy as a steel ship can float: the hollow steel hull of the ship sinks into the water until the weight of the water it has displaced is as great as its own weight. The upthrust from the water will then equal the ship's weight and the ship will float.

flower the reproductive unit of an angiosperm (flowering plant), typically consisting of four whorls of modified leaves: ◊sepals, ◊petals, ◊stamens, and ◊carpels. These are borne on a central axis or receptacle. The many variations in size, colour, number and arrangement of parts are closely related to the method of pollination. Flowers adapted for wind pollination typically have reduced or absent petals and sepals and long, feathery ◊stigmas that hang outside the flower to trap airborne pollen. In contrast, the petals of insect-pollinated flowers are usually conspicuous and brightly coloured.

The sepals and petals are collectively known as the calyx and corolla respectively and together comprise the perianth, with the function of protecting the reproductive organs and attracting pollinators. The stamens lie

flower

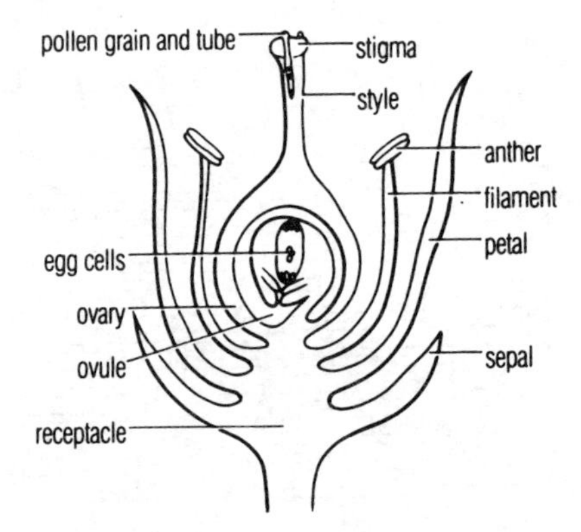

within the corolla, each having a slender stalk, or filament, bearing the pollen-containing anther at the top. Collectively they are known as the androecium. The inner whorl of the flower comprises the carpels, each usually consisting of an ◊ovary in which are borne the ◊ovules, and a stigma borne at the top of a slender stalk, or style. Collectively the carpels are known as the gynoecium. The stalk of a flower is called a pedicel.

Flowers may either be borne singly or grouped together in inflorescences. A flower is called hermaphrodite when it contains both male and female reproductive organs. When male and female organs are carried in separate flowers, they are termed monoecious; when male and female flowers are on separate plants, the term dioecious is used.

fluid any substance, either liquid or gas, in which the molecules are relatively mobile and can 'flow'.

fluoridation addition of small amounts of fluoride salts to drinking water by certain water authorities to help prevent tooth decay. In areas where fluoride ions are naturally present in the water, research found that the incidence of tooth decay in children from those areas was reduced by more than 50%. A concentration of one part per million is sufficient to produce this beneficial effect.

fluorine chemical element, symbol F, atomic number 9, relative atomic mass 19. It occurs naturally as the minerals fluorspar (CaF_2) and cryolite (Na_3AlF_6), and is the first member of the halogen group of elements. At ordinary temperatures it is a pale yellow, highly poisonous, and reactive gas, and it unites directly with nearly all the elements. Hydrogen fluoride is used in etching glass, and the freons, which all contain fluorine, are widely used as refrigerants.

Minute quantities of sodium fluoride are added to some water supplies to help prevent tooth decay (see ◊fluoridation).

fluorocarbon compound formed by replacing the hydrogen atoms of a hydrocarbon with fluorine. Fluorocarbons are used as inert coatings, refrigerants, synthetic resins, and as propellants in aerosols.

There is concern because their release into the atmosphere depletes the ozone layer (see ◊ozone), allowing more ultraviolet light from the Sun to penetrate the Earth's atmosphere, increasing the incidence of skin cancer.

foam mixture of gas and liquid. The volume of air is far greater than the volume of liquid, which is expanded so that it becomes nothing more than bubble walls. For a foam to be stable, it may be necessary to add small amounts of oil or soap, which strengthen the bubbles. The foam used by firefighters consists basically of carbon dioxide; by incorporating the gas into bubbles it is much easier to ensure that the carbon dioxide is directed at the fire, and does not diffuse away.

focal length the distance from the centre of a spherical mirror or lens to the focal point. For a concave mirror or convex lens, it is the distance at which parallel rays of light are brought to a focus to form a real image (for a mirror, this is half the radius of curvature). For a convex mirror or concave lens, it is the distance from the centre to the point at which a virtual image (an image produced by diverging rays of light) is formed.

In the case of lenses, the focal length is the reciprocal of the power (in dioptres—the units used to measure this) of the lens: the greater its power the shorter its focal length.

focus or ***focal point*** in optics, the point at which light rays converge, or from which they appear to diverge, to form a sharp image. Other electromagnetic rays, such as microwaves, and sound waves may also be brought together at a focus. Rays parallel to the principal axis of a lens or mirror are converged at, or appear to diverge from, the ◊principal focus.

food anything eaten by human beings and other animals to sustain life and health. The building blocks of food are called nutrients, and humans can utilize the following nutrients:
carbohydrate, as starch found in bread, potatoes, and pasta; as simple sugars in sucrose (table sugar) and honey; and as cellulose (dietary fibre) in cereals, fruit, and vegetables;
protein, of which good sources are nuts, fish, meat, eggs, milk, and some vegetables;
fat, found in most meat products, fish, dairy products (butter, milk, cheese), margarine, nuts, vegetable oils, and lard;
vitamins, found in a wide variety of foods, except for vitamin B_{12}, which is only found in animal products;
minerals, found in a wide variety of foods; a good source of calcium is milk, of iodine is seafood, and of iron is liver or green vegetables;
water, found everywhere in nature;
alcohol, found in alcoholic beverages, from 40% in spirits to 0.01% in low-alcohol lagers and beers.

Food is needed for both its calorific energy, measured in kilojoules, and

focal length

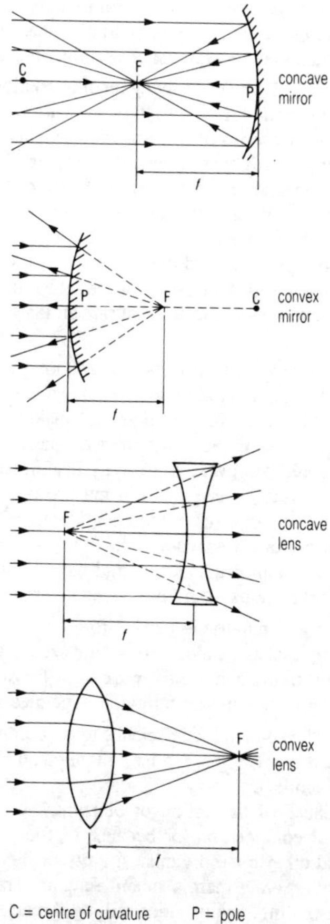

nutrients, such as proteins, that are converted to body tissues. Some nutrients mainly provide energy, such as fat, carbohydrate, and alcohol; other nutrients are important in other ways, such as aiding metabolism.

food chain in ecology, a way of showing feeding relations between plants and animals. Energy in the form of food is shown to be transferred from producer (the plants) to a series of consumers (first a herbivore, then a carnivore). In reality, organisms have varied diets, relying on different kinds of foods, so that the food chain is an over-simplification. The more complex ***food web*** shows a greater variety of relationships, but again emphasizes that energy passes from plants to herbivores to carnivores.

Environmental groups have used the concept of the food chain to show how poisons and other forms of pollution can pass from one animal to another, eventually resulting in the death of rare animals such as the golden eagle.

food irradiation a development in ◊food technology, whereby food is exposed to low-level radiation to kill microorganisms.

Irradiation is highly effective, and does not make the food any more radioactive than it is naturally. Some vitamins are partly destroyed, such as vitamin C, and it would be unwise to eat only irradiated fruit and vegetables. The main cause for concern is that it may be used by unscrupulous traders to 'clean up' consignments of food, particularly shellfish, with high bacterial counts. Bacterial toxins would remain in the food, so that it could still cause illness, although irradiation would have removed signs of live bacteria. Stringent regulations would be needed to prevent this happening.

food store place in which nutrients can be stored; examples include starch grains within chloroplasts and endosperm within seeds. Among higher animals, adipose (fatty) tissue and the liver are used for storage. Organisms tend to have short-term and long-term food-storage measures.

food technology the application of science to the commercial processing of foodstuffs. Food is processed to render it more palatable or digestible, or to preserve it from spoilage.

Food spoils because of the action of ◊enzymes within the food that change its chemical composition, or because of the growth of bacteria, moulds, yeasts, and other microorganisms. Fatty or oily foods also suffer oxidation of the fats, giving them a rancid flavour. Traditional forms of processing include boiling, frying, flour-milling, bread- making, yoghurt- and cheese-making, brewing, and various methods of ***food preservation***,

food chain

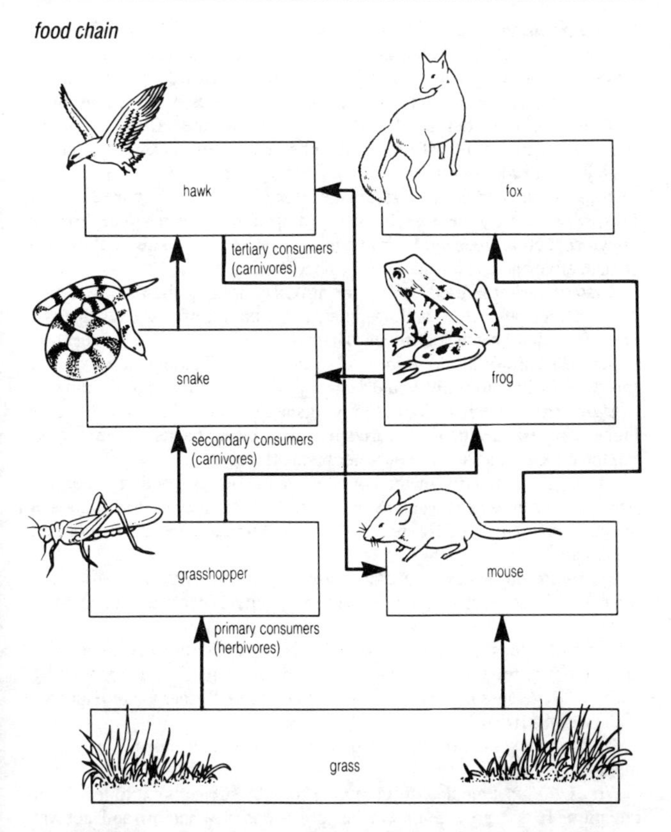

such as salting, smoking, pickling, drying, bottling, and preserving in sugar. Modern food technology still uses traditional methods but also many novel processes and ◊additives, which allow a wider range of foodstuffs to be preserved.

Refrigeration below 5°C (or below 3°C for cooked foods) slows the process of spoilage. Although a convenient form of preservation, this process cannot kill microorganisms, nor stop their growth completely, and a failure to realize its limitations causes many cases of food poisoning. Refrigerator temperatures should be checked as the efficiency of the machinery can decline with age, and higher temperatures are dangerous.

Deep freezing (–18°C or below) stops almost all spoilage processes, although there may be some residual enzyme activity in uncooked vegetables, which is why these are blanched (dipped in hot water to destroy the enzymes) before freezing. Microorganisms cannot grow or divide, but most remain alive and can resume activity once defrosted.

Pasteurization is used mainly for milk. By holding the milk at a high temperature, but below boiling point, for a period of time, all disease-causing bacteria can be destroyed. The milk is held at 72°C for 15 seconds. Other, less harmful bacteria survive, so the milk will still go sour within a few days. Boiling the milk would destroy all bacteria, but impair the flavour.

Ultra-heat treatment is used to produce UHT milk. This process uses higher temperatures than pasteurization, and kills all bacteria present, giving the milk a long shelf life but altering the flavour.

Drying is an effective method of preservation because both microorganisms and enzymes need water to be active. Products such as dried milk and instant coffee are made by spraying the liquid into a rising column of dry, heated air.

Freeze-drying is carried out in a vacuum. It is less damaging to food than straight dehydration, and is used for quality instant coffee and dried vegetables.

Canning relies on high temperatures to destroy microorganisms and enzymes. The food is sealed into a can to prevent any recontamination by bacteria. Beverages may also be canned to preserve the carbon dioxide that makes drinks fizzy.

Pickling makes use of acetic acid, found in vinegar, to stop the growth of moulds.

Irradiation is a method of preserving food by subjecting it to low-level radiation. It is highly controversial (see ◊food irradiation) and not yet widely used in the UK.

Chemical treatments are widely used, for example in margarine manufacture, where hydrogen is bubbled through vegetable oils in the presence of a ◊catalyst to produce a more solid, spreadable fat. The catalyst is later removed. Chemicals that are introduced in processing and remain in the food are known as ***food additives*** and include flavourings, preservatives, antioxidants, emulsifiers, and colourings.

food test any of several types of simple test, easily performed in the laboratory, used to identify the main classes of food.
starch—iodine test Food is ground up in distilled water and iodine is added. A dense black colour indicates that starch is present.
sugar—Benedict's test Food is ground up in distilled water and placed in a test tube with Benedict's reagent. The tube is then heated in a boiling water bath. If a ◊reducing sugar, such as glucose, is present the colour changes from blue to brick-red; if a non-reducing sugar, such as sucrose, is present, there is no colour change.
protein—Biuret test Food is ground up in distilled water and a mixture of copper(II) sulphate and sodium hydroxide is added. If protein is present a mauve colour is seen.

force any influence that tends to change the state of rest or the uniform motion in a straight line of a body. The action of an unbalanced or resultant force results in the acceleration of a body in the direction of action of the force or it may, if the body is unable to move freely, result in its deformation (see ◊Hooke's law). Force is a vector quantity, possessing both magnitude and direction; its SI unit is the newton.

According to Newton's second law of motion the magnitude of a resultant force is equal to the rate of change of ◊momentum of the body on which it acts; the force F (in newtons) producing an acceleration a metres per second squared on a body of mass m kilograms is therefore given by:

$$F = ma$$

See also ◊Newton's laws of motion.

formaldehyde alternative name for ◊methanal.

formalin aqueous solution of formaldehyde (methanal) used to preserve animal specimens.

formic acid alternative name for ◊methanoic acid.

formula representation of a molecule, radical, or ion, in which chemical elements are represented by their symbols. An ***empirical formula*** indicates the simplest ratio of the elements in a compound, without indicating how many of them there are or how they are combined. A ***molecular formula*** gives the number of each type of element present in one molecule. A ***structural formula*** shows the relative positions of the atoms and the bonds between them. For example, for ethanoic acid, the empirical formula is CH_2O, the molecular formula is $C_2H_4O_2$, and the structural formula is CH_3COOH.

fossilization the preservation of an organism or of its shape, tracks, eggs, or faeces. Most fossils are found in rocks, but some ancient animals and plants have been found preserved in ice, hardened tree sap (amber), and even tar.

Fossilization will only occur under certain conditions; for instance, when an animal is covered by mud or sand as soon as it dies, and is not immediately attacked by bacteria. In most fossils it is the hard part of the animal that is preserved: after death, the minerals of the skeleton (bones, teeth, or shell) are steadily replaced by other, rocky minerals. Fossils showing details of soft parts, such as the gut, are much rarer because these parts usually decay very quickly.

Studying the fossils of animal and plant species that existed millions of years ago has played a major part in our understanding of evolution and the history of life on Earth. However, large gaps exist in the fossil record—most species are thought to have left no fossil traces—and this has been taken by some people as evidence that evolution may never have taken place.

fossil fuel fuel, such as coal or oil, formed from the fossilized remains of organic matter that existed hundreds of millions of years ago. Fossil fuels are a non-renewable resource and will run out eventually. Extraction of coal causes considerable environmental pollution, and burning coal contributes to problems of ◊acid rain and the ◊greenhouse effect.

fraction group of similar compounds, the boiling points of which fall within a particular range and which are separated during ◊fractionation.

fractionating column device in which many separate ◊distillations can occur so that a liquid mixture can be separated into its components.

Various designs exist but the primary aim is to allow maximum contact between the hot rising vapours and the cooling descending liquid. As the vapours ascend the column the mixture becomes progressively enriched in the lower-boiling components, so they separate out first.

fractionation or ***fractional distillation*** process used to split complex mixtures (such as petroleum, or crude oil) into their components, usually by repeated heating, boiling, and condensation.

fractionating column

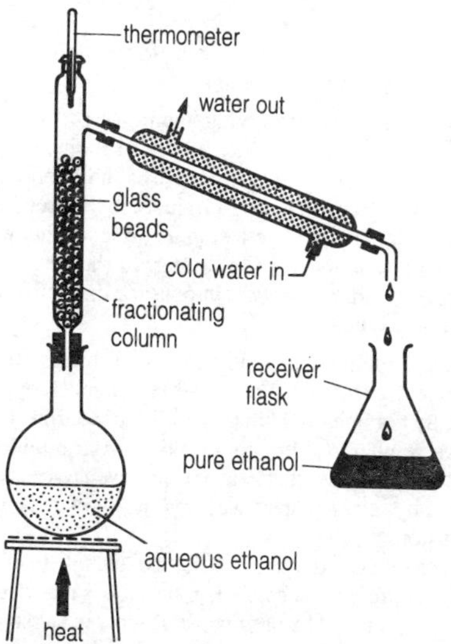

francium metallic element, symbol Fr, atomic number 87, relative atomic mass 223. It is a highly radioactive metal; the most stable isotope has a half-life of only 21 minutes.

Frasch process process used to extract underground deposits of sulphur. Superheated steam is piped to the sulphur deposit and melts it. Compressed air is then pumped down to force the molten sulphur to the surface.

free fall the state in which a body is falling freely under the influence of gravity, as in free-fall parachuting. The term ***weightless*** is normally used to describe a body in free fall in space.

free radical an atom or molecule that has an unpaired electron and is therefore highly reactive. Free radicals are often produced by high temperatures and are found in flames and explosions. A very simple free radical is the methyl radical $CH_3\cdot$ produced by the splitting of the covalent carbon-to-carbon bond in ethane.

$$CH_3CH_3 \leftrightarrow 2CH_3\cdot$$

Most free radicals are very short-lived. If free radicals are produced in living organisms they can be very damaging.

freeze-drying method of preserving food; see ◊food technology. The product to be dried is frozen and then put in a vacuum chamber that forces out the ice as water vapour, a process known as sublimation.

Many of the substances that give products such as coffee their typical flavour are volatile, and would be lost in a normal drying process because they would evaporate along with the water. In the freeze-drying process these volatile compounds do not pass into the ice that is to be sublimed, and are therefore largely retained.

freezing change from liquid to solid state, as when water becomes ice. For a given substance, freezing occurs at a definite temperature, known as the freezing point, that is invariable under similar conditions of pressure, and the temperature remains at this point until all the liquid is frozen. The amount of heat per unit mass that has to be removed to freeze a substance is a constant for any given substance, and is known as the latent heat (enthalpy) of fusion.

Ice is less dense than water since water expands just before its freezing point is reached. If pressure is applied, expansion is retarded and the freezing point will be lowered. The presence of dissolved substances in a liquid also lowers the freezing point (depression of freezing point), the amount of lowering being proportional to the molecular concentration of the solution. Antifreeze mixtures for car radiators and the use of salt to melt ice on roads are common applications of this principle.

freezing-point depression lowering of a solution's freezing point below that of the pure solvent; it depends on the number of molecules of solute dissolved in it. Thus for a single solvent, such as pure water, all substances in the same molecular concentration produce the same lowering of freezing point.

frequency the number of periodic oscillations, vibrations, or waves occurring per unit of time. The unit of frequency is the hertz (Hz), one hertz being equivalent to one cycle per second. Human beings can hear sounds from objects vibrating in the range 20–15,000 Hz.

friction the force that opposes the relative motion of two bodies in contact. The ***coefficient of friction*** is the ratio of the force required to achieve this relative motion to the force pressing the two bodies together.

Friction is greatly reduced by the use of lubricants such as oil, grease, and graphite. Air bearings are now used to minimize friction in high-speed rotational machinery. In other instances friction is deliberately increased by making the surfaces rough—for example, brake linings, driving belts, soles of shoes, and tyres.

fructose ($C_6H_{12}O_6$) monosaccharide sugar that occurs naturally in honey, the nectar of flowers, and many sweet fruits.

fruit structure that develops from the carpel of a flower and encloses one or more seeds. Its function is to protect the seeds during their development and to aid in their dispersal. When fruits are eaten by animals the seeds pass through the alimentary canal unharmed, and are passed out with the faeces.

fuel any source of heat or energy, embracing the entire range of all combustibles and including anything that burns. ***Nuclear fuel*** is any material that produces energy in a nuclear reactor.

functional group small number of atoms in an arrangement that determines the chemical properties of the group and of the molecule to which it is attached (for example the carboxylic acid group $-COOH$, or the amine group $-NH_2$). Organic compounds can be considered as structural skeletons with functional groups attached.

fungus (plural ***fungi***) any of a group of organisms in the kingdom Fungi. Fungi are not considered plants. They lack leaves and roots; they contain no chlorophyll and reproduce by spores. Moulds, yeasts, mildews, and mushrooms are all types of fungus.

Because fungi have no chlorophyll, they must get food from organic substances. They are either ◊parasites, existing on living plants or animals, or ◊saprotrophs, living on dead matter.

fuse in electricity, a wire or strip of metal designed to melt when excessive current passes through. It is a safety device to stop at that point in the circuit surges of current that would otherwise damage equipment and cause fires.

fuse box insulated container housing the electrical fuses that protect the electric circuits and equipment in a building.

fusion the fusing together of atomic nuclei; see ◊nuclear fusion.

G

g symbol for ◊gravitational field strength (the strength of the Earth's gravitational field at any point) and for gravitational acceleration.

gain in electronics, the ratio of the amplitude of the output signal produced by an amplifier to that of the input signal. In a ◊voltage amplifier the voltage gain is the ratio of the output voltage to the input voltage.

galaxy congregation of millions or billions of stars, held together by gravity. Our own galaxy, the Milky Way, is about 100,000 light years in diameter, and contains at least 100 billion stars. It is a member of a small cluster, the Local Group. The Sun lies in one of its spiral arms, about 25,000 light years from the centre.

gall bladder a small muscular sac attached to the underside of the liver and connected to the small intestine by the bile duct. It stores bile from the liver.

galvanizing process for rendering iron rustproof, by plunging it into molten zinc (the dipping method), or by electroplating it with zinc.

gamete sex cell produced by animals and plants as part of the process of sexual reproduction. The function of the gamete is to carry the genetic message of the parent. During ◊fertilization, two gametes fuse to form a zygote, giving rise to a new and unique individual whose genetic code is slightly different from that of either of the parents. In this way genetic variation persists in a population.

Gametes are ◊haploid, containing only half the number of chromosomes found in the parents. They are produced by ◊meiosis, a particular form of cell division. On fertilization, the chromosomes of each gamete pair up, so that the new individual has the original (◊diploid) number of chromosomes.

gamma radiation very high-frequency electromagnetic radiation emitted by the nuclei of radioactive substances during decay. The emission of gamma radiation reduces the energy of the source nucleus, but has no effect on its atomic or mass numbers.

Rays of gamma radiation are stopped only by direct collision with an atom and are therefore very penetrating; they can, however, be stopped by about 4 cm of lead or by a very thick concrete shield. They are less ionizing in their effect than are alpha and beta particles, but are dangerous nevertheless because they can penetrate deeply into body tissues such as bone marrow. They are not deflected by either magnetic or electric fields.

gas form of matter, such as air, in which the molecules move randomly in otherwise empty space, filling any size or shape of container into which the gas is put.

A sugar-lump-sized cube of air at room temperature contains 30 million million million molecules moving at an average speed of 500 metres per second (1,800 kph). Gases can be liquefied by cooling, which lowers the speed of the molecules and enables attractive forces between them to bind them together.

gas collection method used to collect a gas in a laboratory preparation. The properties of the gas, and whether it is required dry, dictate the method used. Dry ammonia is collected by ◊downward displacement of air.

gas exchange movement of gases between an organism and the atmosphere, principally oxygen and carbon dioxide. All aerobic organisms (most animals and plants) take in oxygen in order to burn food and manufacture ATP (adenosine triphosphate). The resultant reactions release carbon dioxide as a waste product to be passed out into the environment. Green plants also absorb carbon dioxide during ◊photosynthesis, and release oxygen as a waste product.

Specialized ◊respiratory surfaces have evolved during evolution to make gas exchange more efficient. In mammals, birds, reptiles, and amphibians, gas exchange occurs in the ◊lungs, aided by the breathing movements of the ribs. Many adult Amphibia and terrestrial invertebrates can absorb oxygen directly through the skin. The bodies of insects and some spiders contain a system of air-filled tubes known as ◊tracheae. Fish have ◊gills as their main respiratory surface. In plants, gas exchange generally takes place via the stomata (see ◊stoma) and the air-filled spaces between the cells in the interior of the leaf.

The process of gas exchange relies upon the ◊diffusion of gases across the respiratory surfaces. For example, oxygen diffuses from the lungs into the blood supply because oxygen molecules are at a higher concentration in the alveoli (air sacs) than they are in the capillaries surrounding the alveoli.

gas exchange

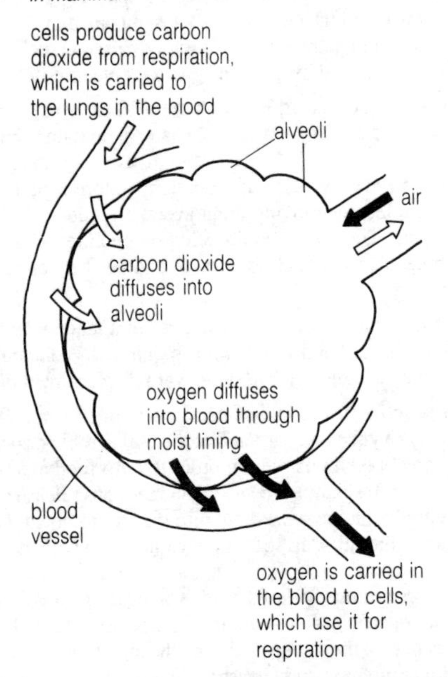

As the oxygen molecules diffuse along the concentration gradient, they will tend to cross the alveolus and capillary walls and pass into the blood. Gas exchange therefore depends on the ability of the organism to maintain a concentration gradient so that oxygen will continue to diffuse across the respiratory surface.

gas laws physical laws concerning the behaviour of gases. They include Boyle's law, Charles's law, and the pressure law, which are concerned with the relationships between the pressure, temperature, and volume of an ideal (hypothetical) gas.

These laws can be combined to give the ***general*** or ***universal gas law***, which may be expressed as

$$\frac{\text{pressure} \times \text{volume}}{\text{temperature}} = \text{constant}$$

or as

$$P_1V_1/T_1 = P_2V_2/T_2$$

gas syringe graduated piece of glass apparatus used to measure accurately volumes of gases produced or consumed in a chemical reaction.

gate see ◊logic gate.

Geiger counter device for detecting and/or counting nuclear radiation and particles. It detects the momentary current that passes between electrodes in a suitable gas when a nuclear particle or a radiation pulse causes ionization in the gas. The electrodes are connected to electronic devices which enable the intensity of radiation or the number of particles passing to be measured. It is named after Hans Geiger.

gel solid produced by the formation of a three-dimensional cage structure, commonly of linked large-molecular-mass polymers, in which a liquid is trapped. A gel may be a jellylike mass (pectin, gelatine) or have a more rigid structure (silica gel).

gelatinous precipitate ◊precipitate that is viscous and jellylike when formed; ◊aluminium hydroxide has this appearance.

gene a unit of inherited material, encoded by a length of ◊DNA. In eukaryotes (organisms other than bacteria), genes are located on the ◊chromosomes. The gene is the inherited factor that consistently affects a particular character in an individual—for example, the gene for eye colour. It occurs at a particular point, or locus, on a particular chromosome and may have several variants, or ◊alleles, each specifying a form of that character—for example, the alleles for blue or brown eyes. Some alleles are ◊dominant. These mask the effect of other, ◊recessive alleles.

Genes produce their visible effects simply by coding for proteins. Precise copies of genes are made by the ◊nucleus during ◊cell division. This ensures that new cells possess the correct genetic material. Variation among individuals and between generations occurs because of new combinations of genes and because of mutation. This is important in evolution.

generator a machine that produces electrical energy from mechanical energy, as opposed to an ◊electric motor, which does the opposite.

The dynamo is a simple generator consisting of a wire-wound coil (armature) that is rotated between the pole pieces of a permanent magnet. The movement (of the wire in the magnetic field) induces a current in the coil by ◊electromagnetic induction, which is converted by means of a commutator into a continuous direct current into an external circuit. Slip rings instead of a commutator produce an alternating current, in which case the generator is called an alternator.

gene therapy a proposed medical technique for curing or alleviating inherited diseases, such as cystic fibrosis. It relies on replacing defective genes with functional ones.

genetic code the way in which instructions for building proteins, the basic structural molecules of living matter, are 'written' in the genetic material ◊DNA. This relationship between the sequence of bases (the subunits in a DNA molecule) and the sequence of ◊amino acids (the subunits of a protein molecule) is the basis of heredity. The code uses sets (codons) of three bases each, and is the same in almost all organisms.

genetic engineering the deliberate manipulation of genetic material by biochemical techniques. It is often achieved by the introduction of new ◊DNA, usually by means of a virus. This can be for pure research or to breed new varieties of plants, animals or bacteria for specific purposes—for example, bacteria may be modified so that they secrete rare drugs. Developments in genetic engineering have already led to the industrial production of human insulin, human growth hormone, and a number of vaccines.

genetic fingerprinting technique used for determining the pattern of certain parts of the genetic material ◊DNA that is unique to each individual. The pattern can be determined from a sample of skin, hair, or semen. Like skin fingerprinting, genetic fingerprinting can accurately distinguish humans from one another, with the exception of identical twins. It is used in paternity testing, forensic medicine, and inbreeding studies.

genetics the study of inheritance and of the units of inheritance (◊genes). The founder of genetics was Gregor Mendel (1822–1884), whose experiments with plants, such as peas, showed that inheritance takes place by means of discrete 'particles', which later came to be called genes.

Before Mendel, it had been assumed that the characteristics of the two parents were blended during inheritance, but Mendel showed that the genes remain intact, although their combinations change. Since Mendel, genetics has advanced greatly, first through ◊breeding experiments and optical-microscope observations (classical genetics), later by means of biochemical and electron-microscope studies (molecular genetics). An advance was the elucidation of the structure of ◊DNA by James D Watson and Francis Crick, and the subsequent discovery of the ◊genetic code. These developments opened up the possibility of deliberately manipulating genes, or ◊genetic engineering. See also ◊genotype, ◊phenotype and ◊monohybrid inheritance.

genitalia the reproductive organs of sexually reproducing animals, particulary the external/visible organs of mammals: in males, the penis and the scrotum, which contains the testes, and in females, the clitoris and vulva.

genome the total information carried by the genetic code of a particular organism.

genotype the particular set of ◊alleles (variants of genes) possessed by a given organism. The term is usually used in conjunction with ◊phenotype.

germanium metallic element, symbol Ge, atomic number 32, relative atomic mass 72.6. It is a grey-white, brittle, crystalline metal in the silicon group, with chemical and physical properties between those of silicon and tin. Germanium is a semiconductor material and is used in the manufacture of transistors and integrated circuits.

germination the initial stages of growth in a seed, spore, or pollen grain. Seeds germinate when they are exposed to favourable external conditions of moisture, light, and temperature.

The process begins with the uptake of water by the seed. The embryonic root is normally the first organ to emerge, followed by the shoot. Food reserves are broken down to nourish the rapidly growing seedling. Germination is considered to have ended with the production of the first true leaves.

gestation the period from the time of implantation of the embryo in the uterus to birth. In humans it is about 266 days.

giant molecular structure or ***macromolecular structure*** solid structure made up of very many atoms joined by covalent bonds in one dimension (long chains, such as polymers), two dimensions (flat sheets, as in graphite), or three dimensions (structures such as diamond and silica).

gill the main respiratory organ of most fishes and immature amphibians, and of many aquatic invertebrates. In all types, water passes over the gills, and oxygen diffuses across the gill membranes into the circulatory system, while carbon dioxide passes from the system out into the water. See ◊gas exchange.

gland a specialized organ of the body that manufactures and secretes enzymes, hormones, or other chemicals. Glands vary in size from the small (for example, tear glands) to the large (for example, the pancreas).

glass brittle, usually transparent or translucent substance that is physically neither a solid nor a liquid. It is made by fusing certain types of sand (silica).

glucose or ***dextrose*** $C_6H_{12}O_6$ monosaccharide sugar present in the blood, and found in honey and fruit juices. It is a source of energy for the body; large quantities of sugar pass into the blood stream when starch is digested in the gut. The sugar travels to the cells, and is burnt by the process of ◊aerobic respiration. In spite of the importance of blood sugar, the actual levels remain fairly constant, a process of control organized by the liver and the pancreas (see ◊insulin).

glycogen or ***animal starch*** carbohydrate stored in the liver cells, made up of glucose molecules. When required as an energy source by the muscles, it is transported to the muscle tissue, where it is converted back to glucose by the hormone ◊insulin. It is then used up to produce energy.

gold heavy, precious, yellow, metallic element; symbol Au, atomic number 79, relative atomic mass 197.0. It is largely unaffected by temperature changes and is highly resistant to acids. For manufacture, gold is alloyed with another strengthening metal (such as copper or silver), its purity being measured in carats on a scale of 24. In 1990 the three leading gold-producing countries were South Africa, 605.4 tonnes; USA, 295 tonnes; and Russia, 260 tonnes.

Gold has long been valued for its durability, malleability, and ductility, and its uses include dentistry, jewellery, and electronic devices.

gonad the part of an animal's body that produces the sperm or ova (eggs) required for sexual reproduction. The sperm-producing gonad is called a ◊testis, and the ovum-producing gonad is called an ◊ovary.

grafting the operation by which a piece of living tissue is removed from one organism and transplanted into the same or a closely related organism where it continues growing.

graphite blackish-grey, laminar, crystalline allotrope of ◊carbon. It is used as a lubricant and as the active component of pencil lead.

The carbon atoms are covalently bonded together in sheets, but the bonds between the sheets are weak so that the sheets are free to slide over one another. Owing to the strong covalent bonds, graphite has a high melting point (3,500°C) and great mechanical strength and is a good conductor of heat and electricity. In its pure form it is used to control the chain reaction in nuclear reactors.

gravitational field the region around a body in which other bodies experience a force due to its gravitational attraction. The gravitational field of a massive object such as the Earth is very strong and easily recognized as the force of gravity, whereas that of an object of much smaller mass is very weak and difficult to detect.

The ***gravitational force of attraction*** F between two masses m_1 and m_2 is given by Newton's universal law of gravitation

$$F = Gm_1m_2/r^2$$

where G is the universal gravitational constant and r is the distance separating the centres of the two masses.

gravitational potential energy energy stored by an object when it is placed in a position from which it can fall under the influence of gravity. The gravitational potential energy E_p of an object of mass m kilograms placed at a height h metres above the ground is given by the formula

$$E_p = mgh$$

where ◊g is the gravitational field strength in newtons per kilogram of the Earth at that place.

If a body possessing gravitational potential energy is released and allowed to fall, then that energy will be converted into ◊kinetic energy. The velocity v of the falling body that has fallen h metres may therefore be calculated by equating the formulae for gravitational potential energy and kinetic energy

$$\text{kinetic energy } E_k = \text{gravitational potential energy } E_p$$

therefore

$$\frac{1}{2}mv^2 = mgh$$

or, assuming that the mass of the body remains the same while falling,

$$\frac{1}{2}v^2 = gh$$

In a ◊hydroelectric power station, gravitational potential energy stored in water held in a high-level reservoir is used to drive turbines to produce electricity.

gravity the force of attraction between objects because of their masses. The force we call gravity on Earth is the force of attraction between any object in the Earth's gravitational field and the Earth itself.

greenhouse effect a phenomenon of the Earth's atmosphere by which solar radiation, absorbed by the Earth and re-emitted from the surface, is prevented from escaping by carbon dioxide and other gases in the air. The result is a warming of the Earth's atmosphere; in a garden greenhouse, the glass walls have the same effect.

Since the Industrial Revolution and particularly since 1950, levels of carbon dioxide in the atmosphere have risen because of the burning of fossil fuels, causing an intensification of the greenhouse effect and a gradual increase in the Earth's temperature. For example, during the 1980s the temperature of the world's oceans rose by about 0.1°C a year; the Arctic ice was 6–7 m thick in 1976 and had reduced to 4–5 m by 1987. Increased concentrations of pollutant gases such as the ◊chlorofluorocarbons (CFCs), methane, and nitrous oxide are also believed to be contributing to global warming. The United Nations Environment Programme estimates that by the year 2025 average world temperatures will have risen by 1.5°C with a consequent rise of 20 cm in sea level. However, predictions about global warming and its possible climatic effects are tentative, and often conflict with each other.

group vertical column of elements in the ◊periodic table. Elements in a group have similar physical and chemical properties; for example, the alkali metals (group I: lithium, sodium, potassium, rubidium, caesium, and francium) are all highly reactive metals that form univalent ions. There is a gradation of properties down any group: in group I, melting and boiling points decrease, and density and reactivity increase.

greenhouse effect

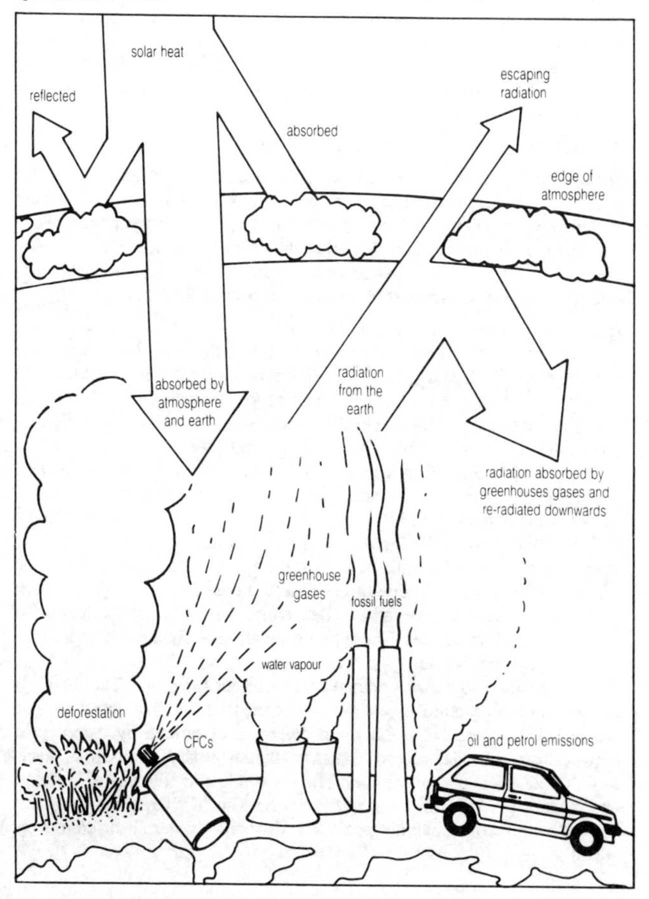

growth the increase in size and dry weight (weight excluding water) that takes place during an organism's development. It is associated with cell division.

Increase of size by expansion, as when a cell enlarges through taking in water, is not usually considered as biological growth because this process does not involve any increase in dry mass.

growth ring another name for ◊annual ring.

guard cell in plants, a specialized cell on the under surface of a leaf for controlling gas exchange and water loss by transpiration. Guard cells occur in pairs and are shaped so that a pore, or ◊stoma, exists between them. They can change shape with the result that the pore disappears. During warm weather, when a plant is in danger of losing excessive water, the guard cells close, cutting down evaporation from the interior of the leaf.

gut or ***alimentary canal*** in animals, a complex specialized tube through which food passes; it extends from the mouth to the anus. It is responsible for digesting food, and for the absorption of nutrients into the blood. In human adults, the gut is about nine metres long, consisting of the mouth, pharynx, oesophagus, stomach, small intestine (duodenum and ileum), large intestine (caecum, colon, and rectum), and anus.

A constant stream of enzymes from the gut wall and from the pancreas assists the breakdown of food molecules into smaller, soluble nutrient molecules, which are absorbed through the gut wall into the blood stream and carried to individual cells for assimilation. The muscles of the gut keep the incoming food moving, mix it with the enzymes and other juices, and slowly push it in the direction of the anus, a process known as ◊peristalsis. The wall of the gut receives an excellent supply of blood and is folded so as to increase its surface area. These two adaptations ensure efficient absorption of nutrient molecules.

Each region of the gut is adapted for different functions. The mouth is adapted for food capture, ingestion, and for the first stages of digestion. The stomach is a storage area, although digestion of protein by the enzyme pepsin starts here; in herbivorous mammals such as sheep and cattle, this is also the site of cellulose digestion. The small intestine follows the stomach and is specialized for digestion and for absorption. The large intestine has a variety of functions, including cellulose digestion, water absorption, storage of faeces, and egestion.

Haber process industrial process in which ammonia is manufactured by direct combination of its elements, nitrogen and hydrogen. The reaction is carried out at 400–500°C and at 200 atmospheres pressure. The two gases, in the proportions of 1:3 by volume, are passed over a ◊catalyst of finely divided iron. Around 10% of the reactants combine, and the unused gases are recycled. The ammonia is separated by either dissolving in water or cooling to liquid.

$$N_2 + 3H_2 \leftrightarrow 2NH_3$$

habitat the localized ◊environment in which an organism lives, and which provides for all (or almost all) of its needs. The diversity of habitats found within the Earth's ecosystem is enormous, and they are changing all the time. Many can be considered inorganic or physical, for example the Arctic ice cap, a cave, or a cliff face. Others are more complex, for instance a woodland or a forest floor. Some habitats are so precise that they are called ***microhabitats***, such as the area under a stone where a particular type of insect lives. Most habitats provide a home for many species.

haemoglobin protein used by all vertebrates and some invertebrates for oxygen transport because the two substances combine reversibly. In vertebrates it occurs in red blood cells (erythrocytes), giving them their colour.

In the lungs or gills where the concentration of oxygen is high, oxygen attaches to haemoglobin forming ***oxyhaemoglobin***. This process effectively increases the amount of oxygen that can be carried in the bloodstream. The oxygen is later released in the body tissues where it is at low concentration, and the deoxygenated blood returned to the lungs or gills. Haemoglobin will also combine with carbon monoxide to form carboxyhaemoglobin, but in this case the reaction is irreversible.

haemophilia inherited disorder in which normal blood clotting is impaired. The sufferer experiences prolonged bleeding from the slightest wound, as well as painful internal bleeding without apparent cause. Cases

haemoglobin

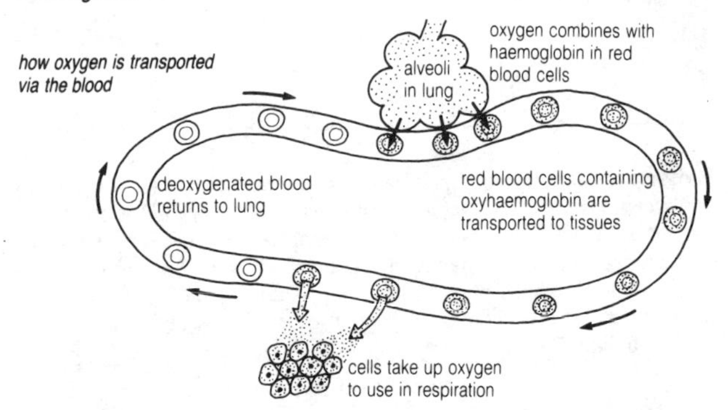

of haemophilia are nearly always sex-linked, transmitted through the female line only to male infants.

hair threadlike structure growing from mammalian skin. Each hair grows from a pit-shaped follicle embedded in the second layer of the skin, the dermis. It consists of dead cells impregnated with the protein keratin.

half-life during ◊radioactive decay, the time in which the strength of a radioactive source decays to half its original value. It may vary from millionths of a second to billions of years.

halogen any of a group of five non-metallic elements with similar chemical bonding properties: fluorine, chlorine, bromine, iodine, and astatine. They form a linked group (VII) in the periodic table of the elements, with fluorine the most reactive and astatine the least reactive. They combine directly with most metals to form salts, for example common salt (NaCl). A more reactive halogen will displace a less reactive halogen from solutions of its salt (see ◊displacement reaction). Each halogen has seven electrons in its valence shell, which accounts for the chemical similarities displayed by the group.

haploid having a single set of ◊chromosomes in each cell. Most higher organisms are ◊diploid—that is, they have two sets—but their gametes (sex cells) are haploid.

hard water water that does not lather easily with soap, and that produces 'fur' or 'scale' in kettles. It is caused by the presence of certain salts of calcium and magnesium.

Temporary hardness is caused by hydrogencarbonates. When water containing these is boiled, they are converted to insoluble carbonates that precipitate as 'scale'. ***Permanent hardness*** is caused by sulphates and silicates, which are not affected by boiling.

Water can be softened by ◊distillation, ◊ion exchange (the principle underlying commercial water softeners), addition of sodium carbonate, addition of large amounts of soap, or boiling (to remove temporary hardness). Liquid detergents readily form a lather with any water without forming a scum.

harmonics vibrations of a stretched string or air column that are multiples of the fundamental vibration. The particular combination of harmonics produced by a musical instrument give it its characteristic sound or ◊timbre.

hazard labels visual system of symbols for indicating the potential dangers of handling certain substances. The symbols used are recognized internationally.

heart muscular organ that rhythmically contracts to force blood around the body. The mammalian heart has four chambers—two thin-walled atria that expand to receive blood, and two thick-walled ventricles that pump it out once more.

heart beat the regular contraction and relaxation of the heart, and the accompanying sounds. As blood passes through the heart a double beat is heard. The first is produced by the sudden closure of the valves between the atria and the ventricles. The second, slightly delayed sound, is caused by the closure of the valves found at the entrance to the major arteries leaving the heart. Diseased valves may make unusual sounds, known as heart murmurs.

heat form of internal energy of a substance due to the kinetic energy in the motion of its molecules or atoms. Its SI unit is the joule (J). The extent to which a body will transfer or absorb heat (its hotness or coldness) is measured by temperature, and is related to the mean kinetic energy of its molecules. Heat energy is transferred by conduction, convection, and radiation. Heat always flows from a region of higher temperature to one of lower temperature unless made to do otherwise, as for example in a refrigerator. Its

hazard labels

effect on a substance may be simply to raise its temperature, cause it to expand, melt it if a solid, vaporize it if a liquid, or increase its pressure if a confined gas.

heart

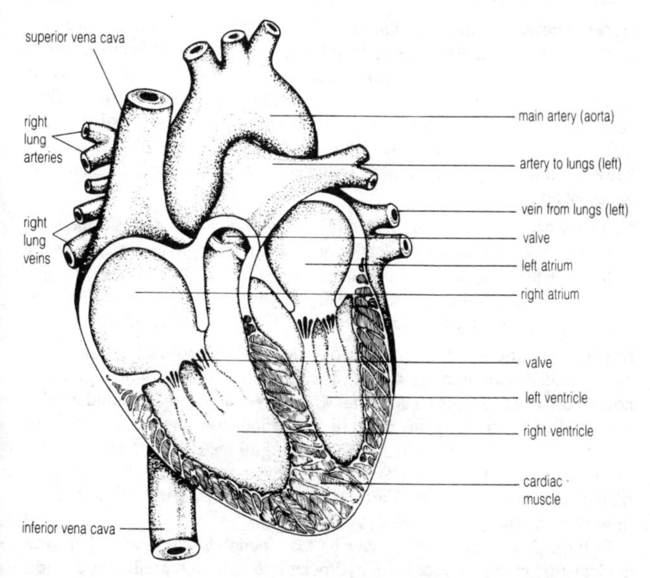

Convection is the transmission of heat through a fluid (liquid or gas) on currents—for example, when the air in a room is warmed by a fire or radiator.

Conduction is the passing of heat along a solid medium to neighbouring parts—for example, when the whole length of a metal rod is heated when one end is held in a fire.

Radiation is heat transfer by infrared rays. It can pass through a vacuum, travels at the same speed as light, can be reflected and refracted, and does not affect the medium through which it passes. For example, heat reaches the Earth from the Sun by radiation.

heat capacity or ***thermal capacity*** the heat energy required to increase the temperature of an object by one degree. It is measured in joules per degree, J/°C or J/K.

heat of reaction alternative term for ◊energy of reaction.

heavy metal metallic element of high relative atomic mass, for instance platinum, gold, and lead. Heavy metals are poisonous and tend to accumulate and persist in living systems, causing, for example, high levels of mercury (from industrial waste and toxic dumping) in shellfish and fish, which are in turn eaten by humans. Treatment of heavy-metal poisoning is difficult because available drugs are not able to distinguish between the heavy metals that are essential to living cells (zinc, copper) and those that are poisonous.

heavy water or ***deuterium oxide*** D_2O water containing the isotope deuterium instead of hydrogen (relative molecular mass 20 as opposed to 18 for ordinary water). Its chemical properties are identical with those of ordinary water, while its physical properties differ slightly. In the nuclear industry, it is used as a moderator to reduce the speed of high-energy neutrons.

helium colourless, odourless, gaseous element, symbol He, atomic number 2, relative atomic mass 4.0026. It is grouped with the ◊noble gases, is non-reactive, and forms no compounds. It is the second most abundant element (after hydrogen) in the universe, and has the lowest boiling (–268.9°C) and melting points (–272.2°C) of all the elements. It is present in small quantities in the Earth's atmosphere from gases issuing from radioactive elements in the Earth's crust; after hydrogen it is the second lightest element.

Helium is a component of most stars, including the Sun, where the nuclear-fusion process converts hydrogen into helium with the production of heat and light.

hematite the principal ore of iron, consisting mainly of iron(III) (ferric) oxide, Fe_2O_3. It occurs as ***specular hematite*** (dark, metallic lustre), ***kidney ore*** (reddish radiating fibres terminating in smooth, rounded surfaces), and as a red earthy deposit.

herbicide or ***weedkiller*** any chemical used to destroy plants or check their growth. Herbicides may be non-selective, killing all plants, or selective, killing only broad-leaved weeds and leaving cereal crops unharmed.

herbivore an animal that feeds on green plants or their products, including seeds, fruit, and nectar. The most numerous type of herbivore is thought to be the zooplankton, tiny invertebrates in the surface waters of the oceans that feed on small photosynthetic algae. Herbivores are more numerous than other animals because their food is the most abundant. They form a link in the ◊food chain between plants and carnivores.

Mammalian herbivores that rely on cellulose as a major part of their diet, for instance cows and sheep, generally possess millions of specialized bacteria in their gut. These are capable of producing the enzyme cellulase, necessary for digesting cellulose; no mammal can manufacture cellulase on its own.

heredity the transmission of traits from parent to offspring. See also ◊genetics.

hermaphrodite an organism that has both male and female sex organs. Hermaphroditism is the norm in species such as earthworms and snails, and is common in flowering plants. Self-fertilization frequently takes place in plant hermaphrodites, but cross-fertilization is the rule among animal hermaphrodites, with the parents functioning as male and female simultaneously, or as one or the other sex at different stages in their development.

hertz SI unit (symbol Hz) of frequency (the number of repetitions of a regular occurrence in one second). Radio waves are often measured in megahertz (MHz), millions of hertz. It is named after Heinrich Hertz.

heterozygous term describing the possession by an organism of two different ◊alleles for a given character. In homozygous organisms, by contrast, the alleles are identical. An individual organism will generally be heterozygous for some genes but homozygous for others.

hibernation a state of ◊dormancy in which certain animals spend the winter. It is associated with a dramatic reduction in all metabolic processes, accompanied by a fall in body temperature, breathing, and heart rate.

hinge joint in vertebrates, a ◊joint where movement occurs in one plane only. Examples are the elbow and knee, which are controlled by pairs of muscles, the ◊flexors and ◊extensors.

HIV (abbreviation for ***human immunodeficiency virus***) the infectious agent believed to cause ◊AIDS.

holdfast an organ found at the base of many seaweeds, attaching them to the sea bed. It may be a flattened, suckerlike structure, or dissected and fingerlike, growing into rock crevices and firmly anchoring the plant.

homeostasis the maintenance of a constant state in an organism's internal environment, particularly with regard to pH, salt concentration, temperature, and blood-sugar levels. Stable conditions are important for the efficient functioning of the ◊enzyme reactions within the cells, which affect the performance of the entire organism.

homologous series any of a number of series of organic chemicals whose members differ by a constant relative molecular mass.

Alkanes (paraffins), alkenes (olefins), and alkynes (acetylenes) form such series whose members differ in mass by 14, 12, and 10 atomic mass units respectively. For example, the alkane homologous series begins with methane (CH_4), ethane (C_2H_6), propane (C_3H_8), butane (C_4H_{10}), and pentane (C_5H_{12}), each member differing from the previous one by a CH_2 group (or 14 atomic mass units).

homozygous term describing the possession by an organism of two identical ◊alleles for a given character. Individuals that are homozygous for a given character always breed true; that is, they produce offspring that resemble them in appearance when crossed with a genetically similar individual. ◊Recessive alleles are only expressed in the homozygous condition. Compare ◊heterozygous.

Hooke's law law stating that the deformation of a body is proportional to the magnitude of the deforming force, provided that the body's elastic limit (see ◊elasticity) is not exceeded. If the elastic limit is not reached, the body will return to its original size once the force is removed.

hormone in animals, a substance produced by the ◊endocrine glands, concerned with control of body functions. Hormones are transported in the blood, and bring about changes in the functions of various target organs according to the body's requirements. The pituitary gland, at the base of the brain, is a centre for overall coordination of hormone secretion; the thyroid hormones determine the rate of general body chemistry; the adrenal hormones prepare the organism during stress for 'fight or flight'; and the sex hormones, such as oestrogen, govern reproductive functions.

host an organism that is parasitized by another.

human reproduction an example of ◊sexual reproduction, where the male produces sperm and the female ova (eggs). These gametes contain only half the normal number of chromosomes, 23 instead of 46, so that on fertilization the resulting cell has the correct genetic complement. Fertilization is internal, which increases the chances of conception; unusually for mammals, copulation and pregnancy can occur at any time of the year.

Human beings are also remarkable for the length of childhood and for the highly complex systems of parental care found in society. The use of contraception and the development of laboratory methods of insemination and fertilization are issues that make human reproduction more than a merely biological phenomenon.

humerus the upper bone of the forelimb of tetrapods. In humans the humerus is the bone above the elbow.

hybrid the offspring from a cross between individuals of two different species, or two inbred lines within a species. In most cases, hybrids between species are infertile and unable to reproduce sexually.

hybridization the production of a ◊hybrid.

hydration the combination of water and another substance to produce a single product. It is the opposite of ◊dehydration. For example, anhydrous copper(II) sulphate reacts with water to give copper(II) sulphate pentahydrate.

$$CuSO_4 + 5H_2O \rightarrow CuSO_4.5H_2O$$

hydrocarbon any of a class of compounds containing only hydrogen and carbon (such as methane). Hydrocarbons are obtained industrially principally from petroleum and coal tar.

hydrochloric acid HCl solution of hydrogen chloride (a colourless, acidic gas) in water. The concentrated acid is about 35% HCl and is corrosive. The acid is a typical strong, monobasic acid forming only one series of salts, the chlorides. When oxidized, for example by manganese(IV) oxide, it releases chlorine.

$$MnO_2 + 4HCl \rightarrow MnCl_2 + Cl_2 + 2H_2O$$

hydroelectric power (HEP) electricity generated by moving water. In a typical hydroelectric power scheme, water stored in a reservoir, often created by damming a river, is piped into water ◊turbines, coupled to electricity generators. In ◊pumped storage plants, water flowing through the turbines is recycled. A ◊tidal power station exploits the rise and fall of the tides. About one-fifth of the world's electricity comes from hydroelectric power.

hydrogen colourless, odourless, gaseous element, symbol H, atomic number 1, relative atomic mass 1.00797. It is the lightest of all the elements and occurs on Earth chiefly in combination with oxygen as water. Hydrogen is the most abundant element in the universe, where it accounts for 93% of the total number of atoms and 76% total mass. It is a component of most stars, including the Sun, whose heat and light are produced through the nuclear-fusion process, which converts hydrogen into helium.

hydrogenation the addition of hydrogen to an unsaturated organic molecule (one that contains ◊double bonds or ◊triple bonds). The process is widely used in the manufacture of margarine and low-fat spreads by the addition of hydrogen to vegetable oils.

hydrogencarbonate compound containing the ion HCO_3^-, an acid salt of carbonic acid (solution of carbon dioxide in water). When heated or treated with dilute acids, they evolve carbon dioxide. The most important compounds are ◊sodium hydrogencarbonate (bicarbonate of soda) and ◊calcium hydrogencarbonate.

hydrophilic ('water-loving') term describing ◊functional groups with a strong affinity for water, such as the carboxylic acid group (–COOH).

If a molecule contains both a hydrophilic and a ◊hydrophobic group, it may have an affinity for both aqueous and non-aqueous molecules. Such compounds are used to stabilize an ◊emulsion or as a ◊detergent.

hydrophobic ('water-hating') term describing ◊functional groups that repel water (compare ◊hydrophilic).

hydroxide inorganic compound containing one or more hydroxide ions (OH^-), combined with a metal. Hydroxides include sodium hydroxide (caustic soda, NaOH), potassium hydroxide (caustic potash, KOH), and calcium hydroxide (slaked lime, $Ca(OH)_2$).

hydroxyl group an atom of hydrogen and an atom of oxygen bonded together and covalently bonded to an organic molecule. Common compounds containing hydroxyl groups are alcohols and phenols. In chemical reactions, the hydroxyl group (–OH) frequently behaves as a single entity.

hygroscopic term used to describe a substance that can absorb moisture from the air without becoming wet. There is a maximum amount of water that any particular hygroscopic substance can absorb.

I

ice solid formed by water when it freezes. It is colourless and its crystals are hexagonal. The water molecules are held together by ◊hydrogen bonds.

The freezing point of water, used as a standard for measuring temperature, is 0° for the Celsius scale and 32° for the Fahrenheit. Ice expands in the act of freezing (leading to, for example, burst pipes), becoming less dense than liquid water (0.9175 at 5°C).

ignition temperature or ***fire point*** the minimum temperature to which a substance must be heated before it will spontaneously burn independently of the source of heat; for example, ethanol has an ignition temperature of 425°C, and a ◊flash point of 12°C.

ileum part of the small intestine of the ◊digestive system, between the duodenum and the colon, that absorbs digested food. Its wall is muscular so that waves of contraction (peristalsis) can mix the food and push it forward. Numerous fingerlike projections, or villi (see ◊villus), point inwards from the wall, increasing the surface area available for absorption. The ileum has an excellent blood supply, which receives the food molecules passing through the wall and transports them to the liver via the hepatic portal vein.

image a picture or appearance of a real object, formed by light that passes through a lens or is reflected from a mirror. If rays of light actually pass through an image, it is called a ***real image***. Real images, such as those produced by a camera or projector lens, can be projected onto a screen. An image that cannot be projected onto a screen, such as that seen in a flat mirror, is known as a ***virtual image***.

immiscible term describing liquids that will not mix with each other, such as oil and water. When two immiscible liquids are shaken together, a turbid mixture will be produced. This normally forms separate layers on standing.

immunity the protection that organisms have against foreign microorganisms, such as bacteria and viruses, and against cancerous cells. The cells that provide this protection are called ◊white blood cells, or leucocytes.

Immunity is also provided by a range of physical and chemical barriers, such as the skin, tear fluid, acid in the stomach, and mucus in the airways. AIDS is one of the many viral diseases in which the immune system is affected.

implantation in mammals, the process by which the developing ◊embryo attaches itself to the wall of the mother's uterus and stimulates the development of the ◊placenta.

impulse in mechanics, the product of a force and the time over which it acts. An impulse applied to a body causes its ◊momentum to change and is equal to that change in momentum. It is measured in newton seconds.

For example, the impulse J given to a football when it is kicked is given by

$$J = Ft$$

where F is the kick force in newtons and t is the time in seconds for which the boot is in contact with the ball.

inbreeding the mating of closely related individuals. It is considered undesirable because it increases the risk that an offspring will inherit copies of rare deleterious recessive ◊alleles from both parents and so suffer from disabilities.

incisor sharp tooth at the front of the mammalian mouth. Incisors are used for biting or nibbling, as when a rabbit or a sheep eats grass. Rodents, such as rats and squirrels, have large, continuously growing incisors, adapted for gnawing. The elephant tusk is a greatly enlarged incisor.

inclined plane or ***ramp*** slope that allows a load to be raised gradually using a smaller effort than would be needed if it were lifted vertically upwards. It is a ◊force multiplier, possessing a ◊mechanical advantage greater than one. Bolts and screws are based on the principle of the inclined plane.

indicator compound that changes its structure and colour in response to its environment. The commonest chemical indicators detect changes in ◊pH, such as ◊litmus, or in the oxidation state of a system (redox indicators).

indicator species a plant or animal whose presence or absence in an area indicates certain environmental conditions, such as soil type, high levels of pollution, or, in rivers, low levels of dissolved oxygen. Many plants show a preference for either alkaline or acid soil conditions, while certain trees require aluminium, and are found only in soils where it is present. Some lichens are sensitive to sulphur dioxide in the air, and absence of these species indicates atmospheric pollution.

induced current electric current that appears when there is relative movement of a conductor (for instance a simple circuit) in a magnetic field. The effect is known as the ***dynamo effect***, and is used in all ◊dynamos and generators to produce electricity. See also ◊electromagnetic induction.

induced current

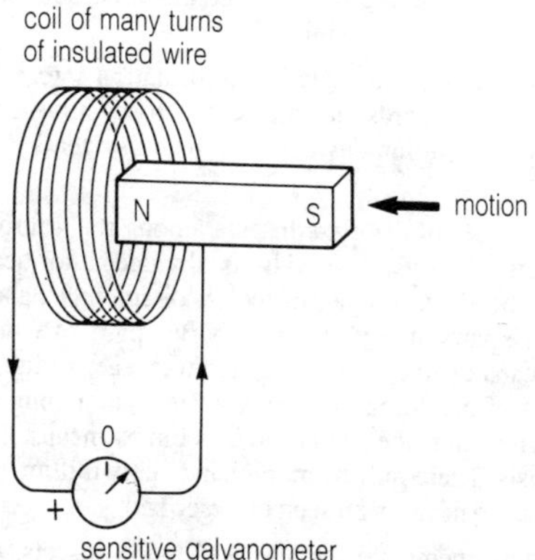

induction alteration in the physical properties of a body that is brought about by the influence of a field. See ◊electromagnetic induction and ◊magnetic induction.

inert gas alternative name for ◊noble gas.

inertia the tendency of an object to remain in a state of rest or uniform motion until an external force is applied, as stated by Newton's first law of motion (see ◊Newton's laws of motion).

infrared radiation invisible electromagnetic radiation of wavelength between about 0.75 micrometres and 1 millimetre, that is, between the limit of the red end of the visible spectrum and the shortest microwaves. All bodies above the ◊absolute zero of temperature absorb and radiate infrared radiation. Infrared radiation is used in medical photography and treatment, and in industry, astronomy, and criminology.

ingestion the process of taking food into the mouth. The method of ◊food capture varies but may involve biting, sucking, or filtering. Many single-celled organisms have a region of their cell wall that acts as a mouth. In these cases surrounding tiny hairs (cilia) sweep food particles together, ready for ingestion.

inorganic chemistry branch of chemistry dealing with the elements and their compounds, excluding the more complex carbon compounds which are considered in ◊organic chemistry.

input device a device for entering information into a computer. Input devices include keyboards, joysticks, touch-sensitive screens, graphics tablets, speech recognition devices, and vision systems. Compare ◊output device.

insect any member of the class Insecta among the arthropods or jointed-legged animals. An insect's body is divided into head, thorax, and abdomen. The head bears a pair of feelers or antennae, and attached to the thorax are three pairs of legs and usually two pairs of wings. Most insects breathe by means of fine air tubes called tracheae, which open to the exterior by a pair of breathing pores (spiracles). The young (larvae) of most insects do not resemble the adults and develop by means of a process called ◊metamorphosis. There may be more than twenty million species of insect; most of these have never been seen or described.

insecticide any chemical pesticide used to kill insects. Among the most effective insecticides are synthetic chlorinated organic chemicals such as DDT; however, these have been shown to accumulate in the environment and to be poisonous to all animal life, including humans, and are consequently banned in many countries.

insulation a method of preventing the flow of heat, electricity or sound.

Electrical insulation makes use of materials such as rubber, PVC, and porcelain, which do not conduct electricity, to prevent a current from leaking from one conductor to another or down to the ground.

Double insulation is a method of constructing electrical appliances that provides extra protection from electric shock, and renders the use of an earth wire unnecessary. In addition to the usual cable insulation, an appliance that meets the double insulation standard is totally enclosed in an insulating plastic body or structure so that there is no direct connection between any external metal parts and the internal electrical components.

Thermal or ***heat insulation*** makes use of insulating materials such as fibreglass to reduce the loss of heat through the roof and walls of buildings. Air trapped between the fibres of clothes acts as a thermal insulator, preventing loss of body warmth.

insulator any poor ◊conductor of heat, sound, or electricity. Most substances lacking free (mobile) ◊electrons, such as non-metals, are electrical or thermal insulators.

insulin hormone, produced by specialized cells in the ◊pancreas, important in controlling blood sugar levels. Insulin is produced in response to rising levels of sugar, and results in glucose being stored in the liver as ◊glycogen. This returns the sugar level to the correct state. People who are diabetic are unable to produce insulin. Without treatment their blood sugar concentrations can frequently approach dangerously high or low levels.

integrated circuit (IC) or ***silicon chip*** a miniaturized electronic circuit produced on a single crystal, or chip, of a semiconducting material such as silicon. It may contain many thousands of components and yet measure

integrated circuit

the packaging of a silicon 'chip'

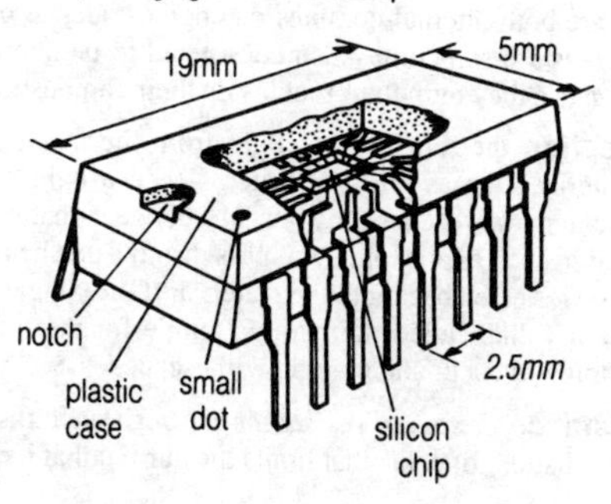

only 5 mm square and 1 mm thick. The IC is encapsulated within a plastic or ceramic case, and linked via gold wires to metal pins with which it is connected to a ◊printed circuit board and the other components that make up electronic devices such as computers and calculators.

intercostal muscle muscle found between the ribs, responsible for producing the rib-cage movements involved in some types of breathing.

When the intercostal muscles contract the ribs move upwards and outwards, enlarging the thorax and causing air to rush into the lungs. On relaxation the ribs move downwards under their own weight, and air is pushed out of the lungs. This type of breathing complements the more gentle contractions of the ◊diaphragm, and occurs for instance during exercise.

interference the phenomenon of two or more wave motions interacting and combining to produce a resultant wave of larger or smaller amplitude (depending on whether the combining waves are in or out of ◊phase with each other).

Interference of white light (multiwavelength) results in spectral coloured fringes, for example, the iridescent colours of oil films seen on water or soap bubbles. Interference of sound waves of similar frequency produces the phenomenon of beats, often used by musicians when tuning an instrument. With monochromatic light (of a single wavelength), interference produces patterns of light and dark bands.

internal-combustion engine engine in which fuel is burned inside the engine itself, contrasting with an external-combustion engine (such as the steam engine) in which fuel is burned in a separate unit. The diesel and petrol engines are both internal-combustion engines. Gas turbines, and jet and rocket engines are sometimes also considered to be internal-combustion engines because they burn their fuel inside their combustion chambers.

internal reflection the reflection of light from the inside surface of a transparent material such as glass or water. When a light ray travelling through a dense material reaches the boundary between that material and a less dense medium such as air, some will pass through and be refracted, but some will also be reflected back into the material. If the ◊angle of incidence exceeds a certain value, called the ◊critical angle for that material, total internal reflection will occur and no light will escape.

internal resistance or ***source resistance*** the resistance inside a power supply, such as a battery of cells, that limits the current that it can supply to a circuit. For example, in order to supply the high current (hundreds of amperes) necessary to work the starter motor in a car, a battery with a very low internal resistance is required. One effect of internal resistance is to cause a drop in the ◊terminal voltage of a supply (the potential difference across its terminals) when that supply is connected in circuit; the difference between its terminal voltage and its electromotive force (emf) increases as current flow increases (see ◊Ohm's law).

intrauterine device (IUD) or ***coil*** a ◊contraceptive device that is inserted into the uterus (womb). It is a tiny plastic object, sometimes containing copper. By causing a mild inflammation of the lining of the uterus it prevents fertilized eggs from becoming implanted.

inverse square law the statement that the magnitude of an effect (usually a force) at a point is inversely proportional to the square of the distance between that point and the point location of its cause.

Light, sound, electrostatic force, gravitational force (Newton's law) and magnetic force (see ◊magnetism) all obey the inverse square law.

invertebrate an animal without a backbone. The invertebrates comprise over 95% of the million or so known animal species and include the sponges, coelenterates (such as jellyfishes and corals), flatworms, nematodes (roundworms), annelid worms (segmented worms), arthropods (such as insects, spiders, millipedes, and crabs), molluscs (such as snails and squid), echinoderms (such as starfishes and urchins) and primitive aquatic chordates (such as sea-squirts and lancelets).

***in vitro* process** an experiment or technique carried out in the laboratory outside a living organism (literally 'in glass', for example in a test tube).

***in vivo* process** an experiment or technique carried out within a living organism.

involuntary muscle or ***smooth muscle*** muscle capable of slow contraction over a period of time. Its presence in the gut wall allows slow rhythmic movements known as ◊peristalsis, which cause food to be mixed and forced along the gut. It is also found in the ◊iris of the eye, where it plays a part in adjusting the diameter of the pupil in response to changing light intensity.

iodide I^- salt of the halide series. When a silver nitrate solution is added to any iodide solution containing dilute nitric acid, a yellow precipitate of silver iodide is formed.

$$KI_{(aq)} + AgNO_{3\,(aq)} \rightarrow AgI_{(s)} + KNO_{3\,(aq)}$$

When chlorine is passed into a solution of an iodide salt, the solution turns brown, then forms a black precipitate as iodine is produced by a ◊displacement reaction.

$$Cl_{2(g)} + 2I^{-}_{(aq)} \rightarrow 2Cl^{-}_{(aq)} + I_{2(s)}$$

iodine greyish-black, non-metallic element, symbol I, atomic number 53, relative atomic mass 126.9044. It is a member of the ◊halogen group. Its crystals give off, when heated, a violet vapour with an irritating odour resembling that of chlorine. It only occurs in combination with other elements. Its salts are known as iodides, which are found in sea water. As a mineral nutrient it is vital to the proper functioning of the thyroid gland, where it occurs in trace amounts as part of the hormone thyroxine.

ion an atom, or group of atoms, which is either positively charged (***cation***) or negatively charged (***anion***), as a result of the loss or gain of electrons during chemical reactions or exposure to certain forms of radiation.

ion exchange process whereby the ions in one compound replace the ions in another. The exchange occurs because one of the products is insoluble in water. For example, when hard water is passed over an ion-exchange resin, the dissolved calcium and magnesium ions are replaced by either sodium or hydrogen ions, so the hardness is removed. Commercial water softeners use ion-exchange resins.

ionic bond or ***electrovalent bond*** bond produced when atoms of one element donate electrons to atoms of another element, forming positively and negatively charged ◊ions respectively. The electrostatic attraction between the oppositely charged ions constitutes the bond. Sodium chloride (Na^+Cl^-) is a typical ionic compound.

Each ion has the electronic structure of a noble gas (see ◊noble-gas structure). The maximum number of electrons that can be gained is usually two and the maximum that can be lost is three.

ionic compound substance composed of oppositely charged ions. All salts, most bases, and some acids are examples of ionic compounds. They possess the following general properties: they are crystalline solids with a high melting point; are soluble in water and insoluble in organic solvents; and always conduct electricity when molten or in aqueous solution. A typical ionic compound is sodium chloride (Na^+Cl^-).

ionic bond

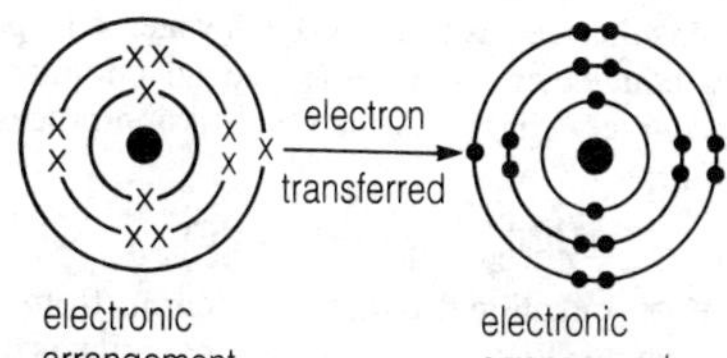

electronic arrangement, 2.8.1 of a sodium atom,

electronic arrangement, 2.8.7 of a chlorine atom,

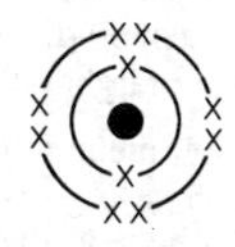

becomes a sodium ion, Na^+, with an electron arrangement 2.8

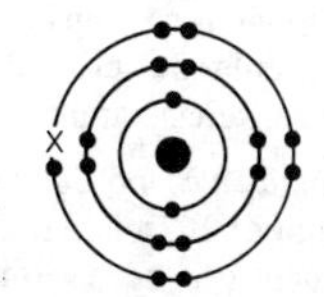

becomes a chloride ion, Cl^-, with an electron arrangement 2.8.8

ionic half equation equation that describes the reactions occurring at the electrodes of a chemical cell or in electrolysis. It indicates which ion is losing electrons (oxidation) or gaining electrons (reduction). Examples are given from the electrolysis of dilute hydrochloric acid (HCl).

positive electrode:

$$2Cl^- - 2e^- \rightarrow Cl_2$$

negative electrode:

$$2H^+ + 2e^- \rightarrow H_2$$

ionization the process of ion formation. It can be achieved in two ways. The first is by the loss or gain of electrons by atoms to form positive or negative ions.

$$Na - e^- \rightarrow Na^+$$

$$^1/_2\,Cl_2 + e^- \rightarrow Cl^-$$

In the second mechanism, ions are formed when a covalent bond breaks, as when hydrogen chloride gas is dissolved in water. One portion of the the molecule retains both electrons, forming a negative ion, and the other portion becomes positively charged. This bond-fission process is sometimes called ***dissociation***.

$$HCl_{(g)} + aq \leftrightarrow H^+_{(aq)} + Cl^-_{(aq)}$$

ionizing radiation radiation that knocks electrons from atoms during its passage, thereby leaving ions in its path. Such radiation is damaging to biological tissue. Alpha and beta particles are far more ionizing than either neutrons or gamma radiation.

iris the coloured muscular diaphragm that controls the size of the pupil in the vertebrate ◊eye. It contains longitudinal muscle that increases the pupil diameter and circular muscle that constricts the pupil diameter. The movement of the iris ensures that the correct amount of light falls on the retina.

iron hard, malleable and ductile, silver-grey, metallic element, symbol Fe, atomic number 26, relative atomic mass 55.847. It is the fourth most abundant element (the second most abundant metal, after aluminium) in the Earth's crust. Iron occurs in concentrated deposits as the ores hematite (Fe_2O_3), spathic ore ($FeCO_3$), and magnetite (Fe_3O_4). It sometimes occurs as a free metal, occasionally as fragments of iron or iron–nickel meteorites.

The metal forms two series of salts: iron(II) (ferrous) and iron(III) (ferric). The metal has the following chemical properties.

with dry air or oxygen When heated in air or oxygen, iron forms the oxide.

$$3Fe + 2O_2 \rightarrow Fe_3O_4$$

with steam When it is heated with steam in the absence of air, iron reduces the steam to hydrogen and forms the oxide.

$$3Fe + 4H_2O \rightarrow Fe_3O_4 + 4H_2$$

with dilute acids Iron forms iron(II) salts with acids, and hydrogen is evolved.

$$Fe + H_2SO_4 \rightarrow FeSO_4 + H_2$$

with air and water In moist air, iron rusts (forms hydrated iron(III) oxide).

$$4Fe + 2H_2O + 3O_2 \rightarrow 2Fe_2O_3.H_2O$$

with chlorine Iron forms iron(III) chloride when reacted with chlorine gas.

$$2Fe + 3Cl_2 \rightarrow 2FeCl_3$$

with other metals in solution Iron displaces less reactive metals when added to a solution of their salts (see ◊reactivity series).

$$Fe_{(s)} + CuSO_{4(aq)} \rightarrow FeSO_{4(aq)} + Cu_{(s)}$$

Iron is the commonest and most useful of all metals; it is strongly magnetic and is the basis for ◊steel, an alloy with carbon and other elements. In electrical equipment it is used in all permanent magnets and electromagnets, and the cores of transformers and magnetic amplifiers. See also ◊cast iron. In the human body, iron is an essential component of haemoglobin,

isomer

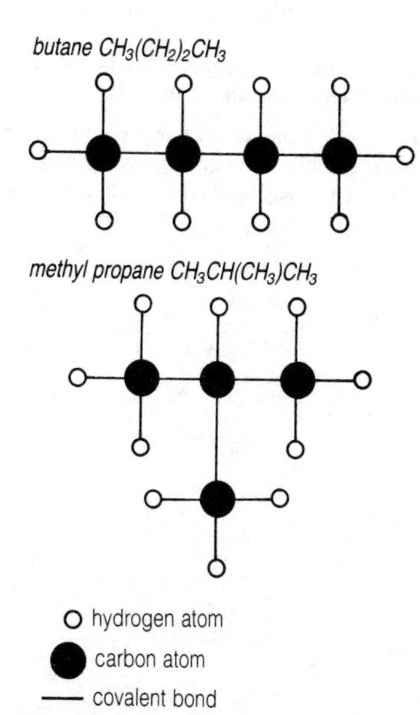

the molecule in red blood cells that transports oxygen to all parts of the body. A deficiency in the diet causes a form of anaemia.

irrigation supplying water to dry agricultural areas by means of artificial dams and channels. Irrigation has been practised for thousands of years.

islets of Langerhans group of cells within the pancreas responsible for the secretion of the hormone ◊insulin. They are sensitive to blood-glucose levels, producing more hormone when glucose levels rise.

isobar line drawn on a synoptic weather chart linking all the places that have the same atmospheric pressure. The distance between isobars on a chart indicates the pressure gradient, rather as the contour lines on a map show the relief of the land.

isomer compound having the same molecular composition and mass as another but with a different arrangement of its atoms. It has the same molecular formula but a different structural formula—for example, butane has two isomers, straight-chain and branched, C_4H_{10} and $CH_3CH(CH_3)$.

isotope one of two or more atoms that have the same proton (atomic) number, but which contain a different number of neutrons, thus differing in their mass numbers. They may be stable or radioactive, naturally occurring or synthetic.

J

joint in vertebrates, the point at which two bones meet. Some joints allow no motion (the sutures of the skull). Most allow a relatively free motion. Of these, some allow a gliding motion (one vertebra of the spine on another, the bones of the wrist), some have a hinge action (elbow and knee), and others allow motion in all directions (hip and shoulder joints), by means of a ball-and-socket arrangement.

The ends of the bones at a moving joint are covered with cartilage for greater elasticity and smoothness, and are enclosed in an envelope (capsule) of tough white fibrous tissue lined with a membrane. This membrane secretes a lubricating and cushioning ◊synovial fluid. The joint is further strengthened by ligaments.

joint

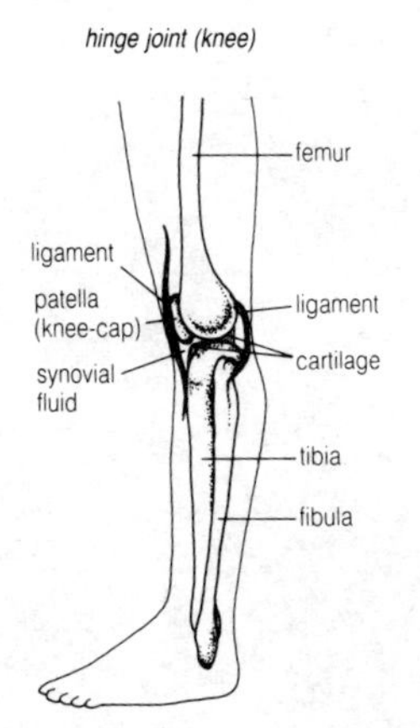

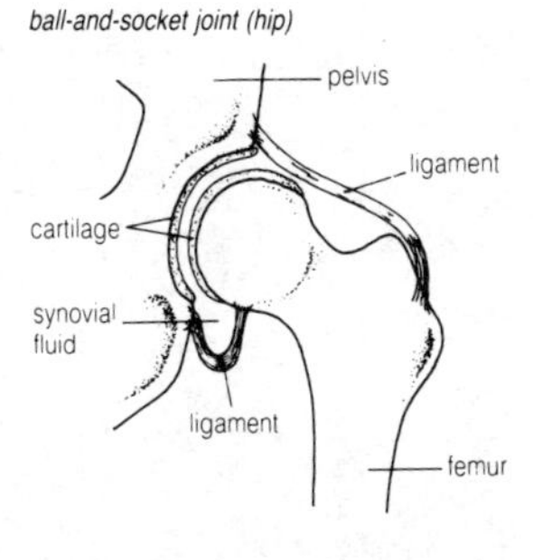

joule SI unit (symbol J) of work and energy. It is defined as the work done (energy transferred) by a force of one newton acting over one metre. It can also be expressed as the work done in one second by a current of one ampere at a potential difference of one volt. One watt is equal to one joule per second.

jugular vein one of two veins in the neck of vertebrates; they return blood from the head to the superior (or anterior) vena cava and from there to the heart.

Kelvin scale temperature scale used by scientists. It begins at ◊absolute zero (–273.16°C) and increases by the same degree intervals as the ◊Celsius scale; that is, 0°C is the same as 273K and 100°C is 373K.

keratin fibrous protein found in the ◊skin of vertebrates and also in hair, nails, claws, hooves, feathers, and the outer coating of horns in animals such as cows and sheep.

kerosene thin oil obtained from the distillation of petroleum; a highly refined form is used in jet aircraft fuel. Kerosene is a mixture of hydrocarbons.

ketone member of the group of organic compounds containing the carbonyl group (–CO–) bonded to two atoms of carbon (instead of one carbon and one hydrogen as in ◊aldehydes). Ketones are liquids or low-melting-point solids, slightly soluble in water. The simplest is propanone (acetone, CH_3COCH_3), used as a solvent.

key a method of identifying an organism. The investigator is presented with sets of statements, for example 'flower has less than five stamens' and 'flower has five or more stamens'. By successively eliminating statements the investigator moves closer to a positive identification. Identification keys assume a good knowledge of the subject under investigation.

kidney in vertebrates, one of a pair of organs responsible for water regulation, excretion of waste products, and maintaining the ionic composition of the blood. The kidneys are situated on the rear wall of the abdomen. Each one consists of a number of long tubules; the outer parts filter the aqueous components of blood, and the inner parts selectively reabsorb vital salts, leaving waste products in the remaining fluid (urine), which is passed through the ureter to the bladder.

The action of the kidneys is vital, although if one is removed, the other enlarges to take over its function. A patient with two defective kidneys may continue near-normal life with the aid of a kidney machine. Kidneys can be successfully transplanted.

kidney

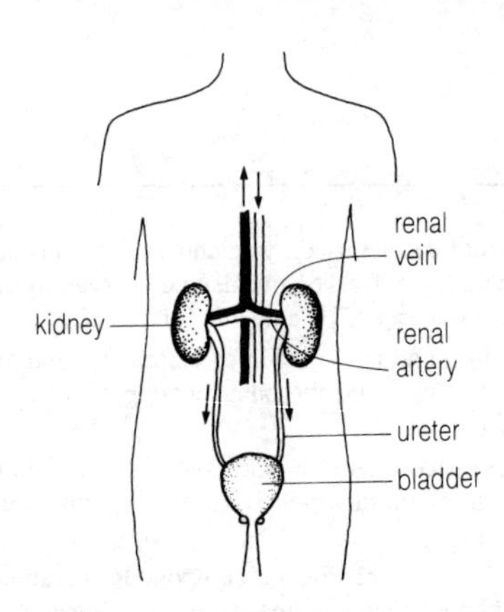

cross section

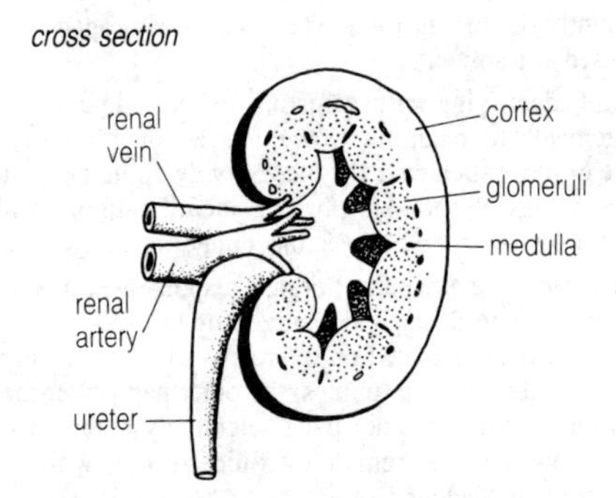

kilowatt-hour commercial unit of electrical energy (symbol kWh), defined as the work done by a power of 1,000 watts in one hour. It is used to calculate the cost of electrical energy taken from the ◊mains electricity supply.

kinetic energy a form of ◊energy possessed by moving bodies. It is contrasted with ◊potential energy.

The kinetic energy of a moving body is equal to the work that would have to be done in bringing that body to rest, and depends up on both the body's mass and speed. The kinetic energy E_k (in joules) of a mass m kilograms travelling with speed v metres per second is given by the formula:

$$E_k = {}^1/_2 mv^2$$

All atoms and molecules possess a certain amount of kinetic energy because they are all in some state of motion (see ◊kinetic theory). Adding heat energy to a substance increases the mean kinetic energy and therefore the mean speed of its constituent molecules—a change that is reflected as a rise in the temperature of that substance.

kinetics the branch of chemistry that investigates the rates of chemical reactions.

kinetic theory theory describing the physical properties of matter in terms of the behaviour—principally movement—of its component atoms or molecules. A gas consists of rapidly moving atoms or molecules and, according to kinetic theory, it is their continual impact on the walls of the containing vessel that accounts for the pressure of the gas.

The slowing of molecular motion as temperature falls, according to kinetic theory, accounts for the physical properties of liquids and solids, culminating in the concept of no molecular motion at ◊absolute zero (0K/–273°C).

krypton colourless, odourless, gaseous, non-metallic element, symbol Kr, atomic number 36, relative atomic mass 83.80. It is grouped with the noble gases and was long believed not to enter into reactions, but it is now known to combine with fluorine under certain conditions; it remains inert to all other reagents. It is present in very small quantities in the air (about 114 parts per million). It is used chiefly in fluorescent lamps, lasers, and gas-filled electronic valves.

kwashiorkor severe protein deficiency in children under five years old, resulting in retarded growth and a swollen abdomen.

L

labelled compound or ***tagged compound*** compound in which a radioactive isotope is substituted for a stable one. Thus labelled, the path taken by the compound through a system can be followed, for example by measuring the radiation emitted. This powerful and sensitive technique is used in medicine, chemistry, biochemistry, and industry.

lactation the secretion of milk from the mammary glands of mammals. In late pregnancy, the cells lining the lobules inside the mammary glands begin extracting substances from the blood to produce milk. The supply of milk starts shortly after birth with the production of colostrum, a clear fluid consisting largely of water, protein, antibodies, and vitamins. The production of milk continues practically as long as the infant continues to suck.

lacteal small vessel responsible for absorbing fat in the small intestine. Occurring in the fingerlike villi (see ◊villus) of the ileum, lacteals have a milky appearance and drain into the lymphatic system.

lactic acid or ***hydroxypropanoic acid*** $CH_3CHOHCOOH$ organic acid, a colourless, almost odourless syrup, produced by certain bacteria during fermentation. It occurs in yoghurt, buttermilk, sour cream, wine, and certain plant extracts; it is present in muscles when they are exercised hard, and also in the stomach. See ◊oxygen debt and ◊anaerobic respiration. It is used in food preservation and in the preparation of pharmaceuticals.

lactose tasteless, white, disaccharide sugar, found in solution in milk; it forms 5% of cow's milk. Each lactose molecule is made up of two monosaccharides glucose and galactose. It is prepared commercially from the whey obtained in cheese-making.

lamp, electric device designed to convert electrical energy into light energy.

In a ***filament lamp***, such as a light bulb, an electric current causes heating of a long thin coil of fine high-resistance wire enclosed at low pressure inside a glass bulb. In order to give out a good light the wire must glow white-hot and therefore must be made of a metal, such as tungsten, that has a high melting point. The efficiency of filament lamps is low because most of the electrical energy is converted to heat.

A ***fluorescent light*** uses an electrical discharge or spark inside a gas-filled tube to produce light. The inner surface of the tube is coated with a fluorescent material that converts the ultraviolet light generated by the discharge into visible light. Although a high voltage is needed to start the discharge, these lamps are far more efficient than filament lamps at producing light.

large intestine the lower ◊gut or bowels, made up of the colon, the caecum, and the rectum. No absorption of food takes place in the large intestine but the colon removes water from the undigested material, which is then stored as faeces in the rectum.

larva the stage between hatching and adulthood in those species in which the young have a different appearance and way of life from the adults. Examples include tadpoles (frogs) and caterpillars (butterflies and moths). The process whereby the larva changes into another stage, such as a pupa (chrysalis) or adult, is known as ◊metamorphosis.

larynx in mammals, a cavity at the upper end of the trachea (windpipe), containing the vocal cords. It is stiffened with cartilage and lined with mucous membrane.

laser (acronym for ***l****ight* ***a****mplification by* ***s****timulated* ***e****mission of* ***r****adiation*) a device for producing a narrow beam of light, capable of travelling over vast distances without dispersion and of being focused to give enormous power intensities (10^8 watts per square centimetre metre for high-energy lasers). Uses of lasers include communications (laser beams can carry far more information than can radio waves), cutting, drilling, welding, satellite tracking, medical and biological research, and surgery.

latent heat heat that changes the state of a substance (for example, from solid to liquid) without changing its temperature.

LCD abbreviation for ◊liquid crystal display.

LDR abbreviation for ◊light-dependent resistor.

leaching process by which substances are washed out of the ◊soil. Fertilizers leached out of the soil find their way into rivers and cause water ◊pollution. In tropical areas, leaching of the soil after ◊deforestation removes scarce nutrients and leads to a dramatic loss of soil fertility.

lead heavy, soft, malleable, grey, metallic element, symbol Pb, atomic number 82, relative atomic mass 207.19. Usually found as an ore (most often in galena), it occasionally occurs as a free metal. It is the final stable product of the decay of uranium. It is the softest and weakest of the commonly used metals, with a low melting point (hence the derivation of its name); it is a poor conductor of electricity and resists acid corrosion. Lead is a cumulative poison that enters the body from lead water pipes, lead-based paints, and leaded petrol. It is an effective shield against radiation and is used in batteries, glass, ceramics, and alloys, such as pewter and solder.

leaf lateral outgrowth on the stem of a plant, and in most species the primary organ of ◊photosynthesis.

Typically leaves are composed of three parts: the sheath or leaf base, the petiole or stalk, and the ◊lamina or blade. The lamina has a network of veins through which water and nutrients are conducted. Structurally the leaf is made up of ◊mesophyll cells surrounded by the epidermis and usually, in addition, a waxy layer, called the ◊cuticle, which prevents excessive evaporation of water from the leaf tissues by ◊transpiration. The epidermis is interrupted by small pores, or stomata (see ◊stoma), through which ◊gas exchange occurs.

A ***simple leaf*** is undivided, as in the beech or oak. A ***compound leaf*** is composed of several leaflets, as in the blackberry, horse chestnut, or ash tree. Leaves that fall in the autumn are called ***deciduous***, while evergreen leaves are called ***persistent***.

Le Chatelier's principle the principle that if a change in conditions is imposed on a system in equilibrium, the system will react to counteract that change and restore the equilibrium.

LED abbreviation for ◊light-emitting diode.

legume plant belonging to the pea family, which has a pod containing dry fruits. Legumes are important in agriculture because of their specialized roots, used to fix nitrogen in the soil (see ◊nitrogen fixation).

lens in optics, a piece of a transparent material, such as glass, possessing two polished surfaces—one concave or convex, and the other plane, concave or convex—to modify rays of light. A convex lens brings rays of light together; a concave lens makes the rays diverge. Lenses are essential to spectacles, microscopes, telescopes, cameras, and almost all optical instruments.

leaf

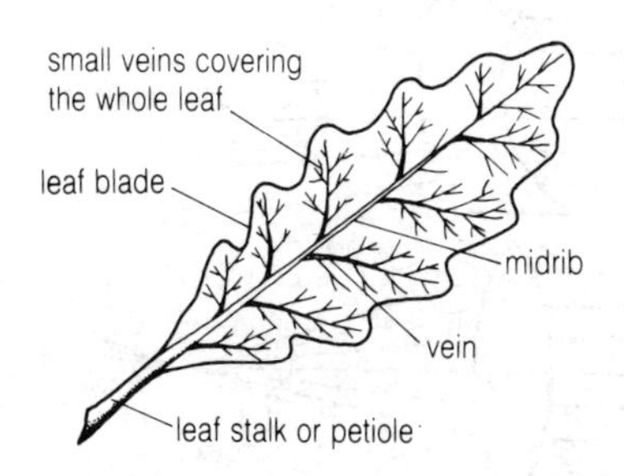

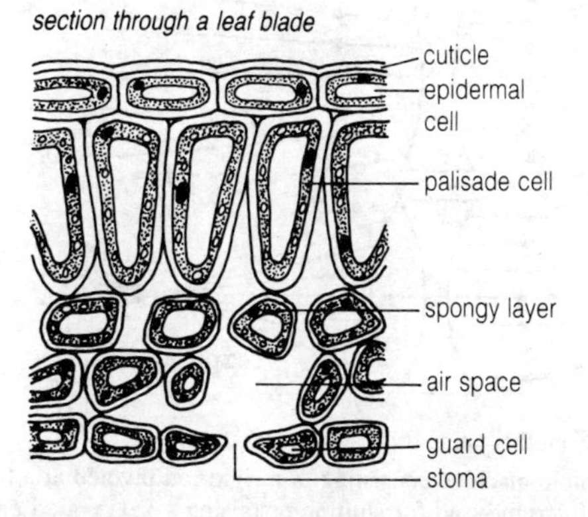

The image formed by a single lens suffers from several defects or aberrations, notably spherical aberration in which a straight line becomes a curved image, and chromatic aberration in which an image in white light tends to have coloured edges. Aberrations are corrected by the use of compound lenses, which are built up from two or more lenses of different refractive index.

lens

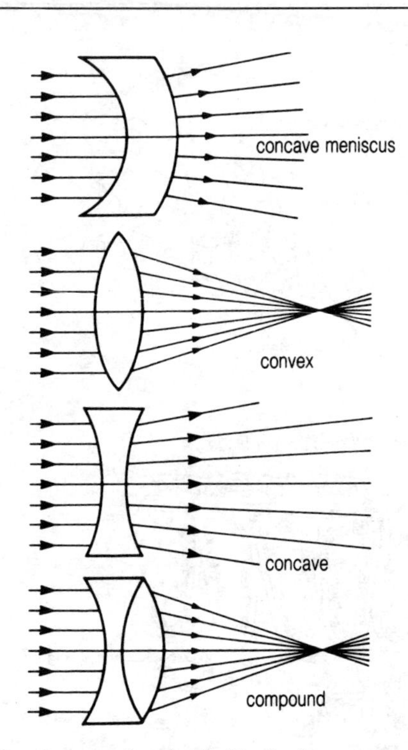

leucocyte another name for a ◊white blood cell.

lever a simple machine consisting of a rigid rod pivoted at a fixed point called the fulcrum, used for shifting or raising a heavy load or applying force in a similar way. Levers are classified into orders according to where the effort is applied, and the load-moving force developed, in relation to the position of the fulcrum.

A ***first-order*** lever has the load and the effort on opposite sides of the fulcrum—for example, a see-saw or pair of scissors.

lever

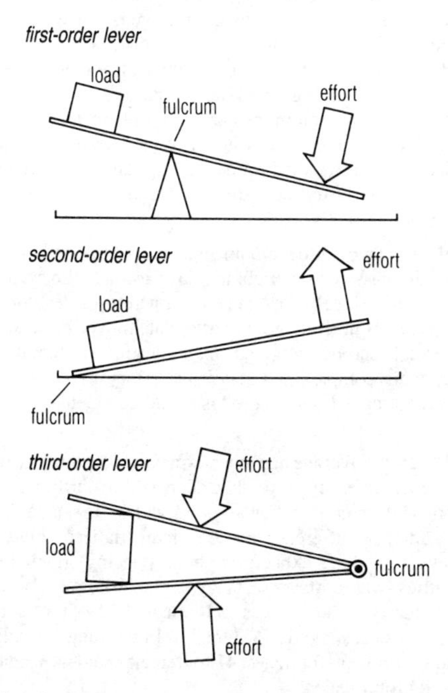

A ***second-order*** lever has the load and the effort on the same side of the fulcrum, with the load nearer the fulcrum—for example, nutcrackers or a wheelbarrow.

A ***third-order*** lever has the effort nearer the fulcrum than the load with both on the same side of it—for example, a pair of tweezers or tongs.

life the ability to grow, reproduce, and respond to such stimuli as light, heat, and sound. It is thought that life on Earth began about 4,000 million years ago. Over time, life has evolved from primitive single-celled organisms to complex multicellular ones.

Life originated in the oceans. The original atmosphere, 4,000 million years ago, consisted of carbon dioxide, nitrogen, and water. It has been shown in the laboratory that more complex organic molecules, such as ◊amino acids and ◊nucleotides, can be produced from these ingredients by passing electric sparks through a mixture. It has been suggested that lightning was extremely common in the early atmosphere, and that this combination of conditions could have resulted in the oceans becoming rich in organic molecules, the so-called primeval soup. These molecules may then have organized into clusters, capable of reproducing and of developing eventually into simple cells.

life cycle the sequence of developmental stages through which members of a given species pass. Most vertebrates have a simple life cycle consisting of ◊fertilization of sex cells or ◊gametes, a period of development as an ◊embryo, a period of juvenile growth after hatching or birth, an adulthood including ◊sexual reproduction, and finally death. Invertebrate life cycles are generally more complex and may involve major reconstitution of the individual's appearance (◊metamorphosis) and completely different styles of life.

life expectancy the average age to which a person, at the time of his or her birth, can expect to live. Life expectancy depends on nutrition, disease control, environmental contaminants, war, stress, and living standards in general.

There is a marked difference between industrialized countries, which generally have a high life expectancy and an ageing population, and the poorest countries, where life expectancy is much shorter. For example, in 1991 life expectancy in the UK was 73 for males and 79 for females, and in Japan was 76 for males and 82 for females. In the same year, life expectancy in Chad was only 39 for males, 41 for females, and in Afghanistan was 43 for males, 44 for females.

lifespan the time between birth and death. Lifespan varies enormously between different species. In humans it is also known as ◊life expectancy and varies according to sex, country, and occupation.

ligament a strong flexible connective tissue, made of the protein collagen, which joins bone to bone at moveable joints. Ligaments prevent bone dislocation (under normal circumstances) but permit joint flexion.

light ◊electromagnetic radiation in the visible range, having a wavelength from about 400 nanometres in the extreme violet to about 770 nanometres in the extreme red. Light is considered to exhibit both particle and wave properties, and the fundamental particle or quantum of light is called the ***photon***. The speed of light (and of all electromagnetic radiation) in a vacuum is approximately 3×10^8 metres per second/186,000 miles per second, and is a universal constant denoted by *c*.

Isaac Newton was the first to discover, in 1666, that sunlight is composed of a mixture of light of different colours in certain proportions and that it could be separated into its components by dispersion. Before his time it was supposed that dispersion of light produced colour instead of separating already existing colours.

light bulb incandescent filament ◊lamp, a thin glass bulb filled with an inert mixture of nitrogen and argon gas. It contains a filament made of fine tungsten wire. When electricity is passed through the wire, it glows white hot, producing light.

light-dependent resistor (LDR) component of electronic circuits whose resistance varies with the level of illumination on its surface. Usually resistance decreases as illumination rises. LDRs are used in light-measuring or light-sensing instruments (for example, in the exposure-meter circuit of an automatic camera) and in switches (such as those that switch on street lights at dusk).

light-emitting diode (LED) means of displaying symbols in electronic instruments and devices. An LED is made of ◊semiconductor material, such as gallium arsenide phosphide, that glows when electricity is passed through it. The first digital watches and calculators had LED displays, but many later models use ◊liquid crystal displays.

lightning high-voltage electrical discharge between two charged rainclouds or between a cloud and the Earth, caused by the build-up of electrical charges. Air in the path of lightning ionizes (becomes conducting), and expands; the accompanying noise is heard as thunder. Currents of 20,000 amperes and temperatures of 30,000°C are common.

lightning conductor device that protects a tall building from lightning strike by providing an easier path for current to flow to earth than through the building. It consists of a thick copper strip of very low resistance that connects a metal rod projecting above the building to the ground below. A good connection to the ground is essential and is made by burying a large metal plate deep in the damp earth. In the event of a direct lightning strike, the current in the conductor may be so great as to melt or even vaporize the metal, but the damage to the building will nevertheless be limited.

light reaction the first stage of ◊photosynthesis, in which light energy splits water into oxygen and hydrogen ions. The second, dark, stage does not require light and results in the formation of carbohydrates.

lime or ***quicklime*** common name for calcium oxide (CaO), a white powdery substance used in making mortar and cement and to reduce soil acidity. It is made commercially by heating calcium carbonate ($CaCO_3$) obtained from limestone or chalk. Lime readily absorbs water to become calcium hydroxide ($Ca(OH)_2$), known as slaked lime.

lime kiln oven used to make quicklime (calcium oxide, CaO) by heating limestone (calcium carbonate, $CaCO_3$) in the absence of air.

$$CaCO_3 \leftrightarrow CaO + CO_2$$

limestone sedimentary rock composed chiefly of calcium carbonate $CaCO_3$, either derived from the shells of marine organisms or precipitated from solution, mostly in the ocean. Various types of limestone are used as building stone. ◊Marble is metamorphosed limestone.

limewater dilute solution of calcium hydroxide used in the laboratory to detect the presence of carbon dioxide. If a gas containing carbon dioxide is bubbled through limewater, the solution turns milky owing to the formation of calcium carbonate. Continued bubbling of the gas causes the limewater to clear again as the calcium carbonate is converted to the more soluble calcium hydrogencarbonate.

limiting factor any factor affecting the rate of a metabolic reaction. Levels of light or of carbon dioxide are limiting factors in ◊photosynthesis because both are necessary for the production of carbohydrates. In experiments, photosynthesis is observed to slow down and eventually stop as the levels of light decrease.

liquefaction the process of converting a gas to a liquid. Liquefaction is normally associated with low temperatures and high pressures (see ◊condensation).

liquid state of matter between a ◊solid and a ◊gas. A liquid forms a level surface and assumes the shape of its container. Its atoms do not occupy fixed positions as in a crystalline solid, nor do they have freedom of movement as in a gas (see ◊change of state). Unlike a gas, a liquid is difficult to compress since pressure applied at one point is equally transmitted throughout (Pascal's principle).

liquid air air that has been cooled so much that it has liquefied. This happens at temperatures below about –196°C. The various constituent gases, including nitrogen, oxygen, argon, and neon, can be separated from liquid air by the technique of ◊fractionation.

liquid crystal display (LCD) display of numbers (for example, in a calculator) or picture (such as on a pocket television screen) produced by molecules of a substance in a semiliquid state with some crystalline properties, in that clusters of molecules align in parallel formations. The display is a blank until the application of an electric field, which 'twists' the molecules so that they reflect or transmit light falling on them.

lithium soft, ductile, silver-white, metallic element, symbol Li, atomic number 3, relative atomic mass 6.941. It is one of the ◊alkali metals, has a very low density (far less than most woods), and floats on water (specific gravity 0.57); it is the lightest of all metals. Lithium is used to harden alloys, and in batteries; its compounds are used in medicine to treat manic depression.

litmus dye obtained from various lichens and used as an indicator to test the acidic or alkaline nature of aqueous solutions; it turns red in the presence of acid, and blue in the presence of alkali.

liver a large organ of vertebrates, which has many regulatory and storage functions. The human liver is situated in the upper abdomen, and weighs about 2kg. It receives the products of digestion, converts glucose to glycogen (a long-chain carbohydrate used for storage), and breaks down fats. It removes excess amino acids from the blood, converting them to urea, which is excreted by the kidneys. The liver also synthesizes vitamins, produces bile and blood-clotting factors, and removes damaged red cells and toxins such as alcohol from the blood.

locomotion the ability to move independently from one place to another, occurring in most animals but not in plants. The development of locomotion as a feature of animal life is closely linked to another vital animal feature, that of nutrition. Animals cannot make their food, as can plants, but must find it first; often the food must be captured and killed, which may require great powers of speed. Locomotion is also important in finding a mate, in avoiding predators, and in migrating to favourable areas.

logic gate

truth tables

inputs		output
0	0	0
0	1	1
1	0	1
1	1	1

OR gate

inputs		output
0	0	0
0	1	0
1	0	0
1	1	1

AND gate

input	output
0	1
1	0

NOT gate

inputs		output
0	0	1
0	1	0
1	0	0
1	1	0

NOR gate

inputs		output
0	0	1
0	1	1
1	0	1
1	1	0

NAND gate

logic gate or ***logic circuit*** in electronics, one of the basic components used in building ◊integrated circuits. The five basic types of gate make logical decisions based on the functions NOT, AND, OR, NAND (NOT AND), and NOR (NOT OR). With the exception of the NOT gate, each has two or more inputs.

Information is fed to a gate in the form of binary-coded input signals (logic value 0 stands for 'off' or 'low-voltage pulse', logic 1 for 'on' or 'high-voltage'), and each combination of input signals yields a specific output (logic 0 or 1). An ***OR*** gate will give a logic 1 output if one or more of its inputs receives a logic 1 signal; however, an ***AND*** gate will yield a logic 1 output only if it receives a logic 1 signal through both its inputs. The output of a ***NOT*** or ***inverter*** gate is the opposite of the signal received through its single input, and a ***NOR*** or ***NAND*** gate produces an output signal that is the opposite of the signal that would have been produced by an OR or AND gate respectively.

The properties of a logic gate, or of a combination of gates, may be defined and presented in the form of a ***truth table***, which lists the output that will be triggered by each of the possible combinations of input signals.

lone pair pair of electrons in the outermost shell of an atom that are not used in bonding. In certain circumstances they will form a coordinate covalent bond with electron-deficient species such as H^+ or BF_3 by providing the pair of electrons.

longitudinal muscle type of involuntary muscle found in the gut (alimentary canal). The action of longitudinal muscles causes the mixing of food and results in its movement along the gut.

longitudinal wave

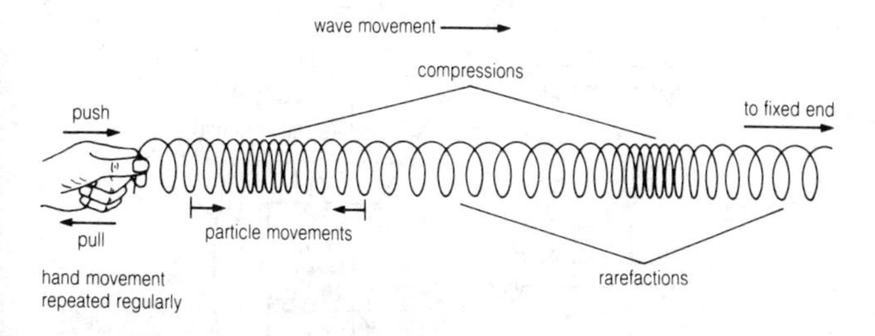

longitudinal wave wave in which the displacement of the medium's particles is in line with or parallel to the direction of travel of the wave motion. It is characterized by its alternating compressions and rarefactions. In the ***compressions*** the particles are pushed together (in a gas the pressure rises); in the ***rarefactions*** they are pulled apart (in a gas the pressure falls). Sound travels through air as a longitudinal wave.

long-sightedness or ***hypermetropia*** defect of vision in which a person is able to focus on objects in the distance, but not on close objects. It is caused by the failure of the lens to return to its normal rounded shape, or by the eyeball being too short, with the result that the image is focused on a point behind the retina. Long-sightedness is corrected by wearing spectacles fitted with ◊converging lenses (convex lenses), each of which acts like a magnifying glass.

loudness subjective judgement of the level or power of sound reaching the ear. The human ear cannot give an absolute value to the loudness of a single sound, but can only make comparisons between two different sounds. Loudness is related to the amplitude of a sound wave and also, because the ear is not equally sensitive to all frequencies or pitches of sound, to frequency. The precise measure of the power of a sound wave at a particular point is called its intensity. Accurate comparisons of sound levels may be made by using sound-level meters, in units called decibels.

loudspeaker device that converts electrical signals into sound waves that are radiated into the air. It is used in all sound-reproducing systems such as tape recorders and televisions. The most common type is the ***moving-coil***

loudspeaker

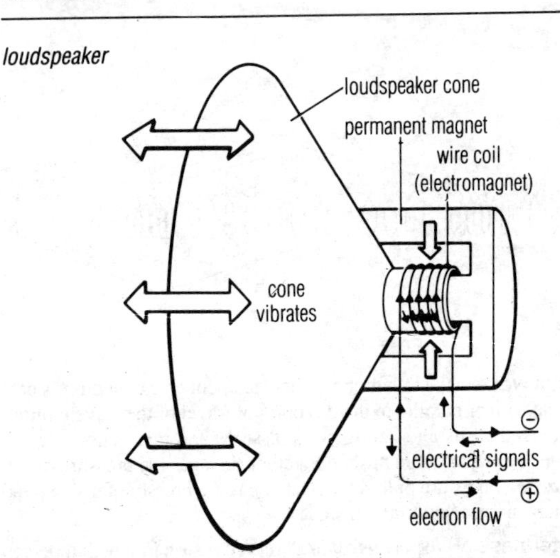

speaker. Electrical signals from, for example, a radio are fed to a coil of fine wire wound around the top of a cone. The coil is surrounded by a magnet. When signals pass through it, the coil becomes an electromagnet, which by moving causes the cone to vibrate, setting up sound waves.

lubricant substance used between moving surfaces to reduce friction. Carbon-based (organic) lubricants, commonly called grease and oil, are recovered from petroleum distillation.

Extensive research has been carried out on chemical additives to lubricants, which can reduce corrosive wear, prevent the accumulation of 'cold sludge' (often the result of stop-start driving in city traffic jams), keep pace with the higher working temperatures of aviation gas turbines, or provide radiation-resistant greases for nuclear power plants. Silicon-based spray-on lubricants are also used; they tend to attract dust and dirt less than carbon-based ones.

A solid lubricant is graphite, an allotropic form of carbon, either flaked or emulsified (colloidal) in water or oil.

lung large cavity of the body used for ◊gas exchange. It has the form of a large sheet that is folded so as to occupy less space. The lung tissue, consists of multitudes of air sacs (alveoli) and blood vessels and is very light and spongy. It brings inhaled air into close contact with the blood so that oxygen can be absorbed and waste carbon dioxide can be passed out. The efficiency of lungs is enhanced by breathing movements, by the thinness and moistness of the surfaces, and by a constant supply of circulating blood.

Air is drawn into the lungs through the trachea and bronchi by the expansion of the ribs and the contraction of the diaphragm. The principal diseases of the lungs are bronchitis, emphysema, and cancer (which occur more frequently in smokers than in non-smokers), and pneumonia.

lung

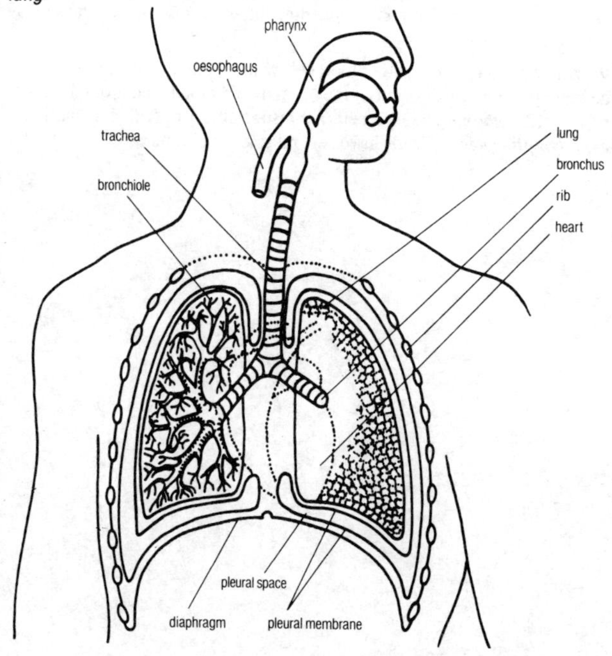

lymph the fluid found in the lymphatic system of vertebrates, which transports nutrients, oxygen, and white blood cells to the tissues, and waste matter away from them. It exudes from ◊capillaries into the tissue spaces between the cells and is made up of blood plasma, plus white cells.

lymph nodes small masses of lymphatic tissue in the body that occur at various points along the major lymphatic vessels. They are chiefly situated in the neck, armpit, groin, thorax, and abdomen. Tonsils and adenoids are large lymph nodes. As the lymph passes through them it is filtered, and bacteria and other microorganisms are engulfed by cells known as macrophages.

Lymph nodes are sometimes mistakenly called 'lymph glands', and the term 'swollen glands' refers to swelling of the lymph nodes caused by infection.

lymphocyte a type of white blood cell with a large nucleus, produced in the bone marrow. Most occur in the ◊lymph and blood, and around sites of infection. ***B-lymphocytes*** or B cells are responsible for producing ◊antibodies. ***T-lymphocytes*** or Tcells have several roles in ◊immunity.

M

MA abbreviation for ◊mechanical advantage.

machine device that allows a small force (the effort) to overcome a larger one (the load). There are three basic machines: the sloping or ◊inclined plane, the ◊lever, and the ◊wheel and axle. All other machines are combinations of these three basic types. Simple machines derived from the inclined plane include the wedge and the screw; the spanner is derived from the lever; the pulley is derived from the wheel.

The two principal features of a machine are its ◊mechanical advantage, which is the ratio load/effort, and its ◊efficiency, which is the work done by the load divided by the work done by the effort; the latter is expressed as a percentage. In a perfect machine, with no friction, the efficiency would be 100%. All practical machines have efficiencies of less than 100%, otherwise perpetual motion would be possible.

macromolecule very large molecule, generally a ◊polymer.

magnesium lightweight, very ductile and malleable, silver-white, metallic element, symbol Mg, atomic number 12, relative atomic mass 24.305. It is one of the ◊alkaline-earth metals, the lightest of the commonly used metals. Magnesium silicate, carbonate, and chloride are widely distributed in nature.

Magnesium is used in light alloys, for example in aircraft, and flash photography. It is a necessary trace element in the human diet, and green plants cannot grow without it because it is an essential constituent of chlorophyll ($C_{55}H_{72}MgN_4O_5$).

magnesium carbonate $MgCO_3$ white solid that occurs in nature as the mineral magnesite. It is a commercial antacid and the anhydrous form is used as a drying agent in table salt. When rainwater containing dissolved carbon dioxide flows over magnesite rocks, the carbonate dissolves to form magnesium hydrogencarbonate, one of the substances that cause temporary hardness in water.

$$H_2O + CO_2 + MgCO_3 \rightarrow Mg(HCO_3)_2$$

magnet any object that forms a magnetic field, either permanently or temporarily through induction, causing it to attract materials such as iron, cobalt, nickel, and alloys of these. It always has two ◊magnetic poles, called north and south.

magnetic compass device for determining the direction of the horizontal component of the Earth's magnetic field. It consists of a magnetized needle with its north-seeking pole clearly indicated, pivoted so that it can turn freely in a plane parallel to the surface of the Earth (in a horizontal circle). The needle will turn so that its north pole points towards the Earth's magnetic north pole.

Walkers, sailors, and other travellers use a magnetic compass to find their direction. The direction of the geographic, or true, North Pole is, however, slightly different from that of the magnetic north pole, and so the readings obtained from a compass of this sort must be adjusted by using tables of magnetic corrections, or information marked on local maps.

magnetic field the physical field or region around a permanent magnet or around a conductor carrying an electric current, in which a force acts on a moving charge or on a magnet placed in that field. The field can be repre-

magnetic field

the Earth's magnetic field

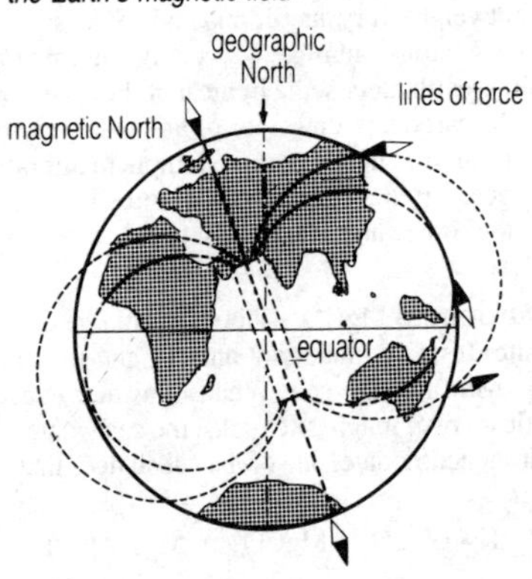

sented by lines of force, which by convention link north and south poles and are parallel to the direction of a small compass needle placed on them. Its magnitude and direction are given by the ◊magnetic flux density, expressed in teslas.

magnetic flux measurement of the strength of the magnetic field around electric currents and magnets.

magnetic induction the production of magnetic properties in unmagnetized iron or other ferromagnetic material when it is brought close to a magnet. The material is influenced by the magnet's magnetic field and the two are attracted. The induced magnetism may be temporary (disappearing as soon as the magnet is removed) or permanent, depending on the nature of the iron and the strength of the magnet. ◊Electromagnets make use of temporary induced magnetism to lift sheets of steel: the magnetism induced in the steel by the approach of the electromagnet enables it to be picked up and transported. To release the sheet, the current supplying the electromagnet is temporarily switched off and the induced magnetism disappears.

magnetic material or ***ferromagnetic material*** one of a number of substances that are strongly attracted by magnets and can be magnetized. These include iron, nickel, and cobalt.

Soft magnetic materials can be magnetized very easily, but the magnetism induced in them (see ◊magnetic induction) is only temporary. They include Stalloy, an alloy of iron with 4% silicon used to make the cores of electromagnets and transformers, and the materials used to make 'iron' nails and paper clips. ***Hard magnetic materials*** can be permanently magnetized by a strong magnetic field. Steel and special alloys such as Alcomax, Alnico, and Ticonal, which contain various amounts of aluminium, nickel, cobalt, and copper, are used to make permanent magnets. The strongest permanent magnets are ceramic, made under high pressure and at high temperature from powders of various metal oxides.

magnetic pole region of a magnet in which its magnetic properties are strongest. Every magnet has two poles, called north and south. The north (or north-seeking) pole is so named because a freely suspended magnet will turn so that this pole points towards the Earth's magnetic north pole. The north pole of one magnet will be attracted to the south pole of another, but will be repelled by its north pole. Like poles may therefore be said to repel, unlike poles to attract.

magnetism branch of physics dealing with the properties of magnets and ◊magnetic fields. Magnetic fields are produced by moving charged particles: in electromagnets, electrons flow through a coil of wire connected to a battery; in magnets, spinning electrons within the atoms generate the field.

magnification measure of the enlargement or reduction of an object in an imaging optical system. ***Linear magnification*** is the ratio of the size (height) of the image to that of the object. ***Angular magnification*** is the ratio of the angle subtended at the observer's eye by the image to the angle subtended by the object when viewed directly. See also ◊magnifying glass.

magnifying glass the simplest optical instrument, a hand-held converging lens used to produce a magnified, erect and virtual image. The image, being virtual, or an illusion created by the ◊refraction of light rays in the lens, can only be seen by looking through the magnifying glass.

The object to be magnified must be placed between the ◊principal focus and the lens; the image produced is best seen if it is at the ◊near point of the eye. The magnification produced by the lens is given by the ratio

$$\text{magnification} = \frac{\text{height of image}}{\text{height of object}}$$

or, where the heights of the image and the object are h_I and h_O respectively,

$$\text{magnification} = h_I/h_O$$

magnifying glass

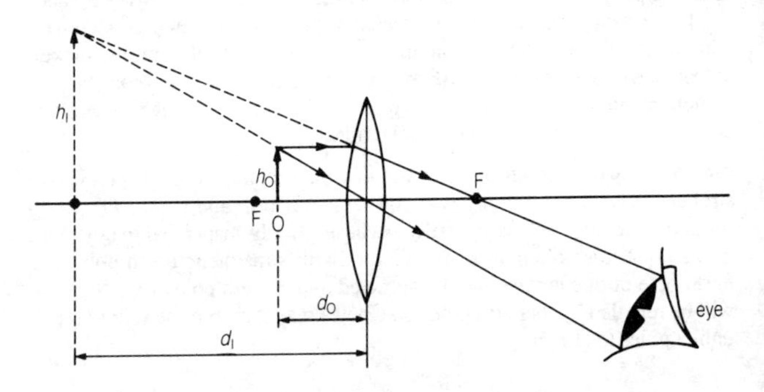

By similar triangles, this ratio can be shown to be equal to the distance ratio, therefore

$$\text{magnification} = \frac{\text{image distance}}{\text{object distance}}$$

or, where the distances from the centre of the lens to the image and to the object are d_I and d_O respectively,

$$\text{magnification} = d_I/d_O$$

It follows that bringing the object closer to the eye increases the magnification.

mains electricity the domestic electricity-supply system. In the UK, electricity is supplied to houses, offices, and most factories as an ◊alternating current at a frequency of 50 hertz and an effective voltage of 240 volts. An advantage of having an alternating supply is that it may easily be changed, by using a ◊transformer, to a lower and safer voltage, such as 9 volts, for operating toys and for recharging batteries.

malnutrition the physical condition resulting from a poor or unbalanced diet. For example, excessive consumption of carbohydrates leads to increasing deposits of fat, and the individual is said to be obese. A lack of protein or of carbohydrate leads to wasting diseases such as ◊kwashiorkor and marasmus, which are eventually fatal. The high global death toll linked to malnutrition arises mostly through diseases killing people weakened by poor diet and an impure water supply.

maltose $C_{12}H_{22}O_{11}$ disaccharide sugar made up of two glucose units. It is produced by the action of enzymes on starch and is a major constituent of malt, produced in the early stages of ◊brewing. It is also used in the production of whisky.

mammal member of the vertebrate class Mammalia. Mammals have hair and feed their young with milk. They are homeothermic (warm-blooded), maintaining a constant body temperature in varied surroundings. Most mammals give birth to live young, but the platypus and echidna lay eggs. There are over 4,000 species, adapted to almost every way of life. The smallest shrew weighs only 2g, the largest whale up to 150 tonnes.

mammary gland in female mammals, milk-producing gland underlying the skin, active only after the production of young. In all but monotremes (egg-laying mammals), the mammary glands terminate in teats, which aid infant suckling. The number of glands and their position vary between species. In humans there are two (called ◊breasts), in cows four, and in pigs between ten and fourteen.

manganese hard, brittle, grey-white, metallic element, symbol Mn, atomic number 25, relative atomic mass 54.9380. It resembles iron (and rusts), but it is not magnetic and is softer. It is used chiefly in making steel alloys, also alloys with aluminium and copper. It is used in fertilizers, paints, and industrial chemicals. It is a necessary trace element in human nutrition.

manganese(IV) oxide or ***manganese dioxide*** MnO_2 brown solid that acts as a ◊catalyst in decomposing hydrogen peroxide to obtain oxygen.

$$2H_2O_{2(aq)} \rightarrow 2H_2O_{(l)} + O_{2(g)}$$

It oxidizes concentrated hydrochloric acid to produce chlorine.

$$MnO_2 + 4HCl \rightarrow MnCl_2 + Cl_2 + 2H_2O$$

It oxidizes the hydrogen gas produced in dry batteries to water; without this process, the performance of the battery is impaired.

manometer instrument for measuring the pressure of liquids (including human blood pressure) or gases. In its basic form, it is a U-tube partly filled with coloured liquid; the pressure of a gas entering at one side is measured by the level to which the liquid rises at the other.

marble ◊limestone that has undergone transformation through the action of heat and pressure. It takes and retains a good polish and is used in building and sculpture. In its pure form it is white and consists almost entirely of calcite $CaCO_3$. Mineral impurities give it various colours and patterns.

marsh gas gas consisting mostly of ◊methane. It is produced in swamps and marshes by the action of bacteria on dead vegetation.

mass the quantity of matter in a body as measured by its inertia. Mass determines the acceleration produced in a body by a given force acting on it, the acceleration being inversely proportional to the mass of the body. The mass also determines the force exerted on a body by ◊gravity on Earth, although this attraction varies slightly from place to place. In the SI system, the base unit of mass is the kilogram (symbol kg).

At a given place, equal masses experience equal gravitational forces, which are known as the weights of the bodies. Masses may, therefore, be compared by comparing the weights of bodies at the same place. The standard unit of mass to which all other masses are compared is a platinum–iridium cylinder of one kilogram.

mass number or ***nucleon number*** the sum (symbol A) of the numbers of protons and neutrons in the nucleus of an atom. It is used along with the ◊atomic number in nuclear notation and in nuclear equations (see ◊nuclear reaction); in symbols that represent nuclear isotopes, such as $^{14}_{6}C$, the lower number is the atomic number, and the upper number is the mass number.

maximum and minimum thermometers thermometers used to measure maximimum and minimum temperature readings over a period of time, usually 24 hours. In the maximum thermometer a pointer rises to the highest reading and stays there; in a minimum thermometer the pointer falls to the minimum reading and stays there. Sometimes the two functions are combined in a ***maximum-and-minimum thermometer***.

mechanical advantage (MA) the number of times the load moved by a machine is greater than the effort applied to that machine. In equation terms:

$$MA = load/effort$$

MA has no units because it is a ratio. ◊Force multipliers have an MA greater than one; ◊distance multipliers, or velocity multipliers, have an MA less than one.

The exact value of a working machine's MA is always less than its predicted value because there will always be some frictional resistance that increases the effort necessary to do the work.

meiosis or ***reduction division*** type of cell division that occurs in the production of ◊gametes, and is therefore important in sexual reproduction. It causes the halving of the chromosome number so that gametes are ◊haploid; this is necessary if the fertilized egg, formed on fusion of the gametes, is to have the same number of chromosomes as the parent. Meiosis is also the stage of the life cycle where genetic variation arises (see ◊recombination); every gamete is slightly different, resulting in non-identical offspring. See also ◊mitosis.

melanism black coloration of animal bodies caused by large amounts of the pigment melanin. Melanin is of significance in insects, because melanic ones warm more rapidly in sunshine than do pale ones, and can be more active in cool weather. A fall in temperature may stimulate such insects to produce more melanin. In industrial areas, dark insects and pigeons match sooty backgrounds and escape predation, but they are at a disadvantage in rural areas where they do not match their backgrounds. This is known as ***industrial melanism***.

melting change of state from a solid to a liquid, associated with an intake of energy (for example, if the temperature is rising).

melting point the temperature at which a substance melts, or changes from a solid to liquid form. A pure substance under standard conditions of pressure (usually one atmosphere) has a definite melting point. If heat is supplied to a solid at its melting point, the temperature does not change until the melting process is complete. The melting point of ice is 0°C.

membrane a continuous layer, made up principally of lipid molecules, that encloses a cell or ◊organelles within a cell. Certain small molecules can pass through the cell membrane, but many must enter or leave the cell via channels in the membrane.

Mendelism in genetics, the theory of inheritance originally outlined by Austrian biologist Gregor Mendel. He suggested that, in sexually reproducing species, all characteristics are inherited through indivisible 'factors' (now identified with ◊genes) contributed by each parent to its offspring. See ◊genetics.

meniscus the curved shape of the surface of a liquid in a thin tube, caused by the cohesive effects of surface tension. Most liquids adopt a concave curvature (viewed from above), although with highly viscous liquids (such as mercury) the meniscus is convex.

menopause or ***change of life*** in women. A period when they lose the ability to reproduce, characterized by menstruation (see ◊menstrual cycle) becoming irregular and eventually stopping completely. This may start at about the age of 50, but varies greatly between women.

menstrual cycle the cycle that occurs in female mammals of reproductive age, in which the body is prepared for pregnancy. At the beginning of the cycle the ovum is released from the ovary, and the lining of the uterus

meniscus

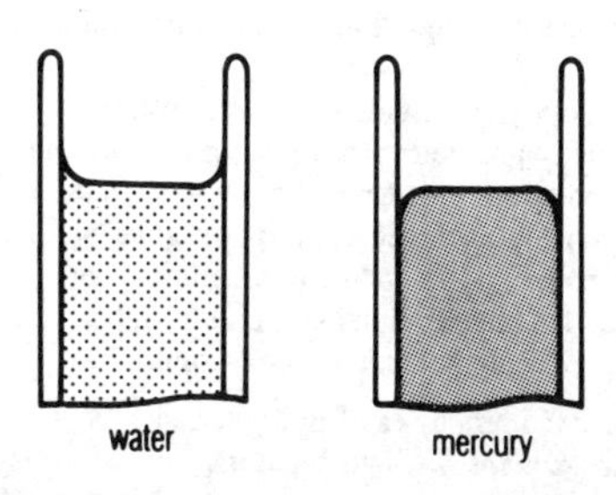

becomes filled with blood vessels. If fertilization does not occur, the corpus luteum (part of the ovary) degenerates, and the uterine lining breaks down, and is shed. This is what causes the loss of blood during menstruation. The cycle then begins again. Human menstruation takes place from puberty to menopause, occurring about every 28 days.

The cycle is controlled by a number of ◊hormones, including ◊oestrogen and ◊progesterone.

mercury or ***quicksilver*** heavy, silver-grey, metallic element, symbol Hg, atomic number 80, relative atomic mass 200.59. It is a dense, mobile liquid with a low melting point (–38.87°C). The chief source is the mineral cinnabar (HgS). Mercury sometimes occurs as a free metal. Its alloys with other metals are called amalgams, and dentistry uses a silver–mercury amalgam for filling teeth, which may be a source of cumulative mercury poisoning to the human body. Industrial uses include drugs and chemicals, mercury-vapour lamps, arc rectifiers, power-control switches, barometers, and thermometers. Industrial dumping of this toxic substance has caused global pollution of land and waters, which has contaminated the food chain.

meristem a region of plant tissue containing cells that are actively dividing to produce new tissues (or have the potential to do so). Meristems found in the tip of roots and stems, the apical meristems, are responsible for the increase in length (◊primary growth) of these organs.

The ◊cambium is a lateral meristem that is responsible for increase in girth (secondary growth) in perennial plants.

mesophyll the tissue between the upper and lower epidermis of a leaf blade, consisting of loosely packed, thin-walled cells containing numerous ◊chloroplasts.

In many plants, mesophyll is divided into two distinct layers. The ***palisade mesophyll*** is usually just below the upper epidermis and is composed of regular layers of elongated cells. Lying below them is the ***spongy mesophyll***, composed of loosely arranged cells of irregular shape. This layer contains fewer chloroplasts and has many intercellular spaces for the diffusion of gases (required for ◊respiration and ◊photosynthesis), linked to the outside by means of pores called stomata (see ◊stoma).

metabolism the chemical processes of living organisms: a constant alternation of building up (***anabolism***) and breaking down (***catabolism***). For example, green plants build up complex organic substances from water, carbon dioxide, and mineral salts (photosynthesis); by digestion animals partly break down complex organic substances, ingested as food, and subsequently resynthesize them in their own bodies.

metal any of a class of elements possessing a specific set of physical and chemical properties. Most elements (about 70) can be classified as metals. They are defined as elements that tend to lose electrons, and are therefore sometimes described as electropositive elements.

physical properties Their physical properties include shiny appearance, a sonorous tone when struck, good conduction of heat and electricity, malleability (can be rolled or pressed into shape), ductility (can be drawn into wire), and the possible emission of electrons when heated (thermionic effect) or when the surface is struck by light (photoelectric effect).

Metals have the following characteristic reactions.

with acids They displace hydrogen from dilute acids.

$$Mg_{(s)} + 2H^{+}_{(aq)} \rightarrow Mg^{2+}_{(aq)} + H_2$$

with oxygen They form basic or amphoteric oxides when reacted with oxygen.

$$2Mg + O_2 \rightarrow 2MgO$$

$$MgO + 2HNO_3 \rightarrow Mg(NO_3)_2 + H_2O$$

with chlorine They form ionic chlorides when reacted with chlorine.

$$Mg + Cl_2 \rightarrow MgCl_2$$

with oxidizing agents They are effective reducing agents; for example, aluminium reduces iron(III) oxide to iron in the ◊thermite reaction.

$$2Al + Fe_2O_3 \rightarrow Al_2O_3 + 2Fe$$

metallic bond the force of attraction operating in a metal that holds the atoms together. In the metal the outer shell electrons are able to move within the crystal and these electrons are said to be delocalized. As they move from one atom to another, they create short-lived, positively charged ions. The electrostatic attraction between the delocalized electrons and the continually forming positively charged ions constitutes the metallic bond.

metallic character chemical properties associated with those elements classed as metals. These properties, which arise from the element's ability to lose electrons, are: the displacement of hydrogen from dilute acids; the formation of ◊basic oxides; the formation of ionic chlorides; and their reducing effect (see ◊reduction).

In the periodic table of the elements, metallic character increases down any group and across a period from right to left.

metalloid or ***weakly metallic element*** element having some but not all of the properties of metals; metalloids are thus usually electrically semiconducting. They comprise the elements germanium, arsenic, antimony, and tellurium.

metamorphic rock a rock that is changed in structure and composition by great heat and pressure. Two examples of metamorphic rocks are marble and slate.

metamorphosis a period during the life cycle of many invertebrates, most amphibians, and some fish, during which the individual's body changes from one form to another through a major reconstitution of its tissues. For example, adult frogs are produced by metamorphosis from tadpoles, and butterflies are produced from caterpillars following metamorphosis within a pupa.

methanal or ***formaldehyde*** HCHO the first of the ◊aldehyde series of organic chemicals. It is a gas at ordinary temperatures, condensing to a liquid at –21°C. It has a powerful penetrating smell. Dissolved in water, it is used as a biological preservative. It is used in the manufacture of plastics, dyes, foam (for example urea–formaldehyde foam, used in insulation), and in medicine.

methane CH_4 the simplest hydrocarbon of the ◊alkane (paraffin) series. Colourless, odourless, and lighter than air, it burns with a bluish flame and explodes when mixed with air or oxygen. It is the chief constituent of natural gas and also occurs in the explosive firedamp of coal mines. In marsh gas, methane forms from rotting vegetation by spontaneous combustion resulting in the pale flame seen over marshland and known as will-o'-the-wisp.

Methane is causing about 38% of the warming of the Earth through the ◊greenhouse effect; the amount of methane in the air is predicted to double over the next 60 years.

methanoic acid common name ***formic acid*** HCOOH the first member of the ◊carboxylic acid series. It is a colourless, slightly fuming liquid that melts at 8°C and boils at 101°C. It occurs in stinging ants, nettles, sweat, and pine needles, and is used in dyeing, tanning, and electroplating.

methanol common name ***methyl alcohol*** or ***wood alcohol*** CH_3OH the simplest of the ◊alcohols. It can be made by the dry distillation of wood, but is usually made from coal or natural gas. When pure, it is a colourless, flammable liquid with a pleasant odour, and is highly poisonous.

methyl orange $C_{14}H_{14}N_3NaO_3S$ orange-yellow powder used as an acid–base indicator in chemical tests, and as a stain in the preparation of slides of biological material. Its colour changes with pH; below pH 3.1 it is red, above pH 4.4 it is yellow.

metric system system of weights and measures developed in France in the 18th century and recognized by other countries in the 19th century. In 1960 an international conference on weights and measures recommended the universal adoption of a revised International System (Système International d'Unités, or SI) for scientific use; see ◊SI units.

microbe another name for ◊microorganism.

microbiology the study of organisms that can only be seen under the microscope, mostly viruses and single-celled organisms such as bacteria, protozoa, and yeasts. The practical applications of microbiology are in medicine (since many microorganisms cause disease); in brewing, baking, and other food and beverage processes, where the microorganisms carry out fermentation; and in genetic engineering, which is creating increasing interest in the field of microbiology.

microclimate the particular climate found in a small area or locality.

microorganism or ***microbe*** a living organism invisible to the naked eye but visible under a microscope. Microorganisms include viruses and single-celled organisms such as bacteria, protozoa, yeasts, and some algae. The study of microorganisms is known as microbiology.

microphone the primary component in a sound-reproducing system, whereby the mechanical energy of sound waves is converted into electrical signals. One of the simplest is the telephone receiver mouthpiece; other types of microphone are used with broadcasting and sound-film apparatus.

microscope instrument for magnification with high resolution for detail. Optical and electron microscopes are the ones chiefly in use. In 1988 a scanning tunnelling microscope was used to photograph a single protein molecule for the first time.

The ***optical*** or ***light microscope*** usually has two sets of glass lenses and an eyepiece. The ***electron microscope*** passes a beam of electrons, instead of a beam of light, through a specimen and, since electrons are not visible, the eyepiece is replaced with a fluorescent screen or photographic plate; far higher magnification and resolution are possible than with the optical microscope.

microwave ◊electromagnetic wave with a wavelength in the range 0.3–30cm, and a frequency of 300–300,000 megahertz (between radio waves and ◊infrared radiation). They are used in radar, as carrier waves in radio broadcasting, and in microwave heating and cooking.

migration the movement, either seasonal or as part of a single life cycle, of certain animals, chiefly birds and fish, to distant breeding or feeding grounds.

milk the secretion of the ◊mammary glands of female mammals, with which they suckle their young (during ◊lactation). Over 85% is water, the remainder comprising protein, fat, lactose (a sugar), calcium, phosphorus, iron, and vitamins. The milk of cows, goats, and sheep is often consumed by humans, but only in Western societies is milk drunk after infancy; for people in most of the world, milk causes flatulence and diarrhoea. Milk composition varies among species, depending on the nutritional requirements of the young; human milk contains less protein and more lactose than that of cows.

milk teeth a child's first set of teeth. The complete group of 20 milk teeth are present after two or three years and begin to be replaced by the second permanent set after the age of six. A normal adult dentition consists of 32 teeth, the extra teeth being molars and wisdom teeth.

mimicry the imitation of one species (or group of species) by another. Frequently, the mimic resembles a model that is poisonous or unpleasant to eat, and has warning coloration; the mimic thus benefits from the fact that predators have learned to avoid the model. Hoverflies that resemble bees or wasps are an example. Appearance is usually the basis for mimicry, but calls, songs, scents, and other signals can also be mimicked.

mineral naturally formed inorganic substance with a particular chemical composition and an ordered internal structure. Either in their perfect crystalline form or otherwise, minerals are the constituents of rocks. In more general usage, a mineral is any substance economically valuable for mining (including coal and oil, despite their organic origins).

mineral oil oil obtained from mineral sources, chiefly ◊petroleum, as distinct from oil obtained from vegetable or animal sources.

mineral salt simple inorganic chemical that is required by living organisms. Plants usually obtain their mineral salts from the soil, while animals get theirs from their food. Important mineral salts include iron salts (needed by both plants and animals), magnesium salts (needed mainly by plants, to make chlorophyll), and calcium salts (needed by animals to make bone or shell).

mirage the illusion seen in hot climates of water on the horizon, or of distant objects being enlarged. The effect is caused by the ◊refraction of light.

Light rays from the sky bend as they pass through the hot layers of air near the ground, so that they appear to come from the horizon. Because the light is from a blue sky, the horizon appears blue and watery. If, during the night, cold air collects near the ground, light can be bent in the opposite direction, so that objects below the horizon appear to float above it. In the same way, objects such as trees or rocks near the horizon can appear enlarged.

mirror any polished surface that reflects light; often made from 'silvered' glass (in practice, a mercury alloy coating of glass). A plane (flat) mirror produces a same-size, erect, 'virtual' image located behind the mirror at the same distance from it as the object is in front of it. A spherical concave mirror produces a reduced, inverted, real image in front or an enlarged, erect, virtual image behind it (as with a shaving mirror), depending on how close the object is to the mirror. A spherical convex mirror produces a reduced, erect, virtual image behind it (as with a car's rear-view mirror).

mitochondria (singular ***mitochondrion***) small bodies or organelles found inside most cells, responsible for aerobic respiration and for providing the organism with ATP (adenosine triphosphate), the essential energy-rich molecule used to drive cellular reactions.

mitosis the process by which a cell divides to give two identical, diploid daughter cells. It takes place during growth and asexual reproduction. See ◊diploid.

mixture substance containing two or more compounds that still retain their separate physical and chemical properties. There is no chemical bonding between them and they can be separated from each other by physical means.

moderator in a nuclear reactor, a material such as graphite or heavy water used to reduce the speed of high-energy neutrons.

molar one of the large teeth found towards the back of the mammalian mouth. The structure of the jaw, and the relation of the muscles, allows a massive force to be applied to molars. In herbivores the molars are flat with sharp ridges of enamel and are used for grinding, an adaptation to a diet of tough plant material. Carnivores have sharp powerful molars called ◊carnassials, which are adapted for cutting meat.

molarity the ◊concentration of a solution expressed as the number of ◊moles of solute per $1000cm^3$ of solution.

molar solution solution that contains one mole of a substance per 1000 cm^3 of solution. For example, a molar solution of sodium hydroxide (NaOH) contains 40g of NaOH in $1000cm^3$ of solution.

molar volume volume occupied by one mole of any gas at standard temperature and pressure, equal to $2.24136 \times 10^{-2}m^3$. Often the more convenient description is that one mole of any gas occupies 24l at room temperature and pressure.

mole SI unit (symbol mol) used in chemistry to indicate the amount of a substance undergoing a physical or chemical change. It is formally defined as the amount of that substance that contains as many particles (atoms, molecules, and so on) as there are atoms in 12g of the carbon–12 isotope.

One mole of an element is present when its relative atomic mass is expressed in grams (one mole of sodium is 23g, and of zinc is 65g). For a molecular substance, one mole is present when the relative molecular mass is expressed in grams (one mole of water is 18g, of sodium hydroxide is 40g, and of sucrose is 342g). For gases at room temperature and pressure, one mole of gas is contained in 24l (see molar volume); one mole of hydrogen is 2g or 24l, one mole of chlorine gas is 71g or 24l.

In a chemical equation it is possible to replace the terms 'atom' and 'molecule' by 'mole' so the precise quantities of substances being reacted or produced can be calculated (in units of mass or volume).

molecular formula ◊formula indicating the actual number of atoms of each element present in a single molecule of a compound. The presence of more than one atom is denoted by a subscript figure—for example, one molecule of the compound water is shown as H_2O, having two atoms of hydrogen and one atom of oxygen.

molecular weight another name for ◊relative molecular mass.

molecule the smallest unit of an ◊element or ◊compound that can exist and still retain the characteristics of the element or compound. A molecule of an element consists of one or more like ◊atoms; a molecule of a compound consists of two or more different atoms bonded together. They vary in size and complexity from the hydrogen molecule (H_2) to the large ◊macromolecules found in polymers. They are held together by covalent bonds, where one electron is contributed by each atom to form an electron pair that is shared by the two contributing atoms, or by coordinate bonds, where the electron pair is donated by one atom only, to be shared between two atoms.

molten state of a solid that has been heated until it melts.

molybdenum heavy, hard, lustrous, silver-white, metallic element, symbol Mo, atomic number 42, relative atomic mass 95.94. The chief ore is the mineral molybdenite. The element is highly resistant to heat and conducts electricity easily. It is used in alloys, often to harden steels. It is a necessary trace element in human nutrition. It has a melting point of 2,620°C, and is not found in the free state.

moment of a force measure of the turning effect, or torque, produced by a force acting on a body. It is equal to the product of the force and the perpendicular distance from its line of action to the point, or pivot, about which the body will turn. Its unit is the newton metre.

If the magnitude of the force is F newtons and the perpendicular distance is d metres then the moment is given by:

$$\text{moment} = Fd$$

momentum the product of the mass of a body and its velocity. If the mass of a body is m kilograms and its velocity is v metres per second, then its momentum is given by:

$$\text{momentum} = mv$$

Its unit is the kilogram metre per second ($kgms^{-1}$) or the newton second.

monoclinic sulphur allotropic form of ◊sulphur, formed from sulphur that crystallizes above 96°C. Once formed, it very slowly changes into the ◊rhombic allotrope.

monocular vision vision in which the eyes are situated on either side of the head, giving a visual field of almost 360°. It enables animals that are vulnerable to predation, such as rabbits and other herbivores, to detect approaching predators. Unlike ◊binocular vision, monocular vision does not allow accurate judgement of distance.

monohybrid inheritance or ***single-factor inheritance*** pattern of inheritance seen in simple genetic experiments, where the two animals (or plants) being crossed are genetically identical except for one gene. Gregor Mendel first carried out experiments of this type, crossing pea plants that differed only in their tallness.

monomer compound composed of simple molecules from which ◊polymers can be made. Under certain conditions the simple molecules (of the monomer) join together (polymerize) to form a very long chain molecule (macromolecule) called a polymer. For example, the polymerization of ethene molecules produces the polymer polyethene.

$$nCH_2{=}CH_2 \rightarrow \!\left[CH_2{-}CH_2\right]\!_n$$

monosaccharide or ***simple sugar*** a ◊carbohydrate that cannot be hydrolysed (split) into smaller carbohydrate units. Examples are glucose and fructose, both of which have the molecular formula $C_6H_{12}O_6$.

moon any small body that orbits a planet. The planet Earth has a moon that has a mass approximately one-eightieth the mass of the Earth. It orbits the Earth every 27.32 days. It has a rocky surface marked with craters, but no atmosphere or water.

motor anything that produces or imparts motion; a machine that provides mechanical power, particularly an ◊electric motor. Machines that burn fuel (petrol, diesel) are usually called ◊engines, but the internal-combustion engine that propels vehicles has long been called a motor, hence 'motorcar'. Strictly speaking, a car's motor is a part of its engine.

motor neuron neuron that transmits impulses from the central nervous system to muscles or body organs (effectors). Motor neurons cause voluntary and involuntary muscle contractions, and stimulate glands to secrete hormones.

mouth the cavity forming the entrance to the gut. In land vertebrates, air from the nostrils enters the mouth cavity to pass down the trachea. The mouth in mammals is enclosed by the jaws, cheeks, and palate.

mp abbreviation for ◊melting point.

mucous membrane thin skin lining all those body cavities and canals that come into contact with the air (for example, eyelids, breathing and digestive passages, genital tract). It secretes mucus, a protective fluid.

mucus a lubricating and protective fluid, secreted by mucous membranes in many different parts of the body. In the gut, mucus smooths the passage of food and keeps potentially damaging digestive enzymes away from the gut lining. In the lungs, it traps airborne particles so that they can be expelled.

multicellular organism a creature consisting of many cells. Although single-celled organisms are common, they can only reach a certain size. In the early stages of evolution, increases in size occurred through cells clumping together as colonies. Over the millions of years organisms became more complex and more efficient by giving different tasks to different groups of cells. In most animals some cells are specialized for nervous control, some for digestion, some for reproduction, and so on.

muscle animal tissue that contracts and expands to produce locomotion and maintain the movement of body substances. Muscle is made of long cells that can contract to between one half and one third of their relaxed length.

Striped muscles are activated by ◊motor neurons under voluntary control; their ends are usually attached via tendons to bones. ***Involuntary*** or ***smooth*** muscles are controlled by motor neurons not under voluntary control, and are located in the gut, blood vessels, iris, and various ducts. ***Cardiac*** muscle occurs only in the heart, and is also controlled by the autonomic nervous system.

mutation change in the genes brought about by a change in the hereditary material ◊DNA . Mutations, the raw material of evolution, result from mistakes during the replication (copying) of DNA molecules. Only a few improve the organism's performance and are therefore favoured by ◊natural selection. Mutation rates are increased by certain chemicals and by radiation.

mycelium an interwoven mass of threadlike filaments or ◊hyphae, forming the main body of most fungi. The reproductive structures, or 'fruiting bodies', grow from the mycelium.

mycorrhiza a mutually beneficial (mutualistic) association between plant roots and a soil fungus. Mycorrhizal roots take up nutrients more efficiently than non-mycorrhizal roots, and the fungus benefits by obtaining carbohydrates from the plant.

myelin sheath in vertebrates, the fatty, insulating layer that surrounds neurons (nerve cells). It acts to speed up the passage of nerve impulses.

NAND gate in electronics, a type of ◊logic gate.

naphtha term originally applied to naturally occurring liquid hydrocarbons, now used for the mixtures of hydrocarbons obtained by destructive distillation of petroleum, coal tar, and shale oil. It is raw material for the petrochemical and plastics industries.

naphthalene $C_{10}H_8$ solid, white, shiny, aromatic hydrocarbon obtained from coal tar. The smell of mothballs is due to their napthalene content. It is used in making indigo and certain azo dyestuffs, as a mild disinfectant, and an insecticide.

narcotic pain-relieving and sleep-inducing drug. The chief narcotics induce dependency, and include opium, its derivatives and synthetic modifications (such as morphine and heroin); alcohols (such as ethanol); and barbiturates.

nasal cavity in vertebrates, the organ of the sense of smell; in vertebrates with lungs, it is the upper entrance of the respiratory tract. The cavity opens to the exterior by means of nostrils, which, in most mammals, are set in a protrusion called a ◊nose. The nasal cavity has a large surface area and is lined with a ◊mucous membrane that warms and moistens the air and ejects dirt. In the upper parts of the cavity the membrane contains ◊olfactory cells (cells sensitive to smell).

national grid the network of cables, carried overhead on pylons or buried under the ground, that connects consumers of electrical power to power stations, and interconnects the power stations. It ensures that power can be made available to all customers at any time, allowing demand to be shared by several power stations, and particular power stations to be shut down for maintenance work from time to time. Britain has the world's largest grid system, with over 140 power stations able to supply up to 55,000 megawatts. See also ◊power transmission.

native metal or ***free metal*** any of the metallic elements that occur in nature in a chemically uncombined or elemental form (in addition to any combined form). They include bismuth, cobalt, copper, gold, iridium, iron, lead, mercury, nickel, osmium, palladium, platinum, ruthenium, rhodium, tin, and silver. Some are commonly found in the free state, such as gold; others, such as mercury, occur almost exclusively in the combined state, but under unusual conditions do occur as native metals.

natural gas mixture of flammable gases found in the Earth's crust (often in association with petroleum), now one of the world's three main fossil fuels (with coal and oil). Natural gas is a mixture of ◊hydrocarbons, chiefly methane, with ethane, butane, and propane.

Before the gas is piped to storage tanks and on to consumers, butane and propane are removed and liquefied to form 'bottled gas'. Natural gas is liquefied for transport and storage, and is therefore often used where other fuels are scarce and expensive.

natural radioactivity radioactivity generated by those radioactive elements that exist in the Earth's crust. All the elements from polonium (atomic number 84) to uranium (atomic number 92) are radioactive. Radioisotopes of some lighter elements are also found in nature (for example potassium-40).

natural selection process whereby some members of a breeding population are by chance genetically advantaged in relation to the environment, and therefore better able to survive and reproduce than other members. As most environments are slowly but constantly changing, natural selection continually discriminates between members of a population, enhancing the reproductive success of those organisms that possess favourable characteristics. The process is slow, relying on random variations thrown up by ◊mutation and the genetic ◊recombination of sexual reproduction; it is believed to be the main cause of ◊evolution.

navel a small indentation in the centre of the abdomen of mammals, the remains of the site of attachment of the ◊umbilical cord, which connects the fetus to the ◊placenta.

near point the closest position to the eye to which an object may be brought and still be seen clearly. For a normal human eye the near point is about 25 cm; however, it gradually moves further away with age, particularly after the age of 40.

nectar a sugary liquid secreted by some plants from a nectary, a specialized gland usually situated near the base of the flower. Nectar attracts insects, birds, bats, and other animals to the flower for ◊pollination and is the raw material used by bees in the production of honey.

neo-Darwinism the modern theory of ◊evolution, built up since the 1930s by integrating Charles Darwin's theory of evolution through natural selection with the theory of genetic inheritance founded on the work of Mendel (see ◊Mendelism).

neon colourless, odourless, gaseous element, symbol Ne, atomic number 10, relative atomic mass 20.183. It is grouped with the ◊noble gases, is non-reactive, and forms no compounds. It occurs in small quantities in the Earth's atmosphere.

Tubes containing neon are used in electric advertising signs, giving off a fiery red glow; it is also used in lasers.

nervous system the system of interconnected neurons (nerve cells) of all vertebrates and most invertebrates. The mammalian nervous system consists of a central nervous system comprising brain and spinal cord, and a peripheral nervous system connecting up with sensory organs, muscles, and glands.

neuron or ***nerve cell*** type of cell found in the nervous system, capable of rapidly transferring information between different parts of an animal. When neurons are collected together in significant numbers (as in the ◊brain), they not only transfer information but also process it. The unit of information is the ***nerve impulse***, a travelling wave of chemical and electrical changes affecting the membrane of the nerve cell. Many neurons have long extensions, or ◊axons, down which the impulse travels before connecting with another neuron. The impulse involves the passage of sodium and potassium ions across the cell membrane. The axon terminates at the ◊synapse, a specialized area closely linked to the next cell. On reaching the synapse, the impulse releases a chemical (neurotransmitter), which diffuses across to the neighbouring neuron and there stimulates another impulse.

neurotransmitter a chemical that diffuses across a ◊synapse, and thus transmits impulses between ◊neurons (nerve cells), or between neurons and effector organs (such as muscles). Nearly 50 different neurotransmitters have been identified.

neutralization process occurring when the excess acid (or excess base) in a substance is reacted with added base (or added acid) in an amount so that

neuron

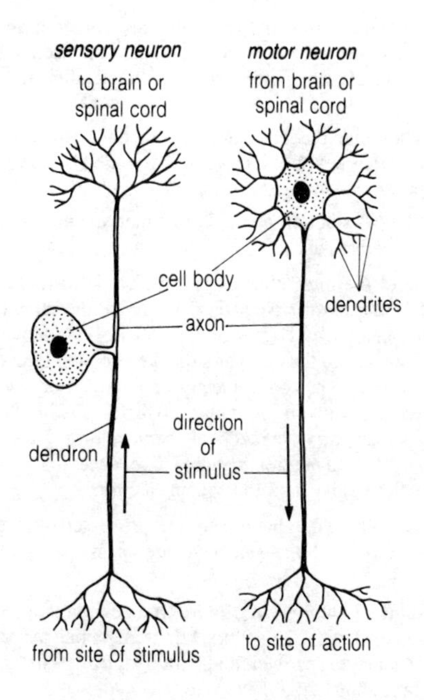

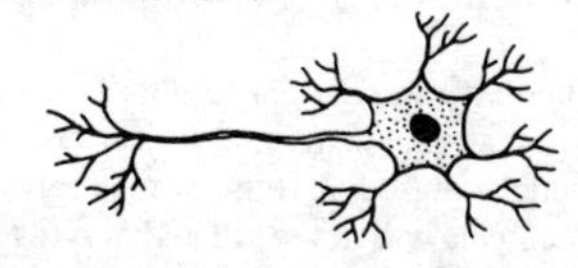

the resulting substance is neither acidic nor basic. This process is represented by the ionic equation

$$H^+_{(aq)} + OH^-_{(aq)} \rightarrow H_2O_{(l)}\ (-\Delta H)$$

In theory neutralization involves adding acid or base as required to achieve ◊pH7. When the colour of an ◊indicator is used to test for neutralization, the final pH may differ from pH7 depending upon the indicator used.

neutral oxide oxide that has neither acidic nor basic properties (see ◊oxide). Neutral oxides are only formed by ◊non-metals. Examples are carbon monoxide, water, and nitrogen(I) oxide.

neutral solution solution of pH 7, in which the concentrations of $H^+_{(aq)}$ and $OH^-_{(aq)}$ ions are equal.

neutron one of the three chief subatomic particles (the others being the ◊proton and the ◊electron). Neutrons have about the same mass as protons but no electric charge, and occur in the nuclei of all ◊atoms except hydrogen. They contribute to the mass of atoms but do not affect their chemistry, which depends on the proton or electron numbers. For instance, ◊isotopes of a single element (with different masses) differ only in the number of neutrons in their nuclei and have identical chemical properties.

Outside a nucleus, a neutron is radioactive, decaying with a ◊half-life of about 12 minutes to give a proton and an electron.

neutron number the number of neutrons possessed by an atomic nucleus. ◊Isotopes are atoms of the same element possessing different neutron numbers.

newton SI unit (symbol N) of ◊force. One newton is the force needed to accelerate an object with mass of one kilogram by one metre per second per second. To accelerate a car weighing 1,000 kg from 0 to 60 mph in 30 seconds would take about 250,000N.

Newton's laws of motion three laws that form the basis of Newtonian mechanics.

(1) Unless acted upon by an external resultant, or unbalanced, force, an object at rest stays at rest, and a moving object continues moving at the same speed in the same straight line. Put more simply, the law says that, if left alone, stationary objects will not move and moving objects will keep on moving at a constant speed in a straight line.

(2) A resultant or unbalanced force applied to an object produces a rate of change of ◊momentum that is directly proportional to the force and is in the direction of the force. For an object of constant mass m, this law may be rephrased as: a resultant or unbalanced force F applied to an object gives it

an acceleration *a* that is directly proportional to, and in the direction of, the force applied and inversely proportional to the mass. This relation is represented by the equation

$$a = F/m$$

which is usually rearranged in the form

$$F = ma$$

(3) When an object A applies a force to an object B, B applies an equal and opposite force to A; that is, to every action there is an equal and opposite reaction.

niacin or ***nicotinic acid*** or ***vitamin B₃*** vitamin of the B complex. Deficiency of this vitamin gives rise to ◊pellagra. Common natural sources are yeast, wheat, and meat. See ◊vitamin.

niche in ecology, the 'place' occupied by a species in its habitat, including all chemical, physical, and biological components, such as what it eats, the time of day at which the species feeds, temperature, moisture, the parts of the habitat that it uses (for example, trees or open grassland), the way it reproduces, and how it behaves. It is believed that no two species can occupy exactly the same niche, because they would be in direct competition for the same resources at every stage of their life cycle.

nickel hard, malleable and ductile, silver-white, metallic element, symbol Ni, atomic number 28, relative atomic mass 58.71. It occurs in igneous rocks, as a free metal (◊native metal), in fragments of iron–nickel meteorites, and in the Earth's core, which consists principally of iron. It has a high melting point, low electrical and thermal conductivity, and can be magnetized. It does not tarnish and is therefore much used for alloys, electroplating, and for coinage.

nicotine $C_{10}H_{14}N_2$ an alkaloid (nitrogenous compound) obtained from the dried leaves of the tobacco plant *Nicotiana tabacum* and used as an insecticide. A colourless oil, soluble in water, it turns brown on exposure to the air.

Nicotine in its pure form is one of the most powerful poisons known. It is the component of cigarette smoke that causes physical addiction.

nitrate inorganic salt containing the NO_3^- ion. Nitrates in the soil can be used by plants to make proteins and nucleic acids. In industry, nitrates are used in the manufacture of explosives and fertilizers and they are an important substance for use in the chemical and pharmaceutical industries. They play a major part in the ◊nitrogen cycle. Run-off from fields results in nitrates polluting rivers and reservoirs and sometimes leading to ◊eutrophication. Excess nitrate in drinking water is considered a health hazard.

nitric acid HNO_3 acid formed when nitrogen dioxide dissolves in water. It is prepared industrially by the oxidation of ammonia over a platinum catalyst.

$$4NH_3 + 5O_2 \rightarrow 4NO + 6H_2O$$

The nitrogen monoxide formed is further oxidized to nitrogen dioxide, which is then reacted with more oxygen and water to give nitric acid.

$$2NO + O_2 \rightarrow 2NO_2$$

$$4NO_2 + 2H_2O + O_2 \rightarrow 4HNO_3$$

In solution it forms fuming (95% HNO_3) or concentrated (68% HNO_3) acid. Both solutions are very corrosive and are strong oxidizing agents. They dissolve most metals, giving off brown fumes of the oxides of nitrogen and forming the nitrate salt.

$$Cu + 4HNO_3 \rightarrow Cu(NO_3)_2 + 2NO_2 + 2H_2O$$

The dilute acid generally gives off brown fumes of oxides of nitrogen when reacted with metals, showing its oxidizing power. In its other reactions it is a strong, monobasic acid.

nitrification a process that takes place in soil when bacteria oxidize ammonia, turning it into nitrates. Nitrates can be absorbed by the roots of plants, so this is a vital stage in the ◊nitrogen cycle.

nitrogen colourless, odourless, tasteless, gaseous element, symbol N, atomic number 7, relative atomic mass 14.0067. It forms about 78% of the Earth's atmosphere by volume and is a necessary part of all plant and animal tissues (in proteins and nucleic acids). For industrial uses it is obtained by liquefaction and fractional distillation of air.

Nitrogen has been recognized as a plant nutrient, found in manures and other organic matter, from early times, long before the complex cycle of ◊nitrogen fixation was understood.

nitrogen cycle in ecology, the process by which nitrogen passes through the ecosystem. Nitrogen, in the form of inorganic compounds (such as

nitrogen cycle

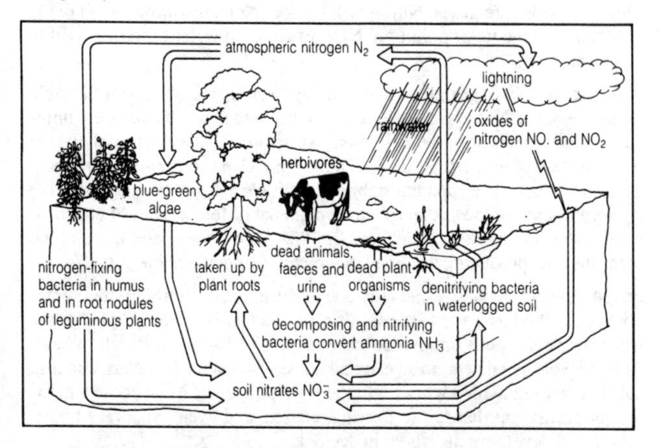

nitrates) in the soil, is absorbed by plants and turned into organic compounds (such as proteins) in plant tissue. A proportion of this nitrogen is eaten by ◊herbivores and used for their own biological processes, with some of this in turn being passed on to the carnivores, which feed on the herbivores. The nitrogen is ultimately returned to the soil as excrement and when organisms die and are converted back to inorganic form by bacterial and fungal ◊decomposers.

Although about 78% of the atmosphere is nitrogen, this cannot be used directly by most organisms. However, certain bacteria are capable of ◊nitrogen fixation; that is, they can extract nitrogen directly from the atmosphere and convert it to compounds such as nitrates that other organisms can use.

nitrogen fixation the process by which nitrogen in the atmosphere is converted into nitrates by the action of bacteria. Some nitrogen-fixing bacteria live mutually with legumes (peas and beans) or other plants (for example, alder), where they form nodules on the roots. Such plants are often cultivated to increase the fertility of soil. Several chemical processes duplicate nitrogen fixation to produce fertilizers; see ◊nitrogen cycle.

nitrogen oxide compound that contains only nitrogen and oxygen. All nitrogen oxides are gases. Nitrogen(II) oxide (NO) and nitrogen(IV) oxide (NO_2) contribute to air pollution. NO_2 dissolves in water to form a weakly acidic solution.

noble gas or ***inert gas*** element belonging to group 0 in the ◊periodic table of the elements. Noble gases are so called because they are extremely unreactive; this is because of their electronic structure. All the electron shells are full and, except for helium, they all have eight electrons in their outermost (◊valency) shell. The apparent stability of this electronic arrangement led to the formulation of the ◊octet rule to explain the different types of chemical bond found in simple compounds. In 1962 xenon was made to combine with fluorine, producing the first known compound of a noble gas.

noise unwanted sound, especially one that is loud or disturbing. It is sometimes described as a form of pollution and this is particularly appropriate when the noise spoils another sound or makes that sound difficult to hear. At low levels, persistent noise can make people irritable, less alert, and less able to carry out skilled work accurately. At high levels it can cause temporary or permanent damage to hearing, nausea, and even temporary blindness. Noise levels are measured in decibels.

Electronic noise takes the form of unwanted signals generated in electronic circuits and in recording processes by stray electrical or magnetic fields, or by temperature variations. In electronic recording and communication systems, 'white noise' frequently appears in the form of high frequencies, or hiss. The main advantages of digital systems are their relative freedom from such noise and their ability to recover and improve noise-affected signals.

non-conductor substance that does not conduct electricity (see ◊insulator) because of a lack of free electrons or ions. Most non-metals are non-conductors.

non-electrolyte compound that does not conduct electricity when molten or in aqueous solution. Sucrose and ethanol are examples of non-electrolytes.

non-metal one of about 20 elements with certain physical and chemical properties opposite to those of ◊metals. Non-metals accept electrons to form negatively charged ions.

Their typical reactions are as follows.

with acids and alkalis Non-metals do not react with dilute acids but may react with alkalis.

$$2NaOH + Cl_2 \rightarrow NaCl + NaOCl$$

with air or oxygen They form acidic or neutral oxides.

$$S + O_2 \rightarrow SO_2$$

with chlorine They react with chlorine gas to form covalent chlorides.

$$2P_{(s)} + 3Cl_{2\,(g)} \rightarrow 2PCl_{3\,(l)}$$

with reducing agents Non-metals act as oxidizing agents.

$$2FeCl_2 + Cl_2 \rightarrow 2FeCl_3$$

non-reducing sugar disaccharide sugar such as sucrose that does not produce a positive result (change in colour) when tested with Benedict's reagent (see ◊food test). Monosaccharides, such as glucose, and the disaccharides maltose and lactose are ◊reducing sugars.

NOR gate in electronics, a type of ◊logic gate.

normal distribution curve the distinctive bell-shaped curve obtained when continuous variation within a population is expressed graphically. When a statistician studies height or intelligence, most people have an intermediate or 'normal' score, with a few individuals scoring either high or low.

nose in humans and most other mammals, it juts out from the front of the face and contains the nostrils, the external openings of the nasal cavity. It is divided down the middle with ◊cartilage. The nostrils also contain plates of cartilage that can be moved by muscles and have a growth of stiff hairs at the margin to prevent dust and dirt from entering.

nostril in vertebrates, the external opening of the ◊nasal cavity. In vertebrates with lungs, the nostrils take in air. In humans, and most other mammals, the nostrils are located on a ◊nose. The passages by which the nasal cavity is connected with the respiratory tract are often called ***internal nostrils***.

NOT gate or ***inverter gate*** in electronics, a type of ◊logic gate.

NPK initials for the symbols of the three elements nitrogen, phosphorus, and potassium. These elements are essential soil nutrients for healthy crop growth. Fertilizers are made up with different amounts of these three elements to suit particular soils and crops. These initials, followed by three numbers, are seen on bags of fertilizer to indicate the relative proportion of each element in that fertilizer.

nuclear energy energy from the inner core or ◊nucleus of the atom, as opposed to energy released in chemical processes, which is derived from the electrons surrounding the nucleus. See ◊nuclear reaction.

nuclear fission process whereby an atomic nucleus breaks up into two or more major fragments with the emission of two or three ◊neutrons. It is accompanied by the release of energy in the form of gamma radiation and the kinetic ◊energy of the emitted particles.

Fission occurs spontaneously in nuclei of uranium-235, the main fuel used in nuclear reactors. However, the process can also be induced by bombarding nuclei with neutrons because a nucleus that has absorbed a neutron becomes unstable and soon splits. The neutrons released spontaneously by the fission of uranium nuclei may therefore be used in turn to induce further fissions, setting up a chain reaction that must be controlled if it is not to result in a nuclear explosion.

nuclear fission

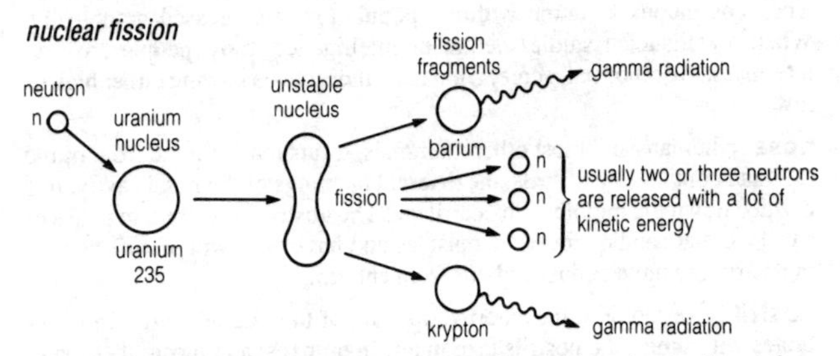

nuclear fusion process whereby two atomic nuclei are 'melted' together, or fused, with the release of a large amount of energy. Very high temperatures and pressures are thought to be required for the process to happen. Under these conditions the atoms involved are stripped of all their electrons so that the remaining particles can come close together at very high speeds and overcome the mutual repulsion of the positive charges on the atomic nuclei. At very close range nuclear forces will come into play, fusing the particles together to form a larger nucleus. As fusion is accompanied by the release of large amounts of energy, the process might one day be harnessed to form the basis of commercial energy production. Methods of achieving controlled fusion are therefore the subject of research around the world.

Fusion is the process by which the Sun and the other stars produce their energy.

nuclear notation method used for labelling an atom according to the composition of its nucleus. The atoms or isotopes of a particular element are represented by the symbol ${}^{A}_{Z}\mathrm{X}$ where A is the mass number of their nuclei, Z is their atomic number, and X is the chemical symbol for that element.

nuclear reaction reaction involving the nuclei of atoms. Atomic nuclei can undergo changes either as a result of radioactive decay, as in the decay of radium to radon (with the emission of an alpha particle) or as a result of particle bombardment in a machine or device, as in the production of cobalt-60 by the bombardment of cobalt-59 with neutrons.

$${}^{226}_{88}\mathrm{Ra} \rightarrow {}^{222}_{86}\mathrm{Rn} + {}^{4}_{2}\mathrm{He}$$

$${}^{59}_{27}\mathrm{Co} + {}^{1}_{0}\mathrm{n} \rightarrow {}^{60}_{27}\mathrm{C0} + \gamma \text{ (gamma radiation)}$$

◊Nuclear fission and ◊nuclear fusion are examples of nuclear reactions. The enormous amounts of energy released may be explained by the mass–energy relation put forward by Einstein, stating that $E = mc^2$ (where E is energy, m is mass, and c is the velocity of light).

The sum of the masses of the products of a nuclear reaction is less than the sum of the masses of the reacting particles. This lost mass is converted to energy according to Einstein's equation.

nuclear reactor central component of a nuclear power station that generates ◊nuclear energy under controlled conditions for use as a source of electrical power. The nuclei of uranium-235 atoms undergo induced ◊nuclear fission in the reactor, and release energy in many forms, one of which is heat. The heat is removed from the core of the reactor by circulating gas or water, and is used to produce the steam that, under high pressure, drives turbines and alternators to produce electrical power.

nuclear safety the use of nuclear energy has given rise to concern over safety. Anxiety has been heightened by accidents such as Windscale (UK), Five Mile Island (USA) and Chernobyl (Ukraine). There has also been mounting concern about the production and disposal of ***nuclear waste***, the radioactive and toxic by-products of the nuclear-energy and nuclear-weapons industries. Burial on land or at sea raises problems of safety, environmental pollution, and security. Nuclear waste may have an active life of several thousand years and there are no guarantees of the safety of the various methods of disposal.

nuclear waste the radioactive and toxic by-products of the nuclear-energy and nuclear-weapons industries. Reactor waste is of three types: high-level spent fuel, or the residue when nuclear fuel has been removed from a reactor and reprocessed; intermediate, which may be long-or short-lived; and low-level, but bulky, waste from reactors, which has only short-lived radioactivity. Disposal, by burial on land or at sea, has raised problems of safety, environmental pollution, and security.

nucleolus dense round structure found in the cell nucleus. It is involved in the production of nucleic acids.

nucleon any particle present in the atomic nucleus. ◊Protons and ◊neutrons are nucleons.

nucleon number another term for ◊mass number.

nucleus (of a cell) the central, membrane-enclosed part of a cell containing the chromosomes.

nucleus (of an atom) the positively charged central part of an ◊atom, which constitutes almost all its mass. Except for hydrogen nuclei, which have only ◊protons, nuclei are composed of both protons and ◊neutrons. Surrounding the nuclei are ◊electrons, which contain a negative charge equal to the protons, thus giving the atom a neutral charge.

nut common name for a dry, single-seeded fruit that does not split open to release the seed. A nut is formed from more than one carpel, but only one seed becomes fully formed; the remainder is aborted. The wall of the fruit, the pericarp, becomes hard and woody, forming the outer shell. Examples are the acorn, hazelnut, and sweet chestnut. The kernels of most nuts provide a concentrated food with about 50% fat and a protein content of 10–20%. Most nuts are produced by perennial trees and bushes.

nutrient any chemical required by an organism to live, grow, and reproduce. Animals need complex organic nutrients, whereas plants can take simpler nutrients such as nitrate and combine them with the products of photosynthesis to make amino acids and proteins, which in turn become the nutrients of animals.

nutrition the strategy adopted by an organism to obtain the chemicals it needs to live, grow, and reproduce. Green plants make their own food from simple chemicals by photosynthesis and are called producers. Other organisms rely on producers for their food and are called consumers.

nylon synthetic long-chain polymer, a polyamide that is similar in chemical structure to protein. Nylon was the first all-synthesized fibre, made from petroleum, natural gas, air, and water by the Du Pont firm in 1938. It is used in the manufacture of moulded articles, textiles, and medical sutures. Nylon fibres are stronger and more elastic than silk and are relatively insensitive to moisture and mildew. Nylon is used for hosiery and woven goods, simulating other materials such as silks and furs; it is also used for carpets.

nymph the immature form of insects that do not have a pupal stage — for example, grasshoppers and dragonflies. Nymphs generally resemble the adult (unlike larvae), but do not have fully formed reproductive organs or wings.

obesity condition of being overweight (generally, 20% or more above the desirable weight for one's sex, build, and height).

Obesity increases susceptibility to disease, strains the vital organs, and lessens ◊life expectancy. It can often be corrected by healthy diet and exercise, unless it is caused by glandular problems.

octane rating numerical classification of petroleum fuels. The efficient running of an internal combustion engine depends on the ignition of a petrol–air mixture at the correct time during the cycle of the engine. Higher-rated petrol burns faster than lower-rated fuels. The use of the correct grade must be matched to the engine.

octet rule rule stating that elements combine in a way that gives them the electronic structure of the nearest ◊inert gas. All the inert gases except helium have eight electrons in their outermost shell, hence the use of the term octet.

This rule is helpful in understanding how the two principal types of bonding—ionic and covalent—are formed.

oesophagus the muscular tube by which food travels from mouth to stomach. The human oesophagus is about 23cm long.

oestrogen a group of hormones produced by the ◊ovaries of vertebrates; the term is also used for various synthetic hormones that mimic their effects. Oestrogens promote the development of female secondary sexual characteristics; stimulate the production of ova (eggs); and, in mammals, prepare the lining of the uterus for pregnancy.

oestrus in mammals, the period during a female's ◊menstrual cycle when mating is most likely to occur. It usually coincides with ovulation.

ohm SI unit (symbol Ω) of electrical ◊resistance. It is defined as the resistance between two points when a potential difference of one volt between them produces a current of one ampere.

Ohm's law law that states that the current flowing in a metallic conductor maintained at constant temperature is directly proportional to the potential difference (voltage) between its ends.

If a current of I amperes flows between two points in a conductor across which the potential difference is V volts, then V/I is a constant called the ◊resistance R ohms between those two points. Hence

$$V/I = R$$

or

$$V = IR$$

Not all conductors obey Ohm's law; those that do are called ***ohmic conductors***.

oil inflammable substance, usually insoluble in water, and chiefly composed of carbon and hydrogen. Oils may be solids (fats and waxes) or liquids. The three main types are: ***essential oils***, obtained from plants; ***fixed oils***, obtained from animals and plants; and ***mineral oils***, obtained chiefly from the refining of petroleum, or crude oil.

Essential oils are volatile liquids that have the odour of their plant source and are used in perfumes, flavouring essences, and in aromatherapy.

Fixed oils are mixtures of ◊esters of fatty acids, of varying consistency, found in both animals (for example, fish oils) and plants (in nuts and seeds). They are used as food; to make soaps, paints, and varnishes; and for lubrication.

Mineral oils derive from organic material that has been compressed underground for millions of years; see ◊petroleum.

olfactory cell in mammals, receptor cell found high inside the nasal cavity, associated with the sense of smell. Olfactory cells are stimulated by chemicals in the air and enhance the related sense of taste.

Olfactory cells can be extremely sensitive, although in humans the sense of smell is not well developed. Many mammals rely on the sense of smell for marking out their territories. Insects can respond to minute levels of airborne chemicals.

omnivore an animal that feeds on both plant and animal material. Omnivores have digestive adaptations intermediate between those of ◊herbivores and ◊carnivores, with relatively unspecialized digestive systems and gut microorganisms that can digest a variety of foodstuffs. Examples are the chimpanzee, the cockroach, and the rat.

optical fibre very fine, optically pure, glass fibre through which light can be reflected to transmit an image or information from one end to the other. Bundles of such fibres are used in endoscopes to inspect otherwise inaccessible parts of machines or of the living body. Optical fibres are increasingly being used to replace copper wire in telephone cables, the messages being coded as pulses of light rather than a fluctuating electric current.

optical instrument instrument that makes use of one or more lenses or mirrors, or of a combination of these, in order to change the path of light rays and produce an image. Optical instruments such as magnifying glasses, microscopes, and telescopes are used to provide a clear, magnified image of the very small or the very distant. Others, such as cameras, photographic enlargers, and film projectors, may be used to store or reproduce images.

optic nerve large nerve passing from the eye to the brain, carrying visual information. In mammals it may contain up to a million sensory neuron fibres, connecting the receptor cells of the retina to the optical centres in the brain.

optics the branch of physics that deals with the study of light and vision—for example shadows and mirror images, lenses, microscopes, telescopes, and cameras. For all practical purposes light rays travel in straight lines, although Albert Einstein demonstrated that they may be 'bent' by a gravitational field. On striking a surface they are reflected or refracted with some absorption of energy, and the study of this is known as geometrical optics.

ore body of rock, a vein within it, or a deposit of sediment, worth mining for the economically valuable mineral it contains.

The term is usually applied to sources of metals. Occasionally metals are found uncombined (◊native metals), but more often they occur as compounds such as carbonates, sulphides, or oxides. The ores often contain unwanted impurities that must be removed when the metal is extracted. Commercially valuable ores include bauxite (aluminium oxide, Al_2O_3) hematite (iron(III) oxide, Fe_2O_3), zinc blende (zinc sulphide, ZnS), and rutile (titanium dioxide, TiO_2).

organ part of a living body, with a distinctive function. Major organs include the brain, the kidneys, the liver, the heart and the lungs.

organelle a specialized structure inside a cell; organelles include ◊mitochondria, ◊chloroplasts, ribosomes, and the ◊nucleus.

organic chemistry

common organic molecule groupings

formula	*name*	*structural formula*
CH_3	methyl	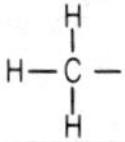
CH_2CH_3	ethyl	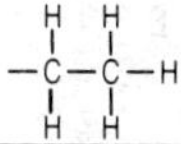
CC	double bond	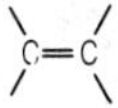
CHO	aldehyde	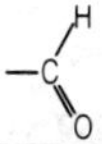
CH_2OH	alcohol	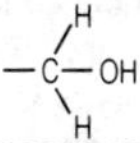
CO	ketone	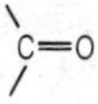
COOH	acid	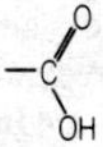
CH_2NH_2	amine	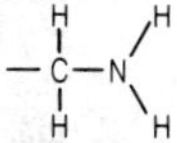
C_6H_6	benzene ring	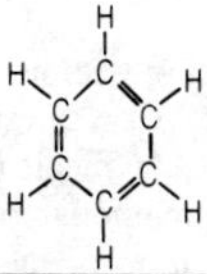

oscillation

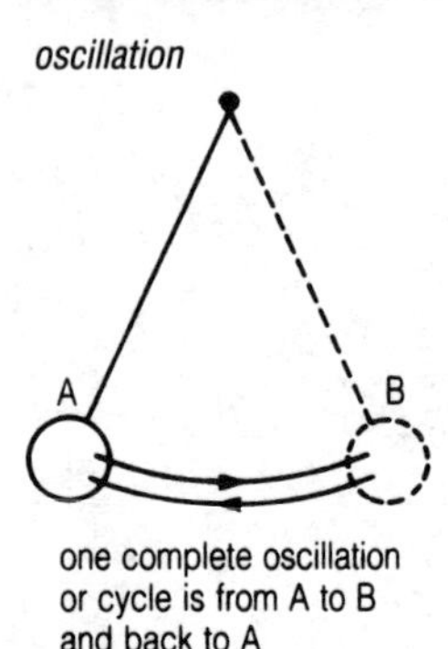

one complete oscillation or cycle is from A to B and back to A

or

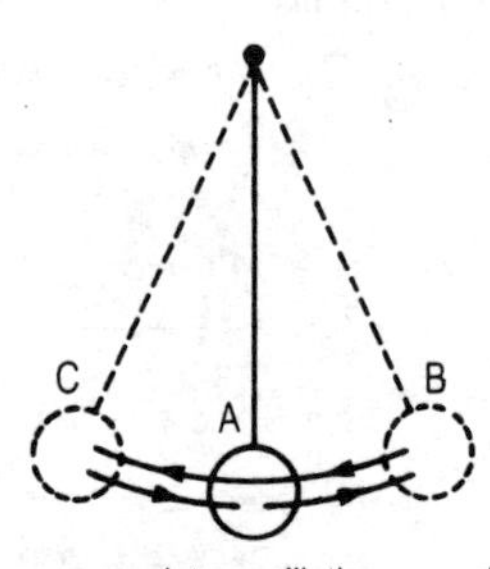

one complete oscillation or cycle is from A to B to C and back to A, moving in the same direction again

organic chemistry branch of chemistry that deals with carbon compounds, in particular the more complex ones. Organic compounds form the chemical basis of life and are more abundant than inorganic compounds. The basis of organic chemistry is the ability of carbon to form long chains of atoms, branching chains, rings, and other complex structures. In a typical organic compound, each carbon atom forms a bond with each of its neighbouring carbon atoms in the chain or ring, and two more with hydrogen atoms (carbon has a ◊valency of four). Other atoms that may be involved in organic molecules include oxygen and nitrogen. Compounds containing only carbon and hydrogen are known as ***hydrocarbons***.

Many organic compounds are made only by living organisms (for example proteins, carbohydrates), and it was once believed that organic compounds could not be made by any other means. This was disproved in the early 19th century when German chemist Friedrich Wöhler synthesized urea, but the name 'organic' (that is 'living') chemistry has remained in use.

organic farming farming without the use of artificial fertilizers (such as ◊nitrates and phosphates) or ◊pesticides (herbicides, insecticides, and fungicides) or other agrochemicals (such as hormones and growth stimulants). Organic farming methods produce food without pesticide residues and greatly reduce pollution of the environment. They are more labour intensive, and therefore more expensive, but use less fossil fuel. Soil structure is greatly improved by organic methods.

In place of artificial fertilizers compost, manure, seaweed, or other substances derived from living things are used (hence the name 'organic'). Growing a crop of a nitrogen-fixing plant, such as alfalfa or clover, and then ploughing it back into the soil also fertilizes the ground.

OR gate in electronics, a type of ◊logic gate.

oscillation one complete to-and-fro movement of a vibrating object or system. For any particular vibration, the time for one oscillation is called its period and the number of oscillations in one second is called its ◊frequency. The maximum displacement of the vibrating object from its rest position is called the ◊amplitude of the oscillation.

oscilloscope see ◊cathode-ray oscilloscope.

osmosis the movement of solvent (liquid) through a semipermeable membrane separating solutions of different concentrations. The solvent

osmosis

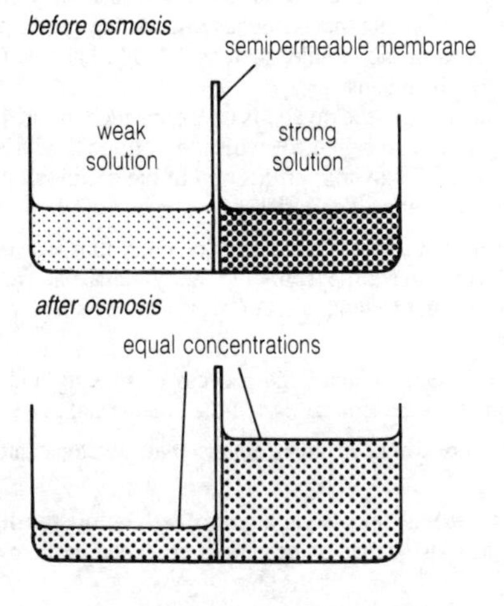

passes from the more dilute solution to the more concentrated solution until the two concentrations are equal. Applying external pressure to the solution on the more concentrated side arrests osmosis, and is a measure of the osmotic pressure of the solution.

Many cell membranes behave as semipermeable membranes, and osmosis is a vital mechanism in the transport of fluids in living organisms—for example in the transport of water from the roots up the stems of plants. Fish have protective mechanisms to counteract osmosis, which would otherwise cause fluid transport between the body of the animal and the surrounding water (outwards in saltwater fish, inwards in freshwater ones).

output device in computing, any device for displaying, in a form intelligible to the user, the results of processing done by a computer. The most common output devices are the VDU (visual display unit, or screen) and the printer.

ovary in female animals, the organ that generates the ◊ovum (egg). In humans, the ovaries are two whitish rounded bodies about 25mm by 35 mm, located in the abdomen near the ends of the ◊Fallopian tubes. Every month, from puberty to the onset of the menopause, an ovum is released from the ovary. This is called ovulation, and forms part of the ◊menstrual cycle. The ovaries secrete the hormones responsible for the secondary sexual characteristics of the female, such as smooth, hairless facial skin and enlarged breasts in humans.

In flowering plants, the ovary is the expanded basal portion of the ◊carpel, containing one or more ◊ovules. It is hollow with a thick wall to protect the ovules. Following fertilization of the ovum within the ovule, it develops into the fruit wall or pericarp.

overtone a note that has a frequency or pitch that is a multiple of the fundamental frequency, the sounding body's natural frequency. Each sound source produces a unique set of overtones, which gives the source its quality or timbre.

ovulation in female animals, the process of making and releasing ova (eggs). In mammals it occurs as part of the ◊menstrual cycle.

ovule a structure found in green plants that develops into a seed after fertilization.

ovum (plural ***ova***) the female gamete (sex cell) before fertilization. In animals (where it is also called an egg), it is produced in the ovary; in plants (where it is also known as an egg cell), the ovum is produced in an ovule. Unlike sperm cells, the ovum is non-motile.

oxidation the loss of ◊electrons, gain of oxygen, or loss of hydrogen by an atom, ion, or molecule during a chemical reaction.

Oxidation may be brought about by reaction with another compound (oxidizing agent), which simultaneously undergoes ◊reduction, or electrically at the anode (positive terminal) of an electric cell.

oxide compound of oxygen and another element, frequently produced by burning the element or a compound of it in air or oxygen.

Oxides of metals are normally ◊bases and will react with an acid to produce a ◊salt. Some of them will also react with a strong alkali to produce a salt in which the metal is part of a complexion, see ◊amphoteric). Most oxides of non-metals are acidic (dissolve in water to form an ◊acid). Some oxides display no pronounced acidic or basic properties.

oxide film thin film of oxide formed on the surface of some metals as soon as they are exposed to the air. This oxide film makes the metal much more resistant to a chemical attack. The considerable lack of reactivity of aluminium to most reagents arises from this property.

oxidizing agent substance that accepts electrons, gives oxygen, or accepts hydrogen (see ◊oxidation). In a redox reaction, the oxidizing agent is the substance that is itself reduced. Common oxidizing agents include oxygen, chlorine, nitric acid, and potassium manganate(VII).

oxygen colourless, odourless, tasteless, gaseous element, symbol O, atomic number 8, relative atomic mass 15.9994. It is the most abundant element in the Earth's crust (almost 50% by mass), forms about 21% by volume of the atmosphere, and is present in combined form in water, carbon dioxide, silicon dioxide (quartz), iron ore, calcium carbonate (limestone) and many other substances. Living things need oxygen, which is a produced by plants through ◊photosynthesis and used up by plants and animals through ◊respiration.

Oxygen is very reactive and combines with all other elements except the ◊inert gases and fluorine. In nature it exists as a molecule composed of two atoms (O_2); single atoms of oxygen are very short-lived owing to their reactivity. They can be produced in electric sparks and by the Sun's ultraviolet radiation in space, where they rapidly combine with molecular oxygen to form ozone.

oxygen debt a physiological state produced by vigorous exercise, in which the lungs cannot supply all the oxygen that the muscles need.

Oxygen is required for the release of energy from food molecules (◊aerobic respiration). Instead of breaking food molecules down fully, muscle cells can switch to a form of partial breakdown that does not require oxygen (◊anaerobic respiration) so that they can continue to generate energy. This partial breakdown produces lactic acid, which results in a sensation of fatigue when it reaches certain levels in the muscles and the blood. Once the vigorous muscle movements cease, the body breaks down the lactic acid, using up extra oxygen to do so. The time required for this is called the ***recovery period***. Panting after exercise is an automatic reaction to 'pay off' the oxygen debt.

oxyhaemoglobin the oxygenated form of ◊haemoglobin, the pigment found in the red blood cells.

ozone (O_3) highly reactive, pale-blue gas with a penetrating odour. It forms a layer in the upper atmosphere, which protects life on Earth from ultraviolet rays, a cause of skin cancer. At lower levels it contributes to the ◊greenhouse effect .

A continent-sized hole has formed over Antarctica as a result of damage to the ozone layer caused by ◊chlorofluorocarbons. In 1989 ozone depletion was 50% over the Antarctic compared with 3% over the Arctic.

At ground level, ozone can cause asthma attacks, stunted growth in plants, and corrosion of certain materials. It is produced by the action of sunlight on car exhaust fumes, and is a major air pollutant in hot summers.

P

palisade cell cylindrical ◊mesophyll cell lying immediately beneath the upper epidermis of a leaf. Palisade cells normally exist as one closely packed row and contain many chloroplasts. During the hours of daylight palisade cells are photosynthetic, using the energy of the Sun to create carbohydrates from water and carbon dioxide.

palladium lightweight, ductile and malleable, silver-white, metallic element, symbol Pd, atomic number 46, relative atomic mass 106.4. It is one of the so-called platinum group of metals, and is resistant to tarnish and corrosion. It often occurs in nature as a free metal in a natural alloy with platinum. Palladium is used as a catalyst, in alloys of gold (to make white gold) and silver, in electroplating, and in dentistry.

pancreas part of the digestive system in vertebrates. It secretes enzymes into the duodenum that digest starches, proteins, and fats. In humans, it is about 18cm long, and lies behind and below the stomach. It contains groups of cells called the ***islets of Langerhans***, which secrete the hormones insulin and glucagon that regulate the blood-sugar level; see ◊blood-glucose regulation.

paraffin common name for ◊alkane, any member of the series of hydrocarbons with the general formula C_nH_{2n+2}. The lower members are gases, such as methane (marsh or natural gas). The middle ones (mainly liquid) form the basis of petrol, kerosene, and lubricating oils, while the higher ones (paraffin waxes) are used in ointment and cosmetic bases.

The fuel commonly sold as paraffin in Britain is more correctly called kerosene.

parallel circuit electrical circuit in which current is split between two or more parallel paths or conductors. The division of the current across each conductor is in the ratio of their resistances. If the currents across two conductors of resistance R_1 and R_2, connected in parallel, are I_1 and I_2 respectively, then the ratio of those currents is given by the equation:

parallel circuit

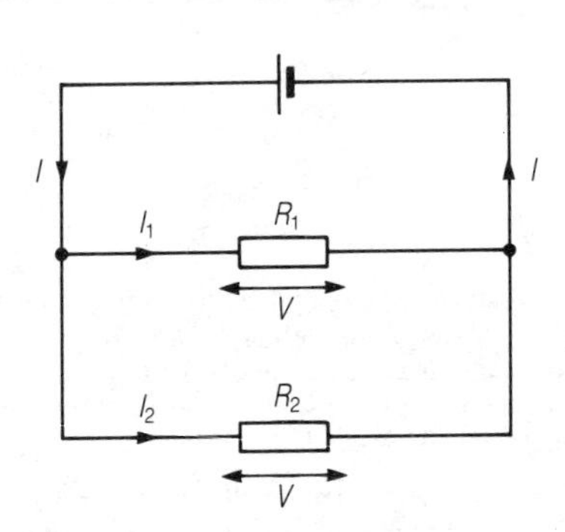

$$I_1/I_2 = R_2/R_1$$

The total resistance R of those conductors is given by:

$$1/R = 1/R_1 + 1/R_2$$

Compare ◊series circuit.

parasite an organism that lives on or in another organism (called the 'host'), and depends on it for nutrition, often at the expense of the host's welfare. Parasites that live inside the host, such as liver flukes and tapeworms, are called ***endoparasites***; those that live on the outside, such as fleas and lice, are called ***ectoparasites***.

pascal SI unit (symbol Pa) of pressure, equal to one newton per square metre.

pathogen microorganism, such as a bacterium or virus, that causes disease. Most pathogens are ◊parasites, and the diseases they cause are incidental to their search for food or shelter inside the host. Non-parasitic organisms, such as soil bacteria or those living in the gut and feeding on waste foodstuffs, can also become pathogenic to a person whose immune system or liver is damaged.

PCB abbreviation for ◊printed circuit board.

pd abbreviation for ◊potential difference.

pellagra chronic disease of subtropical countries in which the staple food is maize, caused by deficiency of niacin (vitamin B_3), which is contained in protein foods, beans and peas, and yeast. Symptoms include digestive disorders, skin eruptions, and mental disturbances.

pelvis in vertebrates, the lower area of the abdomen featuring the bones and muscles used to move the legs or hindlimbs. The ***pelvic girdle*** is a set of bones that allows movement of the legs in relation to the rest of the body and provides sites for the attachment of relevant muscles.

pendulum a weight (called a 'bob') swinging at the end of a cord or rod. When set in motion, it oscillates with a constant frequency that is inversely proportional to the square root of the length of its cord; each oscillation takes the same amount of time regardless of the size of its amplitude.

The regularity of a pendulum's swing was used in making the first really accurate clocks in the 17th century. Pendulums can be used to measure the acceleration due to gravity (an important constant in physics), to measure velocities (ballistic pendulum), and to demonstrate the Earth's rotation (Foucault's pendulum).

penis male reproductive organ, used for internal fertilization; it transfers sperm to the female reproductive tract. Its central passage, through which urine and semen pass, is called the urethra.

pentanol common name ***amyl alcohol*** $C_5H_{11}OH$ clear, colourless, oily liquid, usually having a characteristic choking odour. It is obtained by the fermentation of starches and from the distillation of petroleum.

pepsin enzyme that breaks down proteins during digestion. It requires a strongly acidic environment and is found in the stomach.

peptide a molecule comprising two or more ◊amino acid molecules (not necessarily different) joined by ***peptide bonds***, whereby the acid group of one acid is linked to the amino group of the other (–CO.NH–). The number of amino acid molecules in the peptide is indicated by referring to it as a di-, tri-, or polypeptide (two, three, or many amino acids).

Proteins are built up of interacting polypeptide chains with various types of bond occurring between the chains.

perennating organ in plants, that part of a ◊biennial plant or herbaceous ◊perennial that allows it to survive the winter; usually a root, tuber, rhizome, bulb, or corm.

perennial plant a plant that lives for more than two years. Herbaceous perennials have aerial stems and leaves that die each autumn. They survive the winter by means of an underground perennating organ, such as a bulb or rhizome. Woody perennials, such as trees and shrubs, have stems that persist above ground throughout the year, and may be either ◊deciduous or ◊evergreen. See also ◊annual plant, ◊biennial plant.

periodic table of the elements a table in which the elements are arranged in order of their atomic number. The table summarizes the major properties of the elements and enables predictions to be made about their behaviour.

There are striking similarities in the chemical properties of the elements in each of the vertical columns (called ***groups***), which are numbered I–VII, and a gradation of properties along the horizontal rows (called ***periods***). These features are a direct consequence of the electronic (and nuclear) structure of the atoms of the elements. The periodic table summarizes the major properties of the elements and how they change, and enables predictions to be made. The full version is printed in Appendix I.

period in chemistry, a horizontal row of elements in the periodic table. There is a gradation of properties along each period, from metallic (group I, the alkali metals) to non-metallic (group VII, the halogens).

period in physics, the time taken for one complete oscillation, or cycle of a repeated sequence of events. For example, the time taken for a pendulum to swing from side to side and back again is the period of the pendulum.

peripheral nervous system all parts of the nervous system other than the brain and spinal cord (◊central nervous system).

peristalsis wavelike contractions that pass along tubular organs such as the intestines. They are produced by the alternate contraction and relaxation of antagonistic circular and longitudinal muscles in the organ wall. In the intestines, peristalsis mixes the food and pushes it along.

The same term describes the wavelike motion seen in the locomotion of earthworms and other invertebrates, in which part of the body contracts as another part elongates.

permanent hardness hardness of water that cannot be removed by boiling (see ◊hard water).

perspiration or ***sweating*** the excretion of water and dissolved substances from the ◊sweat glands of the skin of mammals. Perspiration has two main functions: body cooling by the evaporation of water from the skin surface, and excretion of waste products such as salts.

pest any insect, fungus, rodent, or other living organism that has a harmful effect on human beings, other than those that directly cause human diseases. Most pests damage crops or live stock, but the term also covers those that damage buildings, destroy food stores, and spread disease.

pesticide any chemical used in farming, gardening, and indoors to combat pests and the diseases they may carry. Pesticides are of three main types: ◊insecticides (to kill insects), ◊herbicides (to kill plants, mainly those considered weeds), and fungicides (to kill fungal diseases).

petal part of a flower whose function is to attract pollinators such as insects or birds. Petals are frequently large and brightly coloured and may also be scented. Some have a nectary at the base and markings on the petal surface, known as honey guides, to direct pollinators to the source of the ◊nectar. In wind-pollinated plants, however, the petals are usually small and insignificant, and sometimes absent altogether. Petals are derived from modified leaves.

petrochemical chemical derived from the processing of ◊petroleum. The ***petrochemical industry*** is a term embracing those industrial manufacturing processes that obtain their raw materials from the processing of petroleum.

petrol mixture of hydrocarbons derived from petroleum, mainly used as a fuel for internal combustion engines. It is colourless and highly volatile. In the USA, petrol is called gasoline.

Leaded petrol contains antiknock (a mixture of tetraethyl lead and dibromoethane), which improves the combustion of petrol and the performance of a car engine. The lead from the exhaust fumes enters the atmosphere, mostly as simple lead compounds. In recent years the level of lead in the air has risen, and there is strong evidence that it can act as a nerve poison on young children and can cause mental impairment. This has prompted a gradual switch to the use of ***unleaded petrol*** in the UK, which gained momentum owing to a change in the tax on petrol in 1989 that made it cheaper to buy unleaded fuel. Unleaded petrol contains a different mixture of hydrocarbons, and has a lower ◊octane rating than leaded petrol.

petroleum or ***crude oil*** natural mineral oil, a thick greenish-brown flammable liquid found underground in permeable rocks. Petroleum consists of hydrocarbons mixed with oxygen, sulphur, nitrogen, and other elements in varying proportions. It is thought to be derived from the decay of previously

petroleum

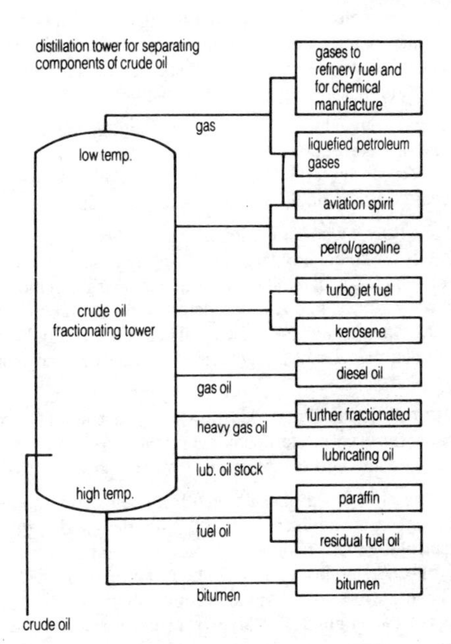

living matter that has been converted by, first, bacterial action, then heat and pressure (but its origin may be chemical also). From crude petroleum, various products are made by distillation and other processes; for example, fuel oil, petrol (gasoline), kerosene, diesel, lubricating oil, paraffin wax, and petroleum jelly.

The organic material in petroleum was laid down millions of years ago (hence the term fossil fuel). Petroleum is often found as large underground lakes floating on water but under a layer of ◊natural gas (mainly methane), trapped below layers of rock that do not allow it to pass through. Oil may flow naturally from wells under gas pressure from above or water pressure from below, causing it to rise up the borehole, but many oil wells require pumping to bring the oil to the surface.

Petroleum products and chemicals are used in large quantities in the manufacture of detergents, artificial fibres, plastics, insecticides, fertilizers, pharmaceuticals, toiletries, and synthetic rubber. Aviation fuel is a more volatile form of petrol.

The burning of fuels derived from petroleum is a major cause of air pollution. Its transport can lead to major catastrophes—for example, the *Torrey Canyon* tanker lost off SW England in 1967, which led to an agreement by the international oil companies in 1968 to pay compensation for massive shore pollution. The 1989 oil spill in Alaska from the *Exxon Valdez* damaged the area's fragile environment, despite clean-up efforts. Drilling for petroleum involves the risks of accidental spillage and drilling-rig accidents such as the 1989 loss of the *Piper Alpha* platform. The problems associated with petroleum have led to the various alternative energy technologies.

pH scale for measuring acidity or alkalinity. A pH of 7.0 (distilled water) indicates neutrality, below 7 is acid, while above 7 is alkaline.

The scale runs from 0 to 14. Strong acids, as used in car batteries, have a pH of about 2; acidic fruits such as citrus fruits are about pH 4. Fertile soils have a pH of about 6.5 to 7, while weak alkalis such as soap are 9 to 10. Corrosive alkalis such as concentrated sodium or potassium hydroxide (lye) are pH 13.

The pH of a solution can be measured by using a broad-range indicator, either in solution or as a paper strip. The colour produced by the indicator is compared with a colour code related to the pH value. An alternative method is to use a pH meter fitted with a glass electrode.

phagocyte a type of ◊white blood cell (leucocyte) that can engulf a bacterium or other invading microorganism. Phagocytes are found in blood, lymph, and other body tissues, where they also ingest foreign matter and dead tissue.

pharynx the interior of the throat, the cavity at the back of the mouth. Its walls are made of muscle strengthened with a fibrous layer and lined with ◊mucous membrane. The internal nostrils lead backwards into the pharynx, which continues downwards into the ◊oesophagus and (through the ◊epiglottis) into the ◊trachea. On each side, a ◊Eustachian tube enters the pharynx from the middle ear cavity.

phase stage in an oscillatory motion, such as a wave motion: two waves are in phase when their peaks and their troughs coincide. Otherwise, there is

pH

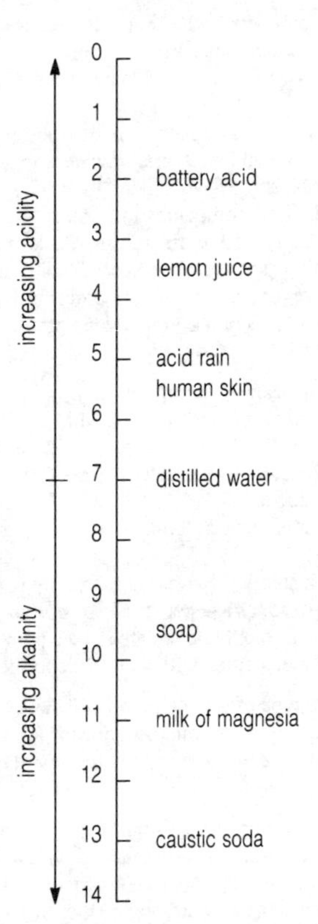

a ***phase difference***, which has consequences in ◊interference phenomena and ◊alternating current electricity.

phenol member of a group of aromatic compounds with weakly acidic properties, which are characterized by a hydroxyl (–OH) group attached directly to an aromatic ring. The simplest of the phenols, derived from benzene, is also known as phenol and has the formula C_6H_5OH. It is also called ***carbolic acid*** and can be extracted from coal tar. Pure phenol consists of colourless, needle-shaped crystals which take up moisture from the atmosphere. It has a strong and characteristic smell and was once used as an antiseptic. It is, however, toxic by absorption through the skin.

phenolphthalein acid–base indicator that is colourless below pH 8 and red above pH 9.6. It is used in titrating weak acids against strong bases.

phenotype in genetics, the traits actually displayed by an organism. The phenotype is not a direct reflection of the ◊genotype because some alleles are masked by the presence of other, dominant alleles (see ◊dominance). The phenotype is further modified by the effects of the environment (for example, poor food stunting growth).

phloem a tissue found in vascular plants whose main function is to conduct sugars and other food materials from the leaves, where they are produced, to all other parts of the plant.

phosphoric acid acid derived from phosphorus and oxygen. Its commonest form (H_3PO_4) is also known as orthophosphoric acid, and is produced by the action of phosphorus pentoxide (P_2O_5) on water. It is used in rust removers and for rust-proofing iron and steel. The 100% acid is known as syrupy phosphoric acid owing to its viscosity. In dilute solution it behaves as a moderately strong, tribasic acid.

Its derivatives are part of many physiological and biochemical processes.

phosphorus highly reactive, non-metallic element, symbol P, atomic number 15, relative atomic mass 30.9738. It occurs in nature as phosphates in the soil, in particular the mineral apatite, and is essential to both plant and animal life. The element has three allotropic forms: a black powder; a white-yellow, waxy solid that ignites spontaneously in air to form the poisonous gas phosphorous pentoxide; and a red-brown powder that neither ignites spontaneously nor is poisonous. Compounds of phosphorus are used in fertilizers, various organic chemicals, for matches and fireworks, and in glass and steel.

photochemical reaction any chemical reaction in which light is produced or light initiates the reaction. Light can initiate reactions by exciting atoms or molecules and making them more reactive: the light energy becomes converted to chemical energy. Many photochemical reactions set up a ◊chain reaction and produce ◊free radicals.

This type of reaction is seen in the bleaching of dyes or the yellowing of paper by sunlight. It is harnessed by plants in ◊photosynthesis and by humans in photography. Chemical reactions that produce light are most commonly seen when materials are burned.

Light-emitting reactions are used by living organisms in bioluminescence. One photochemical reaction is the action of sunlight on car exhaust fumes, which results in the production of ◊ozone. Some large cities, such as Los Angeles, now suffer serious pollution due to photochemical smog.

photodiode semiconductor p–n junction-diode used to detect light or measure its intensity. The photodiode is encapsulated in a transparent plastic case that allows light to fall onto the junction. When this occurs, the reverse-bias resistance (high resistance in the opposite direction to normal current-flow) drops and allows a larger reverse-biased current to flow through the device. The increase in current can then be related to the amount of light falling on the junction. Photodiodes that can detect small changes in light level are used in alarm systems, camera exposure-controls, and optical communication links.

photon the smallest 'package', 'particle', or quantum of energy in which ◊light, or any other form of electromagnetic radiation, is emitted. It has both particle and wave properties; it has no charge, is considered massless, but possesses momentum and energy.

photoperiodism mechanism by which the timing of an organism's activities is altered in response to daylength. The flowering of many plants is initiated in this way, and the breeding seasons of many temperate-zone animals are also triggered by increasing or declining day length.

photosynthesis process by which green plants trap light energy and use it to drive a series of chemical reactions, leading to the formation of carbohydrates. All animals ultimately depend on photosynthesis because it is the method by which the basic food (sugar) is created. For photosynthesis to occur, the plant must possess ◊chlorophyll and must have a supply of carbon dioxide and water. Actively photosynthesizing green plants store excess sugar as starch (this can be tested for in the laboratory by using iodine).

The chemical reactions of photosynthesis occur in two stages. During the ***light reaction*** sunlight is used to split water (H_2O) into oxygen (O_2),

photosynthesis

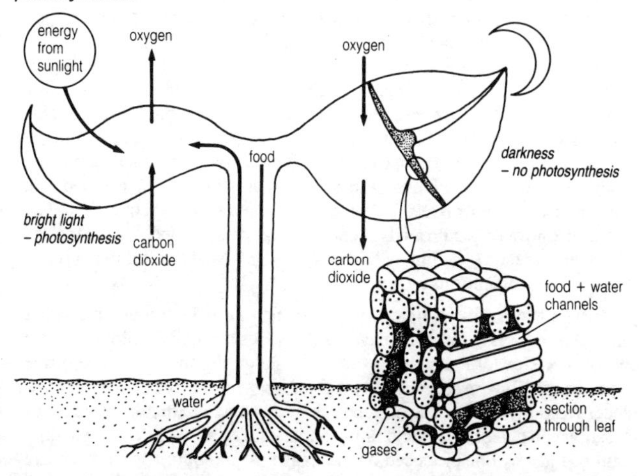

protons (hydrogen ions, H^+), and electrons, and oxygen is given off as a by-product. In the second-stage ***dark reaction***, for which sunlight is not required, the protons and electrons are used to convert carbon dioxide (CO_2) into carbohydrates ($C_m(H_2O)_n$). The reaction can be summarized by the following equation.

$$m\mathrm{CO_2} + n\mathrm{H_2O} + \text{light energy} \rightarrow \mathrm{C}_m(\mathrm{H_2O})_n + m\mathrm{O_2}$$

The reverse equation represents ◊respiration.

phototropism movement of part of a plant towards or away from a source of light. Leaves are positively phototropic, detecting the source of light and orientating themselves to receive the maximum amount.

pig iron or ***cast iron*** the crude, unrefined form of iron produced in a ◊blast furnace. It contains around 4% carbon plus some other impurities. ◊Cast iron is the partly refined form.

Pill, the commonly used term for the contraceptive pill, based on female hormones. The combined pill, which contains oestrogen and progesterone, stops the production of ova (eggs), and makes the mucus produced by the cervix hostile to sperm. It is the most effective form of contraception apart from sterilization, being more than 99% effective.

The ***minipill*** or progesterone-only pill prevents implantation of a fertilized egg into the wall of the uterus. The minipill has a slightly higher failure rate, especially if not taken at the same time each day, but has fewer side effects and is considered safer for long-term use. Possible side effects of the Pill include migraine or headache and high blood pressure. More seriously, oestrogen-containing pills can slightly increase the risk of a blood clot forming in the blood vessels. This risk is increased in women over 35 if they smoke. Controversy surrounds other possible health effects of taking the Pill. The evidence for a link with cancer is slight (and the Pill may protect women from some forms of cancer).

pinhole camera the simplest type of camera, in which a pinhole rather than a lens is used to form an image. Light passes through the pinhole at one end of a box to form a sharp inverted image on the inside surface of the opposite end. The image is equally sharp for objects placed at different distances from the camera because only one ray from a particular distance or direction can enter through the tiny pinhole, and so only one corresponding point of light will be produced on the image. A photographic film or plate fitted inside the box will, if exposed for a long time, record the image.

pinhole camera

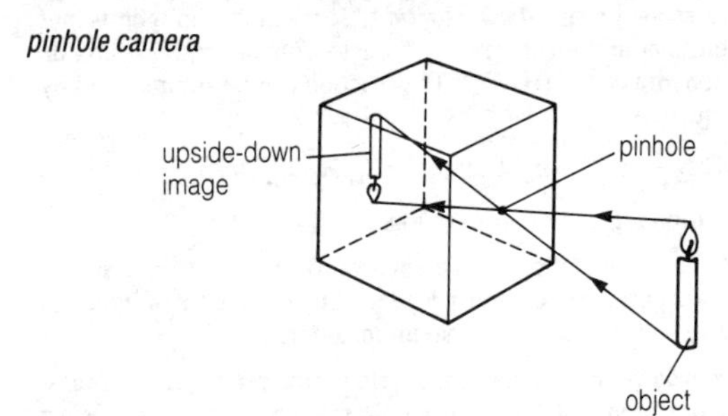

pinna the external part of the ◊ear.

pioneer species in ecology, those species that are the first to colonize and thrive in new areas. Coal tips, recently cleared woodland, and new roadsides are areas where pioneer species will quickly appear. As the habitat matures other species take over, a process known as ***succession***. See also ⟩colonization.

pipette device for the accurate measurement of a known volume of liquid, usually for transfer from one container to another, used in chemistry and biology laboratories.

A conventional pipette is a glass tube, often with an enlarged bulb, which is calibrated in one or more positions. Liquid is drawn into the pipette by suction, to the desired calibration mark. The release of liquid is controlled by careful pressure of the forefinger over the upper end of the tube, or by a plunger or rubber bulb.

pitch black, sticky substance, hard when cold, but liquid when hot, used for waterproofing, roofing, and paving. It is made by the destructive distillation of wood or coal tar, and has been used since antiquity in the waterproofing (caulking) of wooden ships.

pitchblende or ***uraninite*** brownish-black mineral, the major constituent of uranium ore, consisting mainly of uranium oxide (UO_2). It also contains some lead (the final, stable product of uranium decay) and variable amounts of most of the naturally occurring radioactive elements, which are products of either the decay or the fission of uranium isotopes. The uranium yield is 50–80%; it is also a source of radium, polonium, and actinium.

pitfall trap in ecology, a trap consisting of a small container sunk into the earth until its rim is at ground level; invertebrates fall into the container and are unable to escape. The trap is used in surveying the distribution of invertebrates in a particular terrestrial ecosystem.

pituitary gland a major ⟩endocrine gland, situated in the centre of the brain. The anterior lobe secretes hormones, some of which control the activities of other glands (thyroid, gonads, and adrenal cortex); others are direct-acting hormones affecting milk secretion, and controlling growth. Secretions of the posterior lobe control body water balance, and contraction of the uterus. The posterior lobe is regulated by nerves from the hypothalamus, and thus forms a link between the nervous and hormonal systems.

placenta in mammals, the organ that attaches the developing ⟩fetus to the ⟩uterus. It links the blood supply of the fetus to the blood supply of the mother, allowing the exchange of nutrients, wastes, and gases. The two blood systems are not in direct contact, but are separated by thin membranes, with materials diffusing across from one system to the other. The placenta also produces hormones that maintain and regulate pregnancy. It is shed as part of the afterbirth.

It is now understood that a variety of materials, including drugs and viruses, can pass across the placental membrane. For example, HIV, the virus that causes ◊AIDS, can be transmitted in this way.

plant organism that carries out ◊photosynthesis, has cellulose cell walls and complex eukaryotic cells, and is immobile. A few parasitic plants have lost the ability to photosynthesize but are still considered to be plants.

Plants are autotrophs—that is, they make carbohydrates from water and carbon dioxide—and are the primary producers in all food chains, so that all animal life is dependent on them. They play a vital part in the carbon cycle, removing carbon dioxide from the atmosphere and generating oxygen. The study of plants is known as botany.

Originally the plant kingdom included bacteria and fungi, but these are not now thought of as plants. The groups that are usually classified as plants are the multicellular algae (seaweeds and freshwater weeds), bryophytes (mosses and liverworts), pteridophytes (ferns, horsetails, and club mosses), gymnosperms (conifers, yews, cycads, and ginkgos), and angiosperms (flowering plants).

Many of the lower plants (the algae and bryophytes) consist of a simple body, or thallus, on which the organs of reproduction are borne. They are susceptible to drying out and are therefore usually confined to aquatic or damp habitats. The pteridophytes, gymnosperms, and angiosperms have special supportive water-conducting tissues, which identify them as ◊vascular plants.

The seed plants (angiosperms and gymnosperms) are the largest group in the plant kingdom, and structurally the most complex. They are usually divided into three parts: root, stem, and leaves. Stems grow above or below ground. Their cellular structure is designed to carry water and salts from the roots to the leaves in the ◊xylem, and sugars from the leaves to the roots in the ◊phloem. The leaves manufacture the food of the plant by means of photosynthesis, which occurs in the ◊chloroplasts they contain. Flowers and cones are modified leaves arranged in groups, enclosing the reproductive organs from which the fruits and seeds result.

plant hormone a substance produced by a plant that has a marked effect on its growth, flowering, leaf fall, fruit ripening, or some other process. Examples include ◊auxin, ◊giberellin, ethene, and cytokinin.

plant

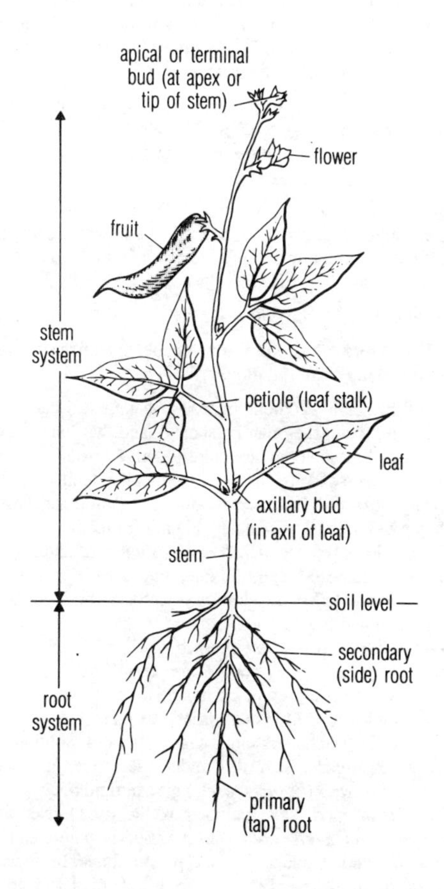
apical or terminal
bud (at apex or
tip of stem)
flower
fruit
stem
system
petiole (leaf stalk)
leaf
axillary bud
(in axil of leaf)
stem
soil level
secondary
(side) root
root
system
primary
(tap) root

Unlike animal hormones, these substances do not always produce their effect in a different site to that in which they were produced, and they may be less specific in their effects. It has therefore been suggested that they should not be described as hormones at all.

plaque a mixture of saliva, food particles, and bacteria. It builds up on teeth and converts sugars to acid, which eats through the hard exterior enamel into the dentine, causing decay. Plaque can be removed by regular brushing.

planet large celestial body in orbit around a star, it consists of rock, metal or gas. Planets do not produce light, but reflect the light of their parent star. There are nine planets in our Solar System; Mercury, Venus, Earth, Mars, Jupiter, Saturn, Uranus, Neptune and Pluto.

plasma the liquid part of the ◊blood.

plaster of Paris form of calcium sulphate, obtained from gypsum, mixed with water for making casts and moulds.

plastic any of the stable synthetic materials that are fluid at some stage in their manufacture, when they can be shaped, and that later set to rigid or semi-rigid solids. Plastics today are chiefly derived from petroleum. Most are polymers, made up of long chains of identical molecules.

Processed by extrusion, injection-moulding, vacuum-forming and compression, they emerge in consistencies ranging from hard and inflexible to soft and rubbery. They replace an increasing number of natural substances, being lightweight, easy to clean, durable, and capable of being rendered very strong, for example by the addition of carbon fibres, for building aircraft and other engineering projects.

Thermosoftening plastics (or ***thermoplastics***) always soften when warmed, then re-harden as they cool. Examples of thermoplastics include polystyrene, a clear plastic used in kitchen utensils or (when expanded into a 'foam' by gas injection) in insulation and ceiling tiles; polyethylene or polythene, used for containers and wrapping; and polyvinyl chloride (PVC), used for drainpipes, floor tiles, audio discs, shoes, and hand bags.

Thermosetting plastics (or ***thermosets***) remain rigid once heated, and do not soften when rewarmed. They include Bakelite, used in electrical insulation and telephone receivers; epoxy resins, used in paints and varnishes, to laminate wood, and as adhesives; and polyurethane, prepared in liquid form as a paint or varnish, and in foam form for upholstery and in lining materials (where it may be a fire hazard). Members of one group of plastics, the silicones, are chemically inert, have good electrical properties, and repel water. Silicones find use in silicone rubber, paints, electrical insulation materials, laminates, waterproofing for walls, stain-resistant textiles, and cosmetics.

Shape-memory polymers are plastics that can be crumpled or flattened and will resume their original shape when heated. They include *trans*-polyisoprene and polynorbornene. The initial shape is determined by heating the polymer to over 35°C and pouring it into a metal mould. The shape can be altered with boiling water and the substance solidifies again when its temperature falls below 35°C.

Biodegradable plastics are increasingly in demand: Biopol was developed in 1990. It is made by microbes that build the plastic in their bodies from carbon dioxide and water (it constitutes 80% of their body). The unused parts of the microbe are dissolved away by heating in water. The discarded plastic can be placed in landfill sites where it breaks back down into carbon dioxide and water. It costs three to five times as much as ordinary plastics to produce.

plate tectonics theory formulated in the 1960s that explains the formation of the major physical features of the Earth's surface. The Earth's outermost layer, or crust, is regarded as a jigsaw of rigid plates up to a hundred kilometres thick, which move relative to each other. Their movement may be due to convection currents within the semisolid mantle beneath.

The world's continents are carried on some of these plates and are therefore slowly shifting their positions (continental drift). Over 200 million years ago, the continents were joined together, forming a single giant continent called Pangaea. However, the gradual movement of the plates caused Pangea to break apart so that by 50 million years ago the continents were approaching their present positions, spread out around the surface of the Earth.

Where two plates are moving apart from each other, molten rock from the Earth's mantle wells up in the space left between the plates and hardens to form new crust, usually in the form of a ridge. Where two plates collide, one of the plates may be forced under the other, eventually melting back into the mantle, or the plates may crumple, forming large mountain ranges such as the Himalayas in Asia, the Andes in South America, and the Rockies in North America. Sometimes two plates will slide past each other—an example is the San Andreas Fault, California, where the movement of the plates sometimes takes the form of sudden jerks, causing the earthquakes common in the San Francisco–Los Angeles area. Most of the earthquake and volcano zones of the world are, in fact, found in regions where two plates meet or are moving apart.

platelet a tiny 'cell' found in the blood, which helps it to clot. Platelets are not true cells, but membrane-bound cell fragments that bud off from large cells in the bone marrow. See ◊blood clotting.

platinum heavy, soft, silver-white, malleable and ductile, metallic element, symbol Pt, atomic number 78, relative atomic mass 195.09. It is the first of a group of six metallic elements (platinum, osmium, iridium, rhodium, ruthenium, and palladium) that possess similar traits, such as resistance to tarnish, corrosion, and attack by acid, and that often occur as free metals (◊native metals). They often occur in natural alloys with each other, the commonest of which is osmiridium. Both pure and as an alloy, platinum is used in dentistry, jewellery, and as a catalyst.

pleural membrane one of a pair of membranes surrounding the lungs, protecting and lubricating them during breathing movements.

plug, three-pin insulated device with three metal projections used to connect the wires in the cable of an electrical appliance with the wires of a mains supply socket. In the UK, plugs have pins of rectangular section, and must comply with the British Standard BS 1363 laid down by the British Standards Institute. A plug must be designed carefully with regard to construction, labelling, clearance between components, accessibility of live parts, earthing (see ◊earth wire), and terminal design. Before being approved, it is tested for the resistance of its insulation, temperature rise while in use, anchorage of cables, mechanical strength, and susceptibility to damage by heat and rust.

plutonium silvery-white, radioactive, metallic element of the ◊actinide series, symbol Pu, atomic number 94, relative atomic mass 239.13. It occurs in nature in minute quantities in ◊pitchblende and other ores, but is produced in large quantities synthetically in nuclear reactors by bombarding uranium-238 with neutrons. Plutonium is one of the three elements capable of ◊nuclear fission (the others are thorium and uranium), and is used as a fuel in fast breeder reactors and in making nuclear weapons. It has a long half-life (24,000 years) during which time it remains highly toxic.

Plutonium is dangerous to handle, difficult to store, and impossible to dispose of.

p–n junction diode in electronics, a two-terminal semiconductor device that allows electric current to flow in only one direction, the ***forward-bias*** direction. A very high resistance prevents current flow in the opposite, or ***reverse-bias***, direction. It is used to convert alternating current (AC) to direct current (DC).

pole, magnetic see ◊magnetic pole.

pollen in flowering plants, the grains that contain the male gametes. They are produced by the ◊anthers and, when mature, have a hard outer wall. Pollen of insect-pollinated plants (see ◊pollination) is often sticky and spiny, and larger than the smooth, light grains produced by wind-pollinated species.

pollination the process by which fertilization occurs in the sexual reproduction of higher plants. The male gametes are contained in pollen grains, which must be transferred from the ◊anther to the ◊stigma in angiosperms (flowering plants), and from the male cone to the female cone in gymnosperms.

Self-pollination occurs when pollen is transferred to a stigma of the same flower, or to another flower on the same plant; ***cross-pollination*** occurs when pollen is transferred to another plant. This involves external pollen-carrying agents, such as wind, water, and animals such as insects, birds, bats, and other small mammals.

Most flowers are adapted for pollination by one particular agent only. Those that rely on animals are generally scented, and have large, brightly coloured petals and sticky pollen grains. They produce nectar, a sugary liquid, or surplus pollen, or both, on which the pollinator feeds. Thus, the relationship between pollinator and plant is an example of mutualism, in which both benefit. In some plants, however, the pollinator receives no benefit, while in others nectar may be removed by animals that do not bring about pollination.

Wind-pollinated, or anemophilous, flowers are usually unscented, have either very reduced petals and sepals or lack them altogether, and do not produce nectar. Their pollen grains are small and smooth walled, and, because air movements are random, are produced in vast amounts. The male flowers have numerous exposed stamens, often on long filaments; the female flowers have long, often branched, feathery stigmas. Many wind-pollinated plants, such as hazel and birch, bear their flowers on catkins to facilitate the free transport of pollen.

pollution the harmful effect on the environment of by-products of human activity, principally industrial and agricultural processes—for example noise, smoke, car emissions, chemical effluents in seas and rivers, pesticides, sewage, and household waste.

Pollution control involves higher production costs for the industries concerned, but failure to implement adequate controls may result in irreversible environmental damage and an increase in the incidence of diseases such as cancer.

vehicle emissions Cars and lorries release a range of pollutants into the atmosphere. Carbon monoxide reduces breathing efficiency by combining irreversibly with ◊haemoglobin in the blood; oxides of nitrogen produce ◊acid rain and can aggravate respiratory conditions such as asthma; particles of carbon react in sunlight to produce low-level ozone, which also causes respiratory problems; and lead compounds impair the development of children. In addition, motor vehicles release large amounts of carbon dioxide, which contributes to the ◊greenhouse effect.

industrial pollution All industry produces waste of some sort; this becomes pollution if not properly managed. For example, most industrial processes involve the use of electricity, apparently a 'clean', non-polluting form of energy. However, almost all power stations, in generating electricity, produce large volumes of warm water, which can alter river ecosystems if released without sufficient cooling. Power stations that burn fossil fuels inevitably release large amounts of carbon dioxide into the air, contributing to climatic change. The smoke produced by coal-fired power stations also contains sulphur dioxide, a major cause of acid rain. Nuclear power stations do not release carbon dioxide or sulphur dioxide, but do produce radioactive waste, which may remain hazardous for tens of thousands of years.

agricultural pollution Intensive farming requires that land be used year after year without interruption. Fast-acting artificial fertilizers, containing high concentrations of nitrates, are therefore necessary. The manufacture of fertilizers requires a considerable amount of electricity, and is therefore in itself an indirect cause of pollution. Once put on the fields, nitrates can pass into rivers and streams, causing ◊eutrophication, and into drinking water, posing a possible hazard to health. High-yield agriculture also makes use of large quantities of herbicides and insecticides to prevent crop damage. Such chemicals can easily pass into the ◊food chain, eventually becoming concentrated in the tissues of predators, causing their death or impairing their ability to reproduce.

Natural disasters may also cause pollution; volcanic eruptions, for example, cause ash to be ejected into the atmosphere and deposited on land surfaces.

polyester synthetic resin formed by the ◊condensation of polyhydric alcohols (alcohols containing more than one hydroxyl group) with dibasic acids (acids containing two replaceable hydrogen atoms). Polyesters are thermosoftening ◊plastics, used in making synthetic fibres, such as Dacron and Terylene, and constructional plastics. With glass fibre added as reinforcement, polyesters are used in car bodies and boat hulls.

polyethene or ***polyethylene*** polymer of the gas ethene (ethylene, C_2H_4). It is a tough, white translucent waxy thermosoftening plastic (which means it can be repeatedly softened by heating). It is used for packaging, bottles, toys, electric cable, pipes, and tubing.

Polyethene is produced in two forms: low-density polyethene, made by high-pressure polymerization of ethene, and high-density poly(ethene), which is made at lower pressure by using catalysts. This form is more rigid at low temperatures and softer at higher temperatures than the low-density type.

In the UK the plastic is better known under the trademark Polythene.

polymer compound made up of large, long-chain molecules composed of many repeated simple units (◊***monomers***). There are many polymers, both natural (cellulose, chitin, lignin) and synthetic (polyethene and nylon, types of plastic). Synthetic polymers belong to two groups: thermosoftening and thermosetting (see ◊plastic).

polymerization the chemical union of two or more (usually small) molecules of the same kind to form a new compound. The process is used extensively in the manufacture of plastics (see ◊thermosetting plastic and ◊thermosoftening plastic).

addition polymerization In this type of polymerization, many molecules of a single compound join together to form long chains. An example is the polymerization of ethene to polyethene.

$$n\mathrm{CH_2{=}CH_2} \rightarrow \mathrm{+\!\!\!\!\;CH_2{-}CH_2\!\!\!\;+}_n$$

polymerization

addition of ethene molecules to form polyethene

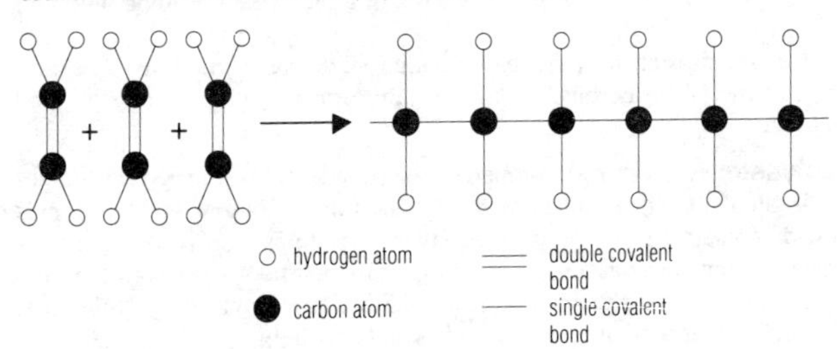

The only product of the reaction is the polymer. Other addition polymers include polyvinyl chloride (PVC).

condensation polymerization In this reaction, a small molecule such as water or hydrogen chloride is given off as a result of the polymerization. An example is the production of polyesters, formed by the polymerization of organic acids and alcohols. Condensation polymerization may involve a single monomer that has two reactive groups (such as an amino acid) or two or more different monomers (such as urea and formaldehyde, which polymerize to form resins).

polypeptide a long-chain ◊peptide.

polysaccharide a long-chain ◊carbohydrate made up of hundreds or thousands of linked simple sugars (monosaccharides) such as glucose and closely related molecules.

The polysaccharides are natural polymers. They either act as energy-rich food stores in plants (starch) and animals (glycogen), or have structural roles in the plant cell wall (cellulose, pectin) or the tough outer skeleton of insects and similar creatures (chitin). See also ◊carbohydrate.

polystyrene polymer made from the monomer styrene (phenylethene, $C_6H_5CH=CH_2$); see ◊plastic.

polytetrafluoroethene (PTFE) polymer made from the monomer tetrafluoroethene (CF_2CF_2). It is a thermosetting plastic with a high melting point that is used to produce 'non-stick' surfaces on pans and to coat bearings. Its trade name is Teflon.

Polythene trade name for a variety of ◊polyethene.

polyunsaturate a type of triglyceride (◊ fat or oil) in which the long carbon chains of the ◊fatty acids contain several double bonds. By contrast, the carbon chains of saturated fats (such as lard) contain only single bonds.

The more double bonds the chains contain, the lower the melting point of the triglyceride. Unsaturated chains with several double bonds produce oils, such as vegetable and fish oils, which are liquids at room temperature. Saturated fats, with no double bonds, are solids at room temperature. The polyunsaturated fats used for margarines are produced by taking a vegetable or fish oil and turning some of the double bonds to single bonds, so that the product is semi-solid at room temperature. Medical evidence suggests that polyunsaturated fats are less likely to contribute to heart disease and arteriosclerosis than are saturated fats.

polyvinyl chloride (PVC) polymer made from the monomer vinyl chloride (chloroethene, CH_2=CHCl); see ◊plastic.

population cycle regular fluctuations in the size of a population, as seen in lemmings, for example. Such cycles are often caused by density-dependent mortality: high mortality due to overcrowding causes a sudden decline in the population, which then gradually builds up again. Population cycles may also result from an interaction between a predator and its prey.

population genetics the branch of genetics that studies the way in which the frequencies of different ◊alleles in populations of organisms change, as a result of natural selection and other processes.

potassium soft, waxlike, silver-white, metallic element, symbol K, atomic number 19, relative atomic mass 39.0983. It is one of the ◊alkali metals and has a very low density—it floats on water, and is the second lightest metal (after lithium). It oxidizes rapidly when exposed to air and reacts violently with water. Of great abundance in the Earth's crust, it is widely distributed with other elements and found in salt and mineral deposits in the form of potassium aluminium silicates.

Potassium, with sodium, plays a role in the transmission of impulses by neurons (nerve cells), and is essential for animals; it is also required by plants for growth.

potassium manganate(VII) or ***potassium permanganate*** $KMnO_4$ dark purple, crystalline solid, soluble in water; it is a strong ◊oxidizing agent in the presence of dilute sulphuric acid. In the process of oxidizing other compounds it is itself reduced to manganese(II) salts (containing the Mn^{2+} ion), which are colourless.

potential difference (pd) measure of the electrical potential energy converted to another form for every unit charge moving between two points in an electric circuit. In equation terms, potential difference V may be defined by

$$V = W/Q$$

where W is the electrical energy converted in joules and Q is the unit charge in coulombs. The unit of potential difference is the volt. See also ◊Ohm's law.

potential energy ◊energy possessed by an object by virtue of its relative position or state (for example, as in a compressed spring). It is contrasted with ◊kinetic energy. See ◊gravitational potential energy.

power, electric the rate at which an electrical machine or component converts ◊electrical energy into other forms of energy—for example, light, heat, or mechanical energy. Its SI unit is the watt (joule per second).

The power P of an electrical machine is given by the formula

$$P = W/t$$

where W is the electrical energy converted by the machine over a time t seconds, or by the formulae

$$P = IV$$

or

$$P = I^2R$$

where I is the current in amperes flowing through the machine, V is the potential difference in volts across it, and R is its resistance in ohms.

For example, an electric lamp that passes a current of 0.25 amperes at 240 volts uses 60 watts of electrical power and converts it into light—in everyday terms, it is a 60-watt lamp. An electric motor that requires 5 amperes at the same voltage consumes 1,200 watts (1.2 kilowatts).

power station building where electrical energy is generated from a fuel or from another form of energy. Fuels used include fossil fuels such as coal, gas, and oil, and the nuclear fuel uranium. Renewable sources of energy include ◊gravitational potential energy, used to produce ◊hydroelectric power, and ◊wind energy. The energy supply is used to turn ◊turbines either directly by means of water or wind pressure, or indirectly by steam pressure, steam being generated by burning fossil fuels or from the heat released by the nuclear fission of uranium nuclei. The turbines in their turn spin alternators, which generate electricity at very high voltage.

power transmission transfer of electrical power from one location, such as a power station, to another. Electricity is conducted along the cables of the ◊national grid at a high voltage (up to 500 kV on the super grid) in order to reduce the current in the wires, and hence minimize the amount of energy wasted from them as heat. ◊Transformers are needed to step down these voltages before power can be supplied to consumers. High voltages require special insulators to prevent current from leaking to the ground and these may clearly be seen on the pylons that carry overhead wires.

ppm abbreviation for ***parts per million***. An alternative (but numerically equivalent) unit used is milligrams per litre (mg l^{-1}).

precipitate (ppt) solid (insoluble) substance sometimes produced when two solutions are mixed or when a gas is passed into a solution. In an equation the precipitate is often represented by the symbol (s).

$$AgNO_{3(aq)} + NaCl_{(s)} \rightarrow AgCl_{(s)} + NaNO_{3(aq)}$$

$$Ca(OH)_{2(aq)} + CO_{2(g)} \rightarrow CaCO_{3(s)} + H_2O_{(l)}$$

predator an animal that hunts and kills other animals for its food.

predict indicate in advance the expected outcomes of experiments or investigations.

pregnancy in humans, the period during which an embryo grows within the womb. It begins at conception and ends at birth, and the normal length is 40 weeks. Menstruation usually stops on conception. Pregnancy in animals is called ◊gestation.

premolar in mammals, one of the large teeth towards the back of the mouth. In herbivores they are adapted for grinding. In carnivores they have a sharp cutting surface. Premolars are present in milk dentition (see ◊milk teeth) as well as permanent dentition.

preservative food ◊additive used to inhibit the growth of bacteria, yeasts, mould, and other microorganisms to extend the shelf-life of foods. All preservatives are potentially damaging to health if eaten in sufficient

pregnancy

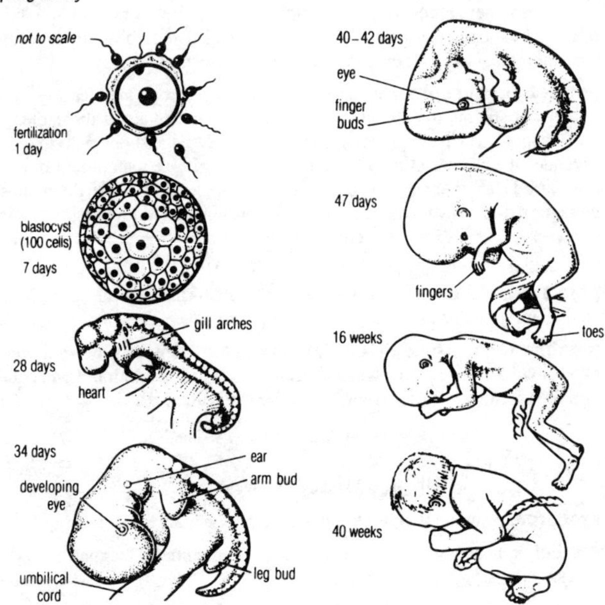

quantity. Both the amount used, and the foods in which they can be used, are restricted by law. See also ◊food technology.

pressure measure of the force acting normally (at right angles) to a body per unit surface area. The pressure *p* exerted by a force *F* newtons acting at right angles over an area of *A* m^2 is given by:

$$p = F/A$$

The SI unit of pressure is the pascal (newton per square metre).

In a fluid (liquid or gas), pressure increases with depth. At the edge of Earth's atmosphere, pressure is zero, whereas at ground level it is about 100 kPa. The pressure *p* at a depth *h* in a fluid of density *d* is given by:

$$p = hdg$$

where g is the gravitational field strength.

pressurized water reactor (PWR) nuclear reactor used in many countries, and in nuclear-powered submarines. In the PWR, water under pressure is the coolant and ◊moderator. It circulates through a steam generator, where its heat boils water to provide steam to drive power ◊turbines.

primeval soup solution of simple organic molecules that gave rise to first forms of life on Earth, about 4,000 million years ago.

principal focus in optics, the point at which incident rays parallel to the principal axis of a ◊lens converge, or appear to diverge, after refraction. The distance from the lens to its principal focus is its ◊focal length.

The principal focus of a converging (convex) lens or of a parabolic concave mirror is the point at which parallel incident rays will converge when refracted or reflected. It is a real focus.

The principal focus of a diverging (concave) lens is the point from which its parallel incident rays appear to diverge after refraction. It is a virtual or imaginary focus at which no light rays actually meet.

printed circuit board (PCB) an electrical circuit created by laying (printing) 'tracks' of a conductor such as copper onto one or both sides of an insulating board.

Components such as integrated circuits (silicon chips), resistors, and capacitors can be soldered to the surface of the board (surface-mounted) or, more commonly, attached by inserting their connecting pins or wires into holes drilled in the board.

prism in optics, a triangular block of transparent material (plastic, glass, silica) commonly used to 'bend' a ray of light or split a beam into its spectral colours. Prisms are used as mirrors to define the optical path in binoculars, camera viewfinders, and periscopes.

productivity, biological in an ecosystem, the amount of material in the food chain produced by the primary producers (plants) that is available for consumption by animals. Plants turn carbon dioxide and water into sugars and other complex carbon compounds by means of photosynthesis. Their net productivity is defined as the quantity of carbon compounds formed, less the quantity used up by the respiration of the plant itself.

progesterone hormone that regulates the menstrual cycle and pregnancy. It is secreted by the corpus luteum (the ruptured Graafian follicle of a discharged ovum).

propagation of plants in horticulture and agriculture, method used to bring about the ◊vegetative reproduction of plants. Its advantage is that all the offspring (daughter plants) are identical; the gardener therefore has control over the features that they show. The various propagation techniques all involve taking tissue from one plant and growing it in another area. Traditional methods include the taking of ◊cuttings, the ◊grafting of shoots from one plant onto the roots of another, and allowing a plant such as the strawberry to grow ◊runners. Modern techniques include the removal of just a few cells from the parent, which can then be grown in a nutrient medium until they form the mature plant.

propane C_3H_8 gaseous hydrocarbon of the ◊alkane series, found in petroleum and used as fuel.

propanol or ***propyl alcohol*** third member of the homologous series of ◊alcohols. Propanol is usually a mixture of two isomeric compounds (see ◊isomer): propan-1-o 1 ($CH_3CH_2CH_2OH$) and propan-2-ol ($CH_3CHOHCH_3$). Both are colourless liquids that can be mixed with water and are used in perfumery.

propanone or ***acetone*** CH_3COCH_3 colourless inflammable liquid used extensively as a solvent, as in nail-varnish remover. It boils at 56.5°C, mixes with water in all proportions, and has a characteristic odour.

propene or ***propylene*** $CH_3CH = CH_2$, second member of the alkene series of hydrocarbons. A colourless, inflammable gas, it is widely used by industry to make organic chemicals, including propylene plastics.

properties the characteristics a substance possesses by virtue of its composition. The ***physical properties*** of a substance can be measured by physical means, for example boiling point, melting point, hardness, elasticity, colour, and physical state. Its ***chemical properties*** are the ways in which it reacts with other substances: whether it is acidic or basic, an oxidizing or a reducing agent, a salt, or stable to heat, for example.

proportion two variable quantities x and y are proportional if, for all values of x,

$$y = kx$$

where k is a constant. This means that if x increases, y increases in a linear fashion. A graph of x against y would be a straight line passing through the origin (the point at which both x and y are equal to zero).

If the graph of y against $1/x$ is a straight line through the origin, y is ***inversely proportional*** to x. The corresponding equation is:

$$y = k/x$$

Many laws of science relate quantities that are proportional (for example, Boyle's law, which relates the volume and pressure of a gas).

prostate gland a gland surrounding, and opening into, the urethra at the base of the bladder in male mammals. The prostate gland produces an alkaline fluid that is released during ejaculation; this fluid activates sperm, and prevents their clumping together.

protease general term for an enzyme capable of splitting proteins. Examples include pepsin, found in the stomach, and trypsin, found in the small intestine.

protein long-chain molecule composed of amino acids joined by ◊peptide bonds. Proteins are essential to all living organisms. As ***enzymes*** they regulate all aspects of metabolism. Structural proteins such as ***keratin*** and ***collagen*** make up the skin, claws, bones, tendons, and ligaments; ***muscle*** proteins produce movement; ***haemoglobin*** transports oxygen; and ***membrane proteins*** regulate the movement of substances into and out of cells.

For humans, protein is an essential part of the diet, and is found in greatest quantity in soya beans and other legumes, meat, eggs, and cheese.

prothrombin precursor enzyme involved in ◊blood clotting. When a blood vessel is damaged, prothrombin is converted to the active enzyme thrombin, which brings about the formation of fibrin, the insoluble protein that forms a mesh over the wound.

proton positively charged subatomic particle, a fundamental constituent of any atomic ◊nucleus. Its lifespan is effectively infinite.

A proton carries a unit positive charge equal to the negative charge of an ◊electron. Its mass is almost 1,836 times that of an electron, 1.67×10^{-24} g. The number of protons in the atom of an ◊element is equal to its atomic, or proton, number.

proton number alternative name for ◊atomic number.

protozoa a group of single-celled organisms without rigid cell walls. Some, such as *Amoeba*, ingest other cells, but most are ◊saprotrophs or parasites.

PTFE abbreviation for ◊polytetrafluoroethene.

puberty stage in human development when the individual becomes sexually mature. It may occur from the age of ten upwards. The sexual organs take on their adult form and pubic hair grows. In girls, menstruation begins, and the breasts develop; in boys, the voice breaks and becomes deeper, and facial hair develops.

pulley simple machine consisting of a fixed, grooved wheel, sometimes in a block, around which a rope or chain can be run. A simple pulley serves only to change the direction of the applied effort (as in a simple hoist for raising loads). The use of more than one pulley results in a ◊mechanical advantage, so that a given effort can raise a heavier load.

The mechanical advantage depends on the arrangement of the pulleys. For instance, a block and tackle arrangement with three ropes supporting the load will lift it with one-third of the effort needed to lift it directly (if friction is ignored), giving a mechanical advantage of 3.

pulse impulse transmitted by the heartbeat throughout the arterial system. When the heart muscle contracts, it forces blood into the aorta, the large artery leading straight out of the left ◊ventricle. Because the arteries are elastic, the sudden rise of pressure causes a throb or sudden swelling through them. In humans, the pulse rate is generally about 70 per minute. The pulse can be felt where an artery is near the surface, such as in the wrist or the neck.

pumped storage hydroelectric plant that uses surplus electricity to pump water back into a high-level reservoir. In normal working the water flows from this reservoir through the ◊turbines to generate power for feeding into the grid. At times of low power demand, electricity is taken from the grid to turn the turbines into pumps that then pump the water back again. This ensures that there is always a maximum 'head' of water in the reservoir to give the maximum output when required.

pupa the non-feeding, largely immobile stage of some insect life cycles, in which larval tissues are broken down, and adult tissues and structures are formed. In butterflies and moths, the pupa is called a chrysalis.

pure-breeding line in genetics, a strain of individuals that when interbred produce genetically identical progeny.

putrefaction the breaking down of organic matter by microorganisms.

PVC abbreviation for ◊polyvinyl chloride.

PWR abbreviation for ◊pressurized water reactor .

pyramid of biomass in ecology, the declining biomass measured at each trophic level of a ◊food chain. In almost all ecosystems, the producers have the highest total biomass, while the primary and secondary consumers have a greatly reduced total. When these data are represented graphically a pyramid of biomass results. The reason for the reduction is that only around 10% of the energy in one trophic level can pass on to the next, the rest being lost as heat or indigestible matter. For instance, 100kg of grass will only support 10 kg of herbivore; in turn, the herbivores will only support 1kg of carnivore.

pyramid of numbers in ecology, the declining numbers of individual organisms that may be counted at each trophic level of a food chain. See ◊pyramid of biomass.

quadrat in ecology, a square structure used to study the distribution of plants or animals in a particular place, for instance a field, rocky shore, or mountainside. The size varies, but is usually 0.5 or 1 metre square, small enough to be carried easily. The quadrat is placed on the ground and the abundance of species estimated. By making such measurements a reliable understanding of species distribution is obtained.

qualitative analysis procedure for determining the identity of the component(s) of a single substance or mixture. A series of simple reactions and tests can be carried out on a compound to determine the elements present.

quantitative analysis procedure for determining the precise amount of a known component present in a single substance or mixture.

quartz crystalline form of ◊silicon(IV) oxide (silica, SiO_2), one of the most abundant minerals of the Earth's crust (12% by volume). Quartz occurs in many different kinds of rock, including sandstone and granite. It is very hard, and is resistant to chemical and mechanical breakdown.

quicklime common name for ◊calcium oxide.

quicksilver former name for the element ◊mercury.

R

radar (acronym for ***ra***dio ***d***irection ***a***nd ***r***anging) device for locating objects in space, direction finding, and navigation by means of transmitted and reflected high-frequency radio waves. The direction of an object is discovered by transmitting a beam of short-wavelength (1–100 cm), short-pulse radio waves, and picking up the reflected beam. Distance is determined by timing the journey of the radio waves (travelling at the speed of light) to the object and back again.

Radar is essential to navigation in darkness, cloud, and fog, and is widely used in warfare to detect enemy aircraft and missiles. It may, however, be thwarted by modifying the shapes of aircraft and missiles in order to reduce their radar cross-section, and by means of devices such as radar-absorbent paints and electronic jamming.

radiant heat energy that is radiated by all warm or hot bodies. It belongs to the infrared part of the electromagnetic spectrum and causes heating when absorbed. Radiant heat is invisible and should not be confused with the red glow associated with very hot objects, which belongs to the visible part of the spectrum.

Infrared radiation can travel through a vacuum and it is in this form that the radiant heat of the Sun travels through space. It is the trapping of this radiation by carbon dioxide and methane in the atmosphere that gives rise to the ◊greenhouse effect.

radiation emission of radiant ◊energy as particles or waves – for example, heat, light, alpha particles, and beta particles (see ◊electromagnetic waves and ◊radioactivity).

radical group of atoms forming part of a molecule, which acts as a unit and takes part in chemical reactions without disintegrating, yet often cannot exist alone; for example, the methyl radical $-CH_3$, or the carboxylic acid radical $-COOH$.

radio transmission and reception of radio waves. In radio transmission a microphone converts ◊sound waves (pressure variations in the air) into

radio

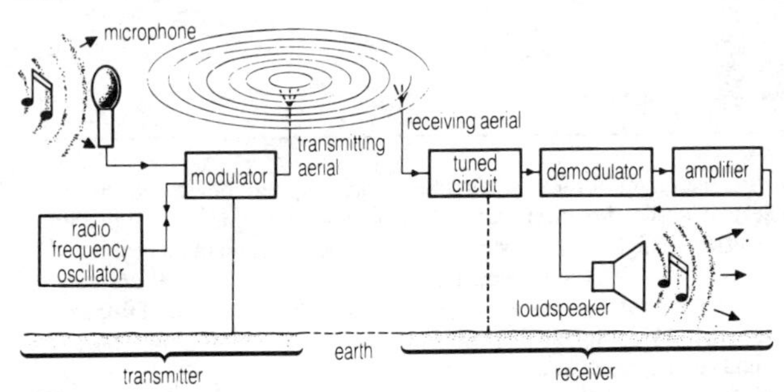

⟩electromagnetic waves that are then picked up by a receiving aerial and fed to a loudspeaker, which converts them back into sound waves.

To carry the transmitted electrical signal, an oscillator produces a carrier wave of high frequency; different stations are allocated different transmitting carrier frequencies. A modulator superimposes the audiofrequency signal on the carrier. There are two main ways of doing this: amplitude modulation (AM), used for long-and medium-wave broadcasts, in which the strength of the carrier is made to fluctuate in time with the audio signal; and frequency modulation (FM), as used for VHF (very high frequency) broadcasts, in which the frequency of the carrier is made to fluctuate. The transmitting aerial emits the modulated electromagnetic waves, which travel outwards from it.

In radio reception a receiving aerial picks up minute voltages in response to the waves sent out by a transmitter. A tuned circuit selects a particular frequency, usually by means of variable ⟩capacitor connected across a coil of wire. A demodulator disentangles the audio signal from the carrier, which is now discarded, having served its purpose. An amplifier boosts the audio signal for feeding to the loudspeaker.

radioactive decay spontaneous alteration of the nucleus of a radioactive atom; this often changes its atomic number, thus changing one element into another, and is accompanied by the emission of radiation. Alpha and beta decay are the most common forms; they may be accompanied by gamma emission. In ***alpha decay*** (the loss of a helium nucleus—two protons and two neutrons) the atomic number decreases by two; in ***beta decay*** (the loss of an electron) the atomic number increases by one.

Decay takes place at a specific rate expressed as the ◊half-life, which is the time taken for half of any mass of that particular isotope to decay completely. The final product in all modes of decay is a stable element (see ◊radioactivity).

radioactivity spontaneous alteration of the nuclei of radioactive atoms, accompanied by the emission of radiation. It is the property exhibited by the radioactive ◊isotopes of stable elements and all isotopes of radioactive elements, and can be either natural or induced. See ◊radioactive decay.

Radioactivity establishes an equilibrium in parts of the nucleus of unstable radioactive substances, ultimately to form a stable arrangement of nucleons (protons and neutrons); that is, a non-radioactive (stable) element. This is most frequently accomplished by the emission of ***alpha particles*** (helium nuclei); ***beta particles*** (electrons); or ***gamma rays*** (electromagnetic waves of very high frequency). It takes place either directly, through a one-step decay, or indirectly, through a number of decays that may change one element into another. This is called a decay series or chain, and sometimes produces an element more reactive than its predecessor.

The instability of the particle arrangements in the nucleus of a radioactive atom (the ratio of neutrons to protons and/or the total number of both) determines the lengths of the ◊half-lives of the isotopes of that atom, which can range from fractions of a second to billions of years. Beta and gamma radiation are both ionizing and are therefore dangerous to body tissues, especially if a radioactive substance is ingested or inhaled.

radioactivity safety precautions taken to ensure the safe handling of radioactive materials. Such materials are hazardous because they emit ◊ionizing radiation, which damages living cells. The consequences of exposure to radiation or of contamination with radioactive materials—for example, by inhaling or ingesting radioactive dust—may be immediate (burns, radiation sickness) or long-term (certain forms of cancer, birth defects due to genetic damage).

Measures taken to protect people who work with radioactive sources include storing and transporting radioactive materials (including contaminated clothing and waste) in sealed, lead-lined containers; the use of thick leaded-glass, lead, or concrete barriers to shield workers from exposed materials; the use of tongs or remote-controlled devices; and the wearing of protective clothing and of monitoring devices, such as film badges, which keep a record of their wearers' exposure history.

radiocarbon dating or ***carbon dating*** method of dating organic materials (for example, bone or wood), used in archaeology. Plants take up carbon dioxide gas from the atmosphere and incorporate it into their tissues, and some of that carbon dioxide inevitably contains a certain amount of the radioactive isotope of carbon, carbon-14. On its death, the plant ceases to take up carbon-14 and the amount already taken up decays at a known rate, the half-life of 5,730 years, so that the time elapsed since the plant died can be measured in a laboratory by detecting how much carbon-14 remains. Animals take carbon-14 into their bodies from eating plant tissues and their remains can be similarly dated. After 120,000 years so little carbon-14 is left that no measurement is possible.

radiochemistry study of radioactive isotopes and their compounds (whether produced from naturally radioactive or irradiated materials) and their use in the study of other chemical processes.

When such isotopes are used in ◊labelled compounds, they enable the biochemical and physiological functioning of parts of the living body to be observed. They can help in the testing of new drugs, showing where the drug goes in the body and how long it stays there. They are also useful in diagnosis—for example, cancer, fetal abnormalities, and heart disease.

radioisotope contraction of ***radioactive ◊isotope***, a radioactive form of an element; it may be naturally occurring or synthetic.

radio wave electromagnetic wave possessing a long wavelength (ranging from about 10^{-3} to 10^{4}m) and a low frequency (from about 10^{5} to 10^{11}Hz). Included in the radio-wave part of the spectrum are ◊microwaves, used for both communications and for cooking; ultra high and very high frequency waves, used for television and frequency-modulated (FM) radio communications; and short, medium, and long waves, used for amplitude-modulated (AM) radio communications. Stars emit radio waves, which may be detected and studied by using radio telescopes.

radium white, radioactive, metallic element, symbol Ra, atomic number 88, relative atomic mass 226.02. It is one of the ◊alkaline-earth metals, found in nature in ◊pitchblende and other uranium ores. Of the 16 isotopes, the commonest, Ra-226, has a half-life of 1.622 years.

Radium decays in successive steps to produce radon (a gas), polonium, and finally a stable isotope of lead.

radon colourless, odourless, gaseous, radioactive element, symbol Rn, atomic number 86, relative atomic mass 222. It is grouped with the ◊noble gases and was formerly considered non-reactive, but is now known to form some compounds with fluorine. Of the 20 known isotopes, only three occur in nature; the longest half-life is 3.82 days.

Radon is the densest gas known and occurs in small amounts in spring water, streams, and the air, being formed from the natural radioactive decay of radium.

rare gas alternative name for ◊noble gas.

rate of reaction the speed at which a chemical reaction proceeds. It is usually expressed in terms of the concentration (usually in ◊moles per litre) of a reactant consumed or product formed in unit time; so the units would be moles per litre per second (mol l^{-1} s^{-1}). The rate of a reaction is affected by the concentration of the reactants, the temperature of the reactants, and the presence of a ◊catalyst. If the reaction is entirely in the gas state, pressure affects the rate, and, for solids, the particle size.

During a reaction at constant temperature the concentration of the reactants decreases and so the rate of reaction decreases. These changes can be represented by drawing graphs.

The rate of reaction is at its greatest at the beginning of the reaction and it gradually slows down. Increasing the temperature produces large increases in the rate of reaction. A 10°C rise can double the rate while a 40°C rise can produce a 50–100-fold increase in the rate. ◊Collision theory is used to explain these effects. Increasing the concentration or the pressure of a gas means there are more particles per unit volume, therefore there are more collisions and more fruitful collisions. Increasing the temperature makes the particles move much faster, resulting in more collisions per unit time and more fruitful collisions; consequently the rate increases.

reaction the coming together of two or more atoms, ions, or molecules resulting in a change in the way these particles are arranged. The nature of the reaction is portrayed by a chemical equation.

reactivity series series produced by arranging the metals in order of their ease of reaction with reagents such as oxygen, water, and acids. This arrangement aids in understanding the properties of metals, helps to explain

rate of reaction

(a) rate of reaction decreases with time

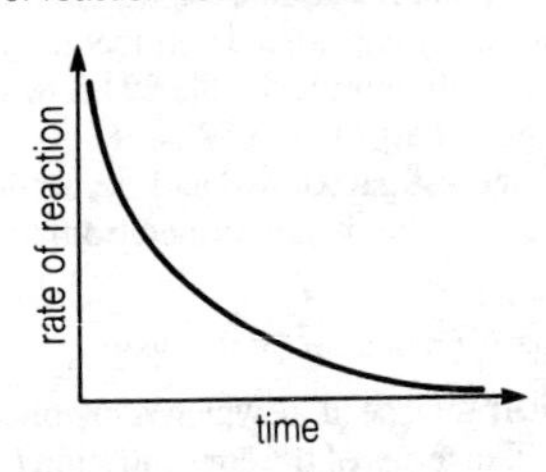

(b) concentration of reactant decreases with time

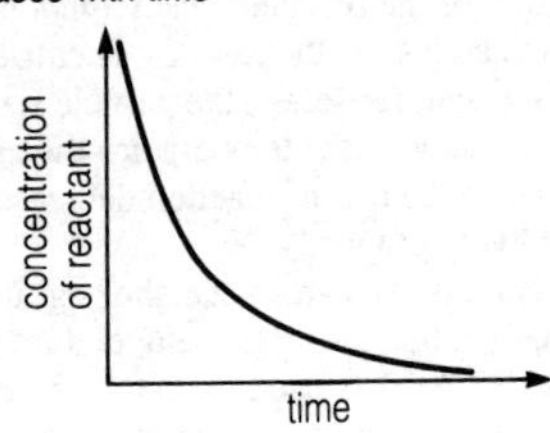

(c) concentration of product increases with time

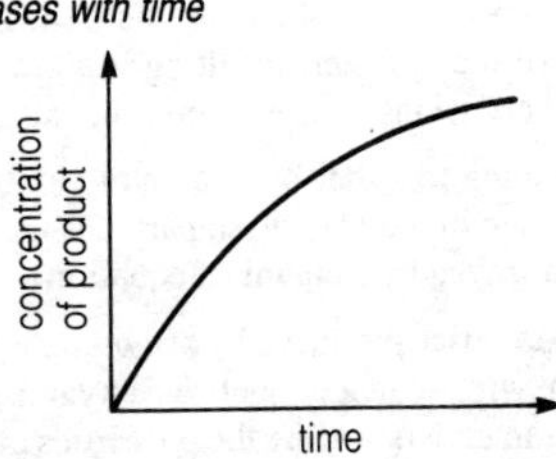

differences, and enables predictions to be made about a certain metal based on a knowledge of its position or properties.

reactor see ◊nuclear reactor.

receptor any cell capable of detecting stimuli. Receptors form part of the nervous system and are used by the body to gather information about the internal or external environment. There are several types, classified according to function. Some respond to light, some to mechanical force, and some to heat. They are essential for ◊homeostasis.

recessive in genetics, term that describes an ◊allele that will show in the ◊phenotype only if its partner allele on the paired chromosome is the same (the organism is ◊homozygous). Such an allele will not show if its partner is ◊dominant, that is if the organism is ◊heterozygous for a particular characteristic. Alleles for blue eyes in humans, and for shortness in pea plants are recessive. Most mutant alleles are recessive and therefore are only rarely expressed (see ◊haemophilia).

recombination in genetics, any process that recombines, or 'shuffles', the genetic material, thus increasing genetic variation in the offspring. The two main processes of recombination both occur during meiosis (reduction division of cells). One is ***crossing over***, in which chromosome pairs exchange segments; the other is the random reassortment of chromosomes that occurs when each gamete (sperm or egg) receives only one of each chromosome pair.

rectification process of converting ◊alternating current (AC) into ◊direct current (DC). It is necessary because almost all electrical power is generated, transmitted, and supplied as alternating current, but many devices, from television sets to electric motors, require direct current. It involves the use of one or more ◊diodes as rectifiers. However, a single diode produces half-wave rectification in which current flows in one direction for one-half of the alternating-current cycle only—an inefficient process in which the power is effectively turned off for half the time. A ***bridge rectifier***, constructed from four diodes, can rectify the alternating supply in such a way that both the positive and negative halves of the alternating cycle can produce a current flowing in the same direction.

rectum lowest part of the gut, which stores faeces before egestion (defecation).

recycling processing of industrial and household waste (such as paper, glass, and some metals) so that it can be reused, thus saving expenditure on

reactivity series

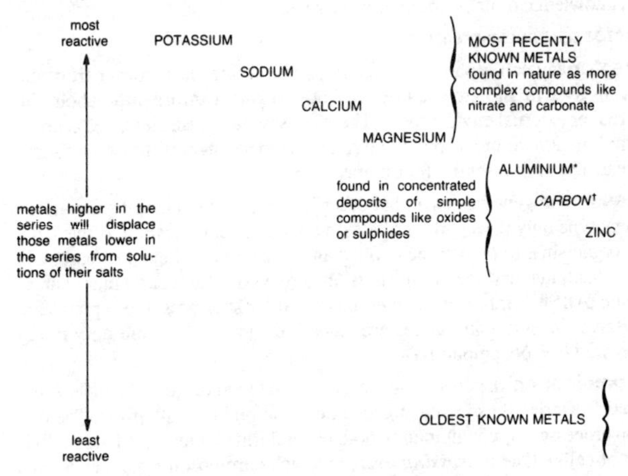

scarce raw materials, slowing down the depletion of non-renewable resources, and helping to reduce pollution.

red blood cell or ***erythrocyte*** the most common type of blood cell, responsible for transporting oxygen around the body. It contains the red protein haemoglobin, which combines with oxygen from the lungs to form oxyhaemoglobin. When transported to the tissues, these cells are able to release the oxygen because the oxyhaemoglobin splits into its original constituents. Mammalian erythrocytes are disclike with a depression in the centre and no nucleus; they are manufactured in the bone marrow and, in humans, last for only four months before being destroyed in the liver and spleen. Those of other vertebrates are oval and possess a nucleus.

redox reaction chemical change where one reactant is reduced and the other reactant oxidized. The reaction can only occur if both are present and

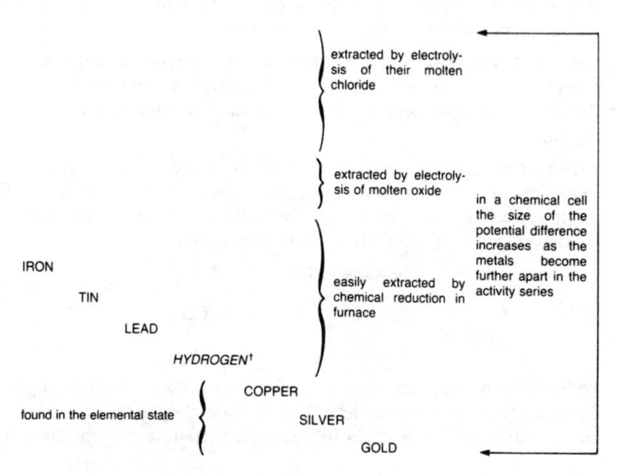

* The position of aluminium may appear anomalous as the oxide on its surface makes it unreactive.

†Carbon and hydrogen, although non-metals, are included to show that they will only reduce metals below them in the reactivity series.

each changes simultaneously. For example, hydrogen reduces copper(II) oxide to copper while it is itself oxidized to water.

$$CuO + H_2 \rightarrow Cu + H_2O$$

Many chemical changes can be classified as redox. Corrosion of iron, the reactions in chemical cells, and electrolysis are just a few instances where redox reactions take place.

red shift in astronomy, the lengthening of the wavelengths of light from an object as a result of the object's motion away from us. It is an example of the ◊Doppler effect. The red shift in light from galaxies is evidence that the universe is expanding.

reducing agent substance that accepts oxygen, donates hydrogen, or donates electrons (see ◊reduction). In a redox reaction, the reducing agent is the substance that is itself oxidized. Strong reducing agents include hydrogen, carbon monoxide, carbon, and metals.

reducing sugar sugar that produces a positive result (change in colour from blue to orange) when tested with Benedict's reagent (see ◊food test). All monosaccharides and the disaccharides maltose and lactose are reducing sugars.

reduction reaction in which an atom, ion, or molecule loses oxygen, gains hydrogen, or gains electrons. Examples include the reduction of iron(III) oxide to iron by carbon monoxide, the hydrogenation of ethene to ethane, and the reduction of a sodium ion to a sodium atom.

$$Fe_2O_3 + 3CO \rightarrow 2Fe + 3CO_2$$

$$CH_2{=}CH_2 + H_2 \rightarrow CH_3{-}CH_3$$

$$Na^+ + e^- \rightarrow Na$$

reflection the throwing back or deflection of waves, such as ◊light or ◊sound waves, when they hit a surface. The ***law of reflection*** states that the angle of incidence (the angle between the ray and a perpendicular line

reflection

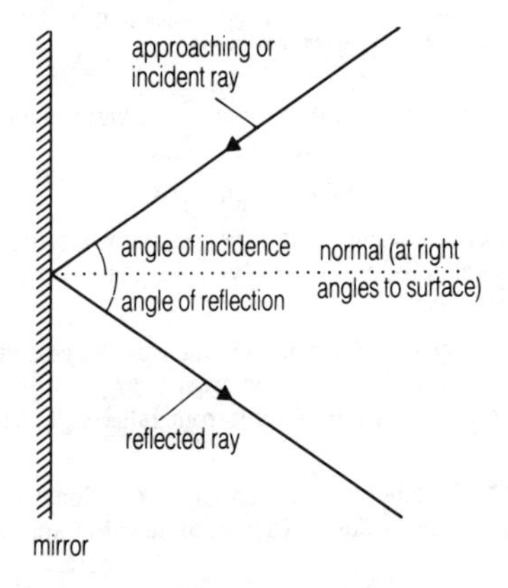

drawn to the surface) is equal to the angle of reflection (the angle between the reflected ray and a perpendicular to the surface).

reflex in animals, a very rapid automatic response to a particular stimulus. It is controlled by the ◊nervous system. A reflex involves only a few neurons, unlike the slower but more complex responses produced by the many processing neurons of the brain.

A ***simple reflex*** is entirely automatic and involves no learning. Examples of such reflexes include the sudden withdrawal of a hand in response to a painful stimulus, or the jerking of a leg when its knee cap is tapped. Sensory cells (receptors) in the knee send signals to the spinal cord along a sensory neuron. Within the spine a ***reflex arc*** switches the signals straight back to the muscles of the leg (effectors) via an intermediate neuron and then a motor neuron; contraction of the leg muscle occurs, and the leg kicks upwards. Only three neurons are involved, and the brain is only aware of

reflex

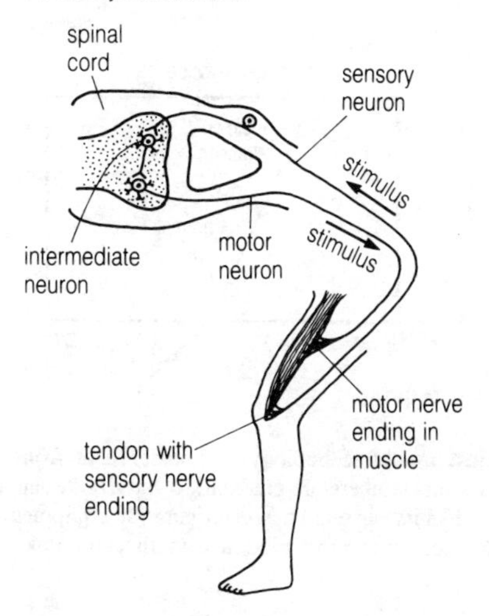

the response after it has taken place. Such reflex arcs are particularly common in lower animals, and have a high survival value, enabling organisms to take rapid action to avoid potential danger. In higher animals (those with a well-developed ▷central nervous system) the simple reflex may be modified by the involvement of the brain—for instance, humans can override the automatic reflex to withdraw a hand from a source of pain.

A ***conditioned reflex*** involves the modification of a reflex action in response to experience (learning). A stimulus that produces a simple reflex response becomes linked with another, possibly unrelated, stimulus. For example, a dog may salivate (a reflex action) when it sees its owner remove a tin-opener from a drawer because it has learned to associate that stimulus with the stimulus of being fed.

refraction the bending of a wave of light, heat, or sound when it passes from one medium to another. Refraction occurs because waves travel at different velocities in different media.

refraction

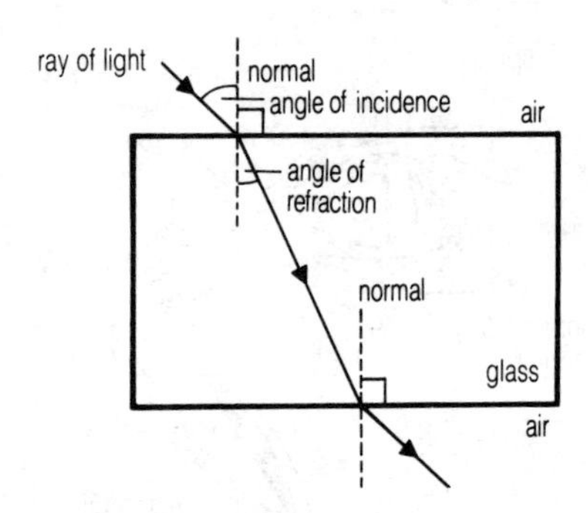

refrigeration use of technology to transfer heat from cold to warm, against the normal temperature gradient, so that a body can remain substantially colder than its surroundings. Refrigeration equipment is used for the chilling and deep freezing of food, and in air conditioners and industrial processes.

Refrigeration is commonly achieved by a vapour-compression cycle, in which a suitable chemical (the refrigerant) travels through a long circuit of tubing, during which it changes from a vapour to a liquid and back again. A compression chamber makes it condense, and thus give out heat. In another part of the circuit, called the evaporator coils, the pressure is much lower, so the refrigerant evaporates, absorbing heat as it does so. The evaporation process takes place near the central part of the refrigerator, which therefore becomes colder, while the compression process takes place near a ventilation grille, transferring the heat to the air outside. The most commonly used refrigerants in modern systems were chlorofluorocarbons (CFCs), but these are now being replaced by coolants that do not damage the ozone layer.

relative atomic mass the mass of an atom. It depends on the number of protons and neutrons in the atom, the electrons having negligible mass. It is calculated relative to one-twelfth the mass of an atom of carbon-12. If more than one ◊isotope of the element is present, the relative atomic mass is calculated by taking an average that takes account of the relative proportions of each isotope, resulting in values that are not whole numbers. The term ***atomic weight***, although commonly used, is strictly speaking incorrect.

relative density or ***specific gravity*** the density (at 20°C) of a solid or liquid relative to (divided by) the maximum density of water (at 4°C). The relative density of a gas is its density divided by the density of hydrogen (or sometimes dry air) at the same temperature and pressure.

relative molecular mass the mass of a molecule, calculated relative to one-twelfth the mass of an atom of carbon-12. It is found by adding the relative atomic masses of the atoms that make up the molecule. The term ***molecular weight*** is often used, but strictly this is incorrect.

relay in electrical engineering, an electromagnetic switch. A small current passing through a coil of wire wound around an iron core attracts an armature, a piece of metal whose movement closes a pair of sprung contacts. This completes a secondary circuit, which may carry a large current or activate other devices.

rennin or ***chymase*** enzyme found in the gastric juice of young mammals, used in the digestion of milk.

replication the production of copies of the genetic material, DNA; it occurs during cell division (◊mitosis and ◊ meiosis). Most mutations are caused by mistakes during replication.

reproduction process by which a living organism produces other organisms similar to itself.

reptile member of the vertebrate class Reptilia. Unlike amphibians, reptiles have hard-shelled, yolk-filled eggs that are laid on land and from which fully formed young are born. Some snakes and lizards retain their eggs and give birth to live young. Reptiles have no internal method of temperature control, unlike mammals and birds. Instead, they adjust their body temperature behaviourally; for instance, by basking or by moving into the shade. Reptiles include snakes, lizards, crocodiles, turtles, tortoises, and the extinct dinosaurs.

residue substance or mixture of substances remaining in the original container after the removal of one or more components by a separation process.

resin substance exuded from pines, firs, and other trees in gummy drops that harden in air. They are solutions of organic molecules, which harden when chain formation occurs, which can be followed by cross-linking between chains. Varnishes are common products of the hard resins, and ointments come from the soft resins.

resistance the property of a substance that restricts the flow of electricity through it, associated with the conversion of electrical energy to heat; also the magnitude of this property. Resistance depends on many factors, such as the nature of the material, its temperature, dimensions, and thermal properties; degree of impurity; the nature and state of illumination of the surface; and the frequency and magnitude of the current. The SI unit of resistance is the ohm (symbol). See also ◊Ohm's law.

resistor any component in an electrical circuit used to introduce ◊resistance to a current. Resistors are often made from wire-wound coils or pieces of carbon. A ◊rheostat is a variable resistor.

resonance rapid and uncontrolled increase in the size of a vibration when the vibrating object is subject to a force varying at its ◊natural frequency. In a trombone, for example, the length of the air column in the instrument is adjusted until it resonates with the note being sounded. Resonance effects are also produced by many electrical circuits. Tuning a radio, for example, is done by adjusting the natural frequency of the receiver circuit until it coincides with the frequency of the radio waves falling on the aerial.

Resonance has many physical applications. Children use it to increase the size of the movement on a swing, by giving a push at the same point during each swing. Soldiers marching across a bridge in step could cause the bridge to vibrate violently if the frequency of their steps coincided with its natural frequency.

respiration biochemical process whereby food molecules are progressively broken down (oxidized) to release energy. In most organisms this requires oxygen, but in some bacteria the oxidant is the nitrate or sulphate ion instead. In all higher organisms, respiration occurs in the ◊mitochondria.

respiratory surface area used by an organism for ◊gas exchange—for example, the lungs, gills or, in plants, the leaf interior. The gases oxygen and carbon dioxide are both usually involved in respiration and photosynthesis. Although organisms have evolved different types of respiratory surface according to need, there are certain features in common. These include thinness and moistness, so that the gas can dissolve in a membrane and then diffuse into the body of the organism. In many animals the gas is then transported away from the surface and towards interior cells by the blood system.

resultant force in mechanics, a single force acting on a particle or body whose effect is equivalent to the combined effects of two or more component forces.

retina the inner layer at the back of the ◊eye, which contains light-sensitive cells and neurons (nerve cells). Light falling on the retina produces chemical changes in the light-sensitive cells, causing them to send electrical signals along the neuron fibres via the optic nerve to the brain.

The light-sensitive cells are of two types, called rod cells and cone cells after the shape of their outer projections. The ***rod cells***, about 120 million in each eye, are distributed throughout the retina. They are sensitive to low levels of light, but do not provide detailed or sharp images, nor can they detect colour. The ***cone cells***, about 6 million in number, are mostly concentrated in a central region of the retina called the ***fovea***, and provide both detailed and colour vision. The cones of the human eye contain three visual pigments, each of which responds to a different primary colour (red, green, or blue). The brain can interpret the varying signal levels from the three types of cone as any of the different colours of the visible spectrum.

reversible reaction chemical reaction that proceeds in both directions at the same time, as the product decomposes back into reactants as it is being produced. Such reactions do not run to completion, providing no substance leaves the system. Examples include the manufacture of ammonia from hy-drogen and nitrogen, and the oxidation of sulphur dioxide to sulphur trioxide.

$$N_2 + 3H_2 \leftrightarrow 2NH_3$$

$$2SO_2 + O_2 \leftrightarrow 2SO_3$$

rheostat variable ◊resistor, usually consisting of a high-resistance wire-wound coil with a sliding contact. It is used to vary electrical resistance without interrupting the current (for example, when dimming lights).

rhesus factor in humans, a protein on the surface of red blood cells that is involved in the rhesus blood-group system. Most individuals possess the main rhesus factor (Rh+), but those with out this factor (Rh–) produce ◊antibodies if they come into contact with it. The name comes from rhesus monkeys, in whose blood rhesus factors were first found.

If a Rh– mother carries a Rh+ fetus, she may produce antibodies if fetal blood crosses the ◊placenta. This is not normally a problem with the first infant because antibodies are only produced slowly. However, the antibod-ies continue to build up after birth, and a second Rh+ child may be attacked by antibodies passing from mother to fetus, causing the child to contract anaemia, heart failure, or brain damage. In such cases, the blood of the infant has to be changed for Rh– blood.

rhizome horizontal underground plant stem. It is a ◊perennating organ in some species, where it is generally thick and fleshy, while in other species it is mainly a means of ◊vegetative reproduction, and is therefore long and slender, with buds all along it that send up new plants. The potato is a rhi-zome that has two distinct parts, the tuber being the swollen end of a long, cordlike rhizome.

rhombic sulphur allotropic form of sulphur. At room temperature, it is the stable allotrope (see ◊allotropy), unlike ◊monoclinic sulphur.

rhythm method method of natural contraception that works by avoiding intercourse when the woman is producing ova (ovulating). The time of ovu-lation can be predicted by the calendar (counting days from the last period), by temperature changes, or by inspection of the cervical mucus. All these methods are unreliable because it is possible for ovulation to occur at any stage of the menstrual cycle.

rib long, usually curved bone that extends laterally from the ◊spine. Most fishes and many reptiles have ribs along most of the spine, but in mammals they are found only in the chest area. In humans, there are 12 pairs of ribs. The ribs protect the lungs and heart, and allow the chest to expand and contract easily.

riboflavin or ***vitamin B_2*** ◊vitamin of the B complex whose absence in the diet causes stunted growth.

rickets defective growth of bone in children because of inadequate calcium deposits. The bones do not harden and are bent out of shape. The condition is usually caused by a lack of calciferol (vitamin D) in the diet and insufficient exposure to sunlight.

ring circuit household electrical circuit in which appliances are connected in series to form a ring with each end of the ring connected to the power supply.

rock hard solid mass composed of mineral particles that have become cemented together. It is the main constituent of the Earth's crust. Rocks may be classified into three main groups: igneous, sedimentary, and metamorphic.

Igneous rock, such as basalt and granite, is made from magma (molten rock) that has solidified on or beneath the Earth's surface. ***Sedimentary rock*** is formed by the compression of deposited particles—for example, sandstone from sand particles, limestone from the remains of sea creatures. such as sand. ***Metamorphic rock*** is formed by changes in existing igneous or sedimentary rocks under high pressure or temperature, or chemical action—for example, marble is formed from limestone.

rod cell a type of light-sensitive cell found in the ◊retina of the eye. Rods are highly sensitive and provide only black-and-white vision. They are used when lighting conditions are poor and are the only type of visual cell found in animals active at night.

root the part of a plant that is usually underground, and whose primary functions are anchorage and the absorption of water and dissolved mineral salts. Roots usually grow downwards and towards water.

The absorptive area of roots is greatly increased by the numerous, slender ◊root hairs formed near the tips. A ***root cap*** protects the tip of the root from abrasion as it grows through the soil.

Symbiotic associations occur between the roots of certain plants, such as clover, and various bacteria that fix nitrogen from the air (see ◊nitrogen fixation).

root hair tiny hairlike outgrowth on the surface cells of plant roots that greatly increases the area available for the absorption of water and other materials. New root hairs are continually being formed near the root tip, one of the places where plants show the most active growth. The hairs are extremely delicate.

roughage or ***dietary fibre*** part of the diet that is indigestible, consisting mostly of cellulose from plant cell walls. It adds bulk to the gut contents, assisting the process of ◊peristalsis, the muscular contractions forcing the food along the intestine. A high roughage content is believed to have several beneficial effects, including reduced cancer risks.

rubidium soft, silver-white, metallic element, symbol Rb, atomic number 37, relative atomic mass 85.47. It is one of the ◊alkali metals, ignites spontaneously in air, and reacts violently with water. It is used in photoelectric cells and vacuum-tube filaments.

runner aerial stem that produces new plants. It is a means of ◊vegetative reproduction.

rust reddish-brown oxide of iron formed by the action of moisture and oxygen on the metal. It consists mainly of hydrated iron(III) oxide ($Fe_2O_3.H_2O$) and iron(III) hydroxide ($Fe(OH)_3$).

rust prevention methods of preventing the formation of ◊rust, the commonest form of ◊corrosion. Ships and bridges use a combination of the two main approaches, barrier methods and sacrificial protection.
barrier methods The most widely used methods introduce a barrier between the metal and the air and moisture, so that reaction is minimized. The metal can be covered with a layer of paint, grease, plastic, or an unreactive metal such as tin, copper, or chromium.
sacrificial protection In this method, iron is either covered with a more reactive metal, such as zinc (galvanization) or connected to a more reactive metal, such as magnesium. In theory, as long as the zinc or magnesium is present it will react (corrode) first, so that corrosion of the iron itself is delayed.

S

saccharide another name for a sugar molecule. See ◊sugar.

saliva in vertebrates, a secretion from the salivary glands that aids the swallowing and digestion of food in the mouth. In mammals, it contains the enzyme amylase, which converts starch to sugar.

salt

		negative ions		
		NO_3^-	SO_4^{2-}	PO_4^{3-}
positive ions	Na^+	$NaNO_3$ sodium nitrate	Na_2SO_4 sodium sulphate	Na_3PO_4 sodium phosphate
	Mg^{2+}	$Mg(NO_3)_2$ magnesium nitrate	$MgSO_4$ magnesium sulphate	$Mg_3(PO_4)_2$ magnesium phosphate
	Fe^{3+}	$Fe(NO_3)_3$ iron (III) nitrate	$Fe_2(SO_4)_3$ iron (III) sulphate	$FePO_4$ iron (III) phosphate

salt any member of a group of compounds containing a positive ion (cation) derived from a ◊metal or ammonia and a negative ion (anion) derived from an ◊acid or ◊non-metal. Salts have the properties typical of ionic compounds.

formula As all salts are electrically neutral, their formulae can be worked out by making sure that the total numbers of positive and negative charges arising from the ions are equal (see table).

preparation Various methods can be used to prepare salts in the laboratory; the choice is dictated by the starting materials available and by whether the required salt is soluble or insoluble. Methods include:

(1) ***acid + metal*** for salts of magnesium, iron, and zinc;
(2) ***acid + base*** for salts of magnesium, iron, zinc, and calcium;
(3) ***acid + carbonate*** for salts of all metals;
(4) ***acid + alkali*** for salts of sodium, potassium, and ammonium;
(5) ***direct combination*** for sulphides and chlorides;
(6) ***double decomposition*** for insoluble salts.

In methods (1)–(3) an excess of the solid reactant is added to the acid to ensure that no acid remains. The excess solid is filtered from the salt solution and the filtrate is boiled to a much smaller volume; it is then allowed to cool and crystallize. The salt crystals are filtered and dried on filter paper.

In method (4) an indicator is used to determine the volume of acid needed to neutralize the alkali (or vice versa). The colour can then be removed by charcoal treatment, or alternatively the experiment can be repeated without the indicator. The solution is boiled to a smaller volume, cooled to crystallize the salt, and the crystals filtered and dried as in (1)–(3) above.

In method (5) the salt is made in one step and does not require drying.

In method (6) the two solutions are mixed and stirred. The precipitated salt is filtered, washed well with water to remove the soluble impurities, and allowed to dry in air or an oven at 60–80°C.

salt, common popular name for ◊sodium chloride (NaCl).

saprotroph organism that feeds on the excrement or the dead bodies or tissues of others. They include most fungi (the rest being parasites); many bacteria and protozoa; animals such as dung beetles and vultures; and a few unusual plants, including several orchids. Saprotrophs cannot make food for themselves, so they are a type of ◊heterotroph. They are useful scavengers, and in sewage farms and refuse dumps break down organic matter into nutrients easily assimilable by green plants.

satellite any small body that orbits a larger one, either natural or artificial. Natural satellites that orbit planets are called ***moons***. The first artificial satellite, *Sputnik 1*, was launched into orbit around the Earth by the USSR in 1957.

The uses to which artificial satellites are put include:

scientific experiments and observation Many astronomical observations are best taken above the atmosphere, and satellites have been used to survey the skies and map the positions of stars with great precision.
reconnaissance and mapping applications Remote-sensing satellites are

used to make observations of the Earth's surface—gathering information about water sources and drainage, snow and ice cover, vegetation, geological structures, oil and mineral locations, land use, and pollution. Some satellites collect information that has military implications, such as the siting of military bases and the movement of armies.

weather monitoring Specialized satellites provide continuous worldwide observation of the atmosphere and its behaviour (for example, wind patterns and cloud cover), and have made it possible to forecast the weather with greater success.

communications Satellites play a major role in the worldwide communications network, being used to link computer systems and relay television, telephone, and other telecommunications signals across long distances. Signals are sent to and from the satellite via ground-based stations.

saturated compound organic compound that contains only single covalent bonds, such as propane. Saturated compounds are less reactive than their unsaturated relatives; for instance, the ◊alkenes.

saturated solution solution obtained when a solvent (liquid) can dissolve no more of a solute (usually a solid) at a particular temperature. Normally, a slight fall in temperature causes some of the solute to crystallize out of solution. If this does not happen the phenomenon is called supercooling, and the solution is said to be ***supersaturated***.

scale ◊calcium carbonate deposits that form on the inside of a kettle or boiler as a result of boiling ◊hard water.

scurvy disease caused by deficiency of ascorbic acid (vitamin C), which is contained in fresh vegetables and fruit. The signs are weakness and aching joints and muscles, progressing to bleeding of the gums and then other organs, and drying-up of the skin and hair. Treatment is by giving the vitamin.

seat belt safety device in a motor vehicle that is designed to reduce the risk of injury to a passenger during a collision or when brakes are applied sharply. In an emergency, it extends the time over which the decelerating force acts on a passenger, thereby reducing that force to a safe level. It also spreads the force over a broad band across the chest and over the hip bone, reducing the pressure applied to the person.

The principle behind the operation of the seat belt is based on Newton's second law of motion (see ◊Newton's laws of motion). The change of momentum (◊impulse) required to stop a passenger is equal to the product of the decelerating force applied and the time over which that force acts. It follows that if the time is increased, the force will be reduced.

secretion any substance (normally a fluid) produced by a cell or specialized gland, for example, sweat, saliva, enzymes, and hormones. The process whereby the substance is discharged from the cell is also known as secretion.

sedative drug that lessens nervousness, excitement, or irritation. Sedatives will induce sleep in larger doses.

sedimentary rocks layers of material laid down by wind, water, ice or gravity, the lower layers becoming compressed as the layer thickens and hardening into sedimentary rocks. Materials such as sand, gravel and clay are produced by erosion of existing rocks and can be moved by winds, rivers or glaciers to form layers of sediment. The remains of dead animals and plants collecting at the bottoms of seas as a result of gravity are also converted into sedimentary rocks.

seed the reproductive structure of higher plants. It develops from a fertilized ovule and consists of an embryo and a food store, surrounded and protected by an outer seed coat, called the testa. In flowering plants the seed is enclosed within a ◊fruit. After ◊germination, the seed develops into a new plant.

Seeds may be dispersed from the parent plant in a number of different ways. Agents of dispersal include animals, as with ◊burs and fleshy edible fruits, and wind, where the seed or fruit may be winged or plumed. Water can disperse seeds or fruits that float, and various mechanical devices may eject seeds from the fruit, as in some pods or legumes.

seismic waves shock waves produced as a result of earthquakes. Scientists studying the patterns of these waves can collect information on the nature of the materials through which the shock waves have passed and this helps them to understand the structure of the Earth. Artificial seismic waves, on a smaller scale, can be generated by explosions and mechanical vibrators and used for oil and mineral exploration.

selenium grey, non-metallic element, symbol Se, atomic number 34, relative atomic mass 78.96. It belongs to the sulphur group and occurs in several allotropic forms that differ in their physical and chemical properties. It is an essential trace element in human nutrition. Obtained from many sulphide ores and selenides, it is used as a red colouring for glass and enamel.

semen fluid containing ◊sperm from the testes and secretions from various sex glands, such as the prostate gland, that is produced by male animals during copulation. The secretions serve to nourish and activate the sperm cells, and prevent their clumping together.

semicircular canal one of three looped tubes that form part of the labyrinth in the inner ◊ear. They are filled with fluid and detect changes in the position of the head, contributing to the sense of balance.

semiconductor crystalline material with an electrical conductivity between that of metals (good) and insulators (poor).

The conductivity of semiconductors can usually be improved by minute additions of different substances or by other factors. Silicon, for example, has poor conductivity at low temperatures, but this is improved by the application of light, heat, or voltage; hence silicon is used in transistors, rectifiers, and integrated circuits (silicon chips).

sense organ any organ that an animal uses to gain information about its surroundings. All sense organs have specialized receptors (such as light receptors in an eye) and some means of translating their response into a nerve impulse that travels along sensory neurons to the brain. The main human sense organs are the ◊eye, which detects light and colour (different wavelengths of light); the ◊ear, which detects sound (vibrations of the air) and gravity; the ◊nasal cavity, which detects some of the chemical molecules in the air, giving a sense of smell; and the ◊tongue, which detects some of the chemicals in food, giving a sense of taste. There are also many small sense organs in the skin, including pain sensors, temperature sensors, and pressure sensors, contributing to our sense of touch.

Some animals can detect small electric discharges, underwater vibrations, minute vibrations of the ground, or sounds that are below (infrasound) or above (ultrasound) the range of hearing in humans. Sensitivity to light varies greatly. Most mammals cannot distinguish different colours, whereas many insects can see light in the ultraviolet range, and snakes can form images of infrared radiation (radiant heat).

sensitivity the ability of an organism, or part of an organism, to detect changes in the environment. Although all living things are capable of some sensitivity, evolution has led to the formation of highly complex mechanisms for detecting light, sound, chemicals, and other stimuli. It is essential to an animal's survival that it can process this type of information and make an appropriate response.

sepal part of a flower, usually green, that surrounds and protects the flower in bud. The sepals are derived from modified leaves and are collectively known as the calyx.

In some plants, such as the marsh marigold, where true ◊petals are absent, the sepals are brightly coloured and petal-like, taking over the role of attracting insect pollinators to the flower.

series circuit an electric circuit in which the components are connected end to end, so that the current flows through them all one after the other. The division of the ◊terminal voltage across each conductor is in the ratio of the resistances of each conductor. If the potential differences across two conductors of resistance R_1 and R_2, connected in series, are V_1 and V_2 respectively, then the ratio of those potential differences is given by the equation:

$$V_1/V_2 = R_1/R_2$$

The total resistance R of those conductors is given by:

$$R = R_1 + R_2$$

Compare ◊parallel circuit.

sex chromosome one of the pair of chromosomes that determines the sex of an organism.

Maleness and femaleness in animals usually depends on which of the sex chromosomes have been inherited. The two types of chromosome, known as the X and Y chromosomes, are paired in the adult organism. Human beings are male when they possess XY and female when they possess XX. An ovum (egg) or a sperm cell is haploid and can only contain one sex chromosome so all ova are X. Sperm, however, may be X or Y; the zygote resulting from fusion of human gametes will be male if the sperm cell contained the Y chromosome.

sex linkage in genetics, the tendency for certain characteristics to occur exclusively, or predominantly, in one sex only. Human examples include red–green colour blindness and haemophilia, both found predominantly in males. In both cases, these characteristics are ◊recessive and are determined by genes on the ◊X chromosome.

sexual reproduction in humans

female reproductive system

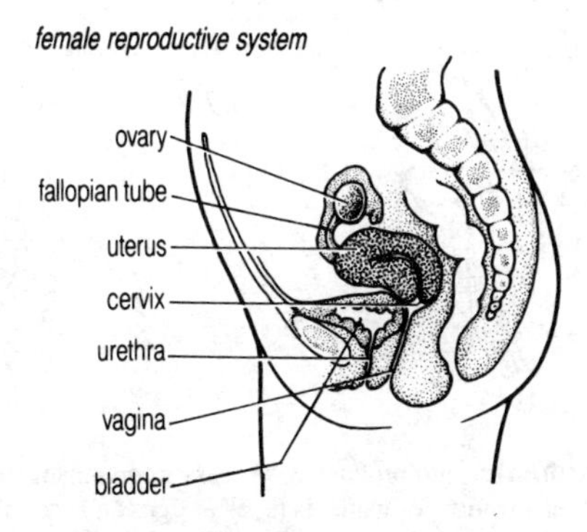

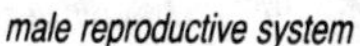
male reproductive system

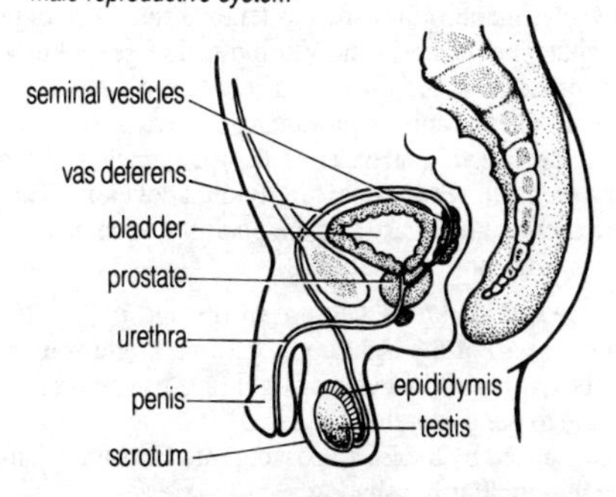

Since females possess two X chromosomes, any such recessive ◊allele on one of them is likely to be masked by the corresponding allele on the other. In males (who have only one X chromosome paired with a largely inert ◊Y chromosome) any gene on the X chromosome will automatically be expressed. Colour blindness and haemophilia can appear in females, but only if they are ◊homozygous for these traits, owing to inbreeding, for example.

sexual reproduction in flowering plants

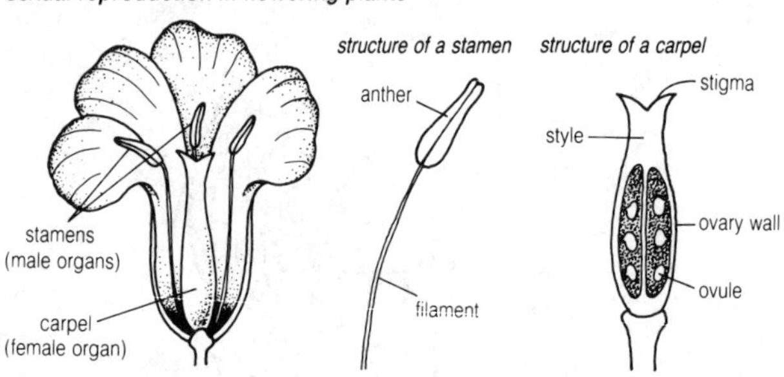

sexual reproduction a reproductive process in organisms that requires the union, or ◊fertilization, of gametes (such as eggs and sperm). These are usually produced by two different individuals, although self-fertilization occurs in a few ◊hermaphrodites such as tapeworms. Most organisms other than bacteria and cyanobacteria show some sort of sexual process. Except in some lower organisms, the gametes are of two distinct types called ova (eggs) and sperm. The organisms producing the ova are called females, and those producing the sperm, males. The fusion of a male and female gamete produces a ***zygote***, from which a new individual develops. The alternatives to sexual reproduction are binary fission, budding, vegetative reproduction, parthenogenesis, and spore formation.

shadow the area of darkness behind an opaque object that cannot be reached by some or all of the light coming from a light source in front. Its presence may be explained in terms of light rays travelling in straight lines and being unable to bend round obstacles.

◊Eclipses are caused by the Earth passing into the Moon's shadow or the Moon passing into the Earth's shadow.

short circuit direct connection between two points in an electrical circuit. Its relatively low resistance means that a large current flows through it, bypassing the rest of the circuit, and this may cause the circuit to overheat dangerously.

short-sightedness or ***myopia*** defect of the eye in which a person can see clearly only those objects that are close up. It is caused by either the eyeball being too long or by the cornea and lens system of the eye being too powerful, both of which cause the images of distant objects to be formed in front of the retina. Short-sightedness can be corrected by wearing spectacles fitted with ◊diverging lenses, or by wearing diverging (concave meniscus) contact lenses.

SI abbreviation for ***Système International [d'Unités]*** (French 'International System [of Metric Units]'); see ◊SI units.

sight the detection of light by an ◊eye, which can form visual images of the outside world.

silica common name for ◊silicon(IV) oxide, SiO_2.

silicate compound containing silicon and oxygen combined together with one or more metal ions.

Common natural silicates are sands (common sand is the oxide of silicon known as silica). Glass is a manufactured complex polysilicate material in which other elements (boron in borosilicate glass) have been incorporated.

silicon brittle, non-metallic element, symbol Si, atomic number 14, relative atomic mass 28.086. It is the second most abundant element (after oxygen) in the Earth's crust and occurs in amorphous and crystalline forms. In nature it is found only in combination with other elements, chiefly with oxygen in silicates and silica (◊silicon(IV) oxide, SiO_2), the main constituents of sands and gravels.

Pottery glazes and glassmaking are based on the use of silica sands and date back into prehistory. Today the crystalline form of silicon is used as a deoxidizing and hardening agent in steel, and has become the basis of the electronics industry because of its ◊semiconductor properties, being used to make silicon chips (integrated circuits) for microprocessors.

silicon chip popular name for ◊integrated circuit.

silicon(IV) oxide or ***silicon dioxide*** SiO_2 colourless or white solid, insoluble in water, that occurs naturally in various forms, the most familiar and pure of which is ◊quartz. It has a ◊giant molecular structure.

silver white, lustrous, extremely malleable and ductile, metallic element, symbol Ag, atomic number 47, relative atomic mass 107.868. It occurs in nature in ores and as a free metal; the chief ores are sulphides, from which the metal is extracted by smelting with lead. It is the best metallic conductor of both heat and electricity; its most useful compounds are the chloride and bromide, which darken on exposure to light and are the basis of photographic emulsions.

Silver is used ornamentally, for jewellery and tableware, for coinage, in electroplating, electrical contacts, and dentistry, and as a solder.

SI units (French ***Système International d'Unités***) standard system of scientific units used by scientists worldwide. Originally proposed in 1960, it is based on seven 'base units': the metre (m) for length, kilogram (kg) for mass, second (s) for time, ampere (A) for electrical current, kelvin (K) for thermodynamic temperature, mole (mol) for amount of substance, and candela (cd) for luminous intensity.

Some non-SI units, well established and internationally recognized, remain in use in conjunction with SI: minute, hour, and day in measuring time; multiples or submultiples of base or derived units which have long-established names, such as tonne for mass, the litre for volume; and specialist measures such as the metric carat for gemstones.

SI units and the prefixes used with them are listed in Appendix II.

skeleton the rigid or semirigid framework that supports an animal's body, protects its internal organs, and provides anchorage points for its muscles. The skeleton may be composed of bone and cartilage (vertebrates), chitin (arthropods), calcium carbonate (molluscs and other invertebrates), or silica (many unicellular eukaryotes).

It may be internal, forming an ◊endoskeleton, or external, forming an ◊exoskeleton. Another type of skeleton, found in invertebrates such as earthworms, is the ***hydrostatic skeleton***. This gains partial rigidity from fluid enclosed within a body cavity. Because the fluid cannot be compressed, contraction of one part of the body results in extension of another part, giving peristaltic motion.

skin in vertebrates, the covering of the body. In mammals, the outer layer (epidermis) is dead and protective, and its cells are constantly being rubbed away and replaced from below. The lower layer (dermis) contains blood vessels, nerves, hair roots, and sweat and sebaceous glands, and is supported by a network of fibrous and elastic cells.

skull in vertebrates, the collection of flat and irregularly shaped bones (or cartilage) that enclose the brain and the organs of sight, hearing, and smell, and provide support for the jaws. In mammals, the skull consists of 22 bones joined by sutures (immovable joints). The floor of the skull is pierced by a large hole for the spinal cord and by several smaller apertures through which other nerves and blood vessels pass.

slag the molten mass of impurities that is produced in the smelting or refining of metals. The slag produced in the manufacture of iron in a ◊blast furnace floats on the surface above the molten iron. It contains mostly silicates, phosphates, and sulphates of calcium. When cooled, the solid is broken up and used as a core material in the foundations of roads and buildings.

slaked lime or ***calcium hydroxide*** $Ca(OH)_2$ substance produced by adding water to lime (calcium oxide, CaO). Much heat is given out and the solid crumbles as it absorbs water. A solution of slaked lime is called ◊limewater.

small intestine the length of gut between the stomach and the large intestine, consisting of the duodenum and the ileum. It is responsible for digesting and absorbing food.

The wall is glandular, producing mucus and enzymes to aid digestion, and muscular so that food can be moved down the gut. Absorption, the passage of small molecules across the wall, is made more efficient by the presence of numerous villi, small fingerlike projections that increase the surface area.

smell sense that responds to chemical molecules in the air. It is used to detect food and to communicate with other animals. It works by having receptors (olfactory receptors) for particular chemical groups, into which the airborne chemicals must fit to trigger a message to the brain. The receptors are in the ◊nasal cavity.

smelting processing a metallic ore in a furnace to produce the metal. Oxide ores such as iron ore are smelted with coke (carbon), which reduces the ore into metal and also provides fuel for the process.

A substance such as ◊limestone is often added during smelting to facilitate the melting process and to form a slag, which dissolves many of the impurities present.

smokeless fuel fuel that does not give off any smoke when burnt, as all the carbon is fully oxidized to carbon dioxide (CO_2). Natural gas, oil, and coke are smokeless fuels.

smoking inhaling the fumes from burning tobacco, usually in the form of cigarettes. The practice can be habit-forming and is dangerous to health, since carbon monoxide and other poisonous materials result from the combustion process. There is a direct link between lung cancer and tobacco smoking; the habit is also linked to respiratory conditions such as bronchitis and emphysema, and to coronary heart diseases. In the West, smoking is now forbidden in many public places because even ***passive smoking***—breathing in fumes from other people's cigarettes—can be harmful.

Manufacturers have attempted to filter out harmful substances such as tar and nicotine, and to use milder tobaccos, and governments have carried out extensive antismoking advertising campaigns. In the UK and the USA all cigarette packaging must carry a government health warning, and television advertising of cigarettes is forbidden.

smooth muscle another name for ◊involuntary muscle.

soap mixture of the sodium salts of various ◊fatty acids: palmitic, stearic, and oleic. It is made by the action of sodium hydroxide (caustic soda) or potassium hydroxide (caustic potash) on fats of animal or vegetable origin (saponification). Soap makes grease and dirt disperse in water in a similar manner to a ◊detergent.

soda lime powdery mixture of calcium hydroxide and sodium hydroxide or potassium hydroxide, used in medicine and as a drying agent.

sodium soft, waxlike, silver-white, metallic element, symbol Na, atomic number 11, relative atomic mass 22.898. It is one of the ◊alkali metals and has a very low density, being light enough to float on water. It is the sixth most abundant element (the fourth most abundant metal) in the Earth's crust. Sodium is highly reactive, oxidizing rapidly when exposed to air and reacting violently with water. Its most familiar compound is sodium chloride (common salt), which occurs naturally in the oceans and in salt deposits left by dried-up ancient seas.

Other sodium compounds used industrially include sodium hydroxide (caustic soda, NaOH), sodium carbonate (washing soda, $Na_2CO_3.10H_2O$) and hydrogencarbonate (sodium bicarbonate, $NaHCO_3$), sodium nitrate (saltpetre, $NaNO_3$, used as a fertilizer), and sodium thiosulphate (hypo, $Na_2S_2O_3$, used as a photographic fixer). Thousands of tons of these are manufactured annually. Sodium metal is used to a limited extent in spectroscopy and in discharge lamps, and is alloyed with potassium as a heat-transfer medium in nuclear reactors.

sodium carbonate or ***soda ash*** Na_2CO_3 anhydrous white solid. The hydrated, crystalline form ($Na_2CO_3.10H_2O$) is also known as washing soda. It is used as a mild alkali, since it is hydrolysed in water to produce hydroxide ions:

$$CO^{2-}_{3(aq)} + H_2O_{(l)} \rightarrow HCO^{-}_{3(aq)} + OH^{-}_{(aq)}$$

It is used to neutralize acids, in glass manufacture, and in water softening.

sodium chloride or ***common salt*** NaCl white, crystalline compound found widely in nature (see ◊crystal). It is a typical ionic solid with a high melting point (801°C); it is soluble in water, insoluble in organic solvents, and is a strong electrolyte when molten or in aqueous solution. It is found dissolved in sea water and as concentrated deposits of rock salt (halite). It is widely used in the food industry as a flavouring and preservative, and in the chemical industry in the manufacture of sodium, chlorine, and sodium carbonate.

While common salt is an essential part of our diet, some medical experts believe that excess salt, largely from processed food, can lead to high blood pressure and increased risk of heart attacks.

sodium hydrogencarbonate or ***bicarbonate of soda*** $NaHCO_3$ mild alkaline substance. It has the following chemical reactions:

with acids It behaves as an alkali in solution, neutralizing acids to form a salt, water, and carbon dioxide.

$$NaHCO_{3(aq)} + HCl_{(aq)} \rightarrow NaCl_{(aq)} + CO_{2(g)} + H_2O_{(l)}$$

with heat On heating it decomposes to give off carbon dioxide and water, leaving the carbonate behind.

$$2NaHCO_{3(s)} \rightarrow Na_2CO_{3(s)} + CO_{2(g)} + H_2O_{(l)}$$

with water It undergoes some hydrolysis in aqueous solution.

$$HCO^{-}_{3(aq)} + H_2O_{(l)} \rightarrow H_2CO_{3(aq)} + OH^{-}_{(aq)}$$

It is used as an antacid in indigestion preparations. Other uses include baking powders and effervescent health drinks.

sodium hydroxide or ***caustic soda*** NaOH the commonest ◊alkali. The solid and the solution are corrosive. It is used to neutralize acids, and in the manufacture of soap and oven cleaners.

It is prepared industrially from sodium chloride by the ◊electrolysis of concentrated brine. The brine contains the ions Na^+, Cl^-, H^+, and OH^- in solution. Two of these ions are discharged at the electrodes. The reactions are as follows.

negative electrode:

$$2H^+ + 2e^- \rightarrow H_2$$

positive electrode:

$$2Cl^- - 2e^- \rightarrow Cl_2$$

This means the Na^+ and OH^- remain in solution, so it becomes sodium hydroxide (containing some sodium chloride). This solution is sometimes called caustic soda liquor and is about 40% NaOH.

soft water water that contains very few dissolved metal ions such as calcium (Ca^{2+}) or magnesium (Mg^{2+}). It lathers easily with soap, and no ◊scale is formed inside kettles or boilers. It has been found that the incidence of heart disease is higher in soft-water areas.

soil loose mixture of finely ground rock and decaying organic material covering the Earth's surface. It provides nutrients for plants, but can lose its fertility if these nutrients are not replaced. The decomposition of dead animals and plants in the upper surface of the soil—the humus layer—continually restores the organic and mineral content. Aeration and mixing is also necessary for a healthy soil, and this is often achieved by the burrowing activities of earthworms.

There are many types, with varying organic and mineral content making some soils more suitable than others for particular types of natural vegetation or for agriculture. Particle size is responsible for one of the major differences between soils. Clay soils have very tiny particles, with only small spaces between each grain, leading to poor drainage qualities. Sandy soils have large particles, but can drain too easily, leading to drying out and, in some cases, desertification. The organic content of soil is widely variable, ranging from zero in some desert soils to almost 100% in peats. Loams are very fertile soils containing a mixture of organic material and particles of different sizes.

soil depletion a decrease in soil quality over time. Causes include loss of nutrients caused by over-farming, erosion by wind, and chemical imbalances caused by acid rain.

soil

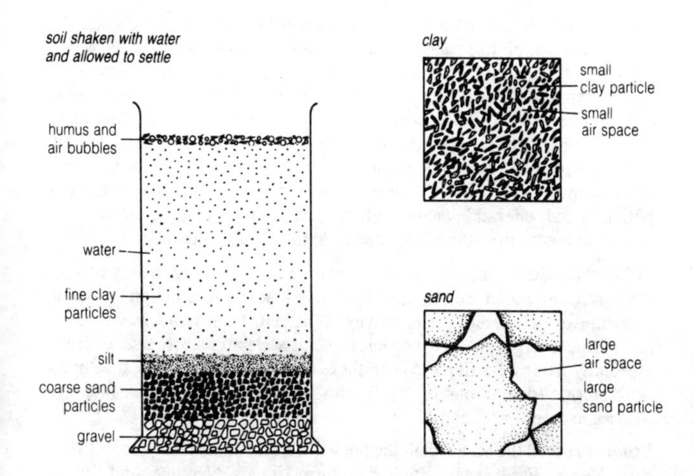

soil erosion the wearing away and redistribution of the Earth's soil layer. It is caused by the action of water, wind, and ice, and also by improper methods of agriculture. If unchecked, soil erosion results in the formation of deserts.

If the rate of erosion is greater than the rate of soil formation (from rock) then the land will decline and eventually become infertile. The removal of forests or other vegetation often leads to serious soil erosion, because plant roots bind soil, and without them the soil is free to wash or blow away. Soil erosion can be countered by erecting windbreaks, such as hedges or strips planted with coarse grass; ◊organic farming can reduce soil erosion by as much as 75%.

sol colloidal suspension of very small solid particles in a liquid that retains the physical properties of a liquid (see ◊colloid).

solar energy energy derived from the Sun's radiation. The amount of energy falling on one square kilometre of the Earth's surface is about 4,000 megawatts, enough to heat and light a small town. In one second the Sun gives off 13 million times more energy than all the electricity used in the USA in one year.

Solar heaters have industrial or domestic uses. They usually consist of a black (heat-absorbing) panel containing pipes through which air or water, heated by the Sun, is circulated, either by thermal ◊convection currents or by a pump. Solar energy may also be harnessed indirectly by using ***solar cells*** (photovoltaic cells) made of panels of ◊semiconductor material (usually silicon), which generate electricity when illuminated by sunlight. Although it is difficult to generate a high output from solar energy compared with sources such as nuclear or fossil-fuel energy, it is a major non-polluting and renewable energy source used as far north as Scandinavia as well as in the southwestern USA and in Mediterranean countries.

solar radiation radiation given off by the Sun, consisting mainly of visible light, ◊ultraviolet radiation, and ◊ infrared radiation, although the whole spectrum of ◊electromagnetic waves is present, from radio waves to X-rays. High-energy charged particles such as electrons are also emitted, especially from solar flares. When these reach the Earth, they cause magnetic storms (disruptions of the Earth's magnetic field), which interfere with radio communications.

Solar System the Sun and all the bodies orbiting it: the nine planets (Mercury, Venus, Earth, Mars, Jupiter, Saturn, Uranus, Neptune, and Pluto), their moons, the asteroids, and the comets. It is thought to have formed from a cloud of gas and dust in space about 4.6 billion years ago. The Sun contains 99% of the mass of the Solar System. The edge of the Solar System is not clearly defined, marked only by the effective limit of the Sun's gravitational influence, which extends about 1.5 light years, about a third of the distance to the nearest star, Alpha Centauri, 4.3 light years away.

solenoid elongated coil of wire. A strong and uniform magnetic field is produced inside it when a current passes through the wire. A solenoid fitted with an iron core forms an ◊electromagnet.

solid state of matter that holds its own shape (as opposed to a liquid, which takes up the shape of its container, or a gas, which totally fills its container). According to ◊kinetic theory , the atoms or molecules in a solid are not free to move but merely vibrate about fixed positions, such as those in ◊crystals.

solidification change of state of a substance from liquid or vapour to solid on cooling. It is the opposite of melting or sublimation.

solubility measure of the amount of solute (usually a solid or gas) that will dissolve in a given amount of solvent (usually a liquid) at a particular temperature. Solubility may be expressed as grams of solute per 100 grams of solvent or, for a gas, in parts per million (ppm) of solvent.

solute substance that is dissolved in another substance (see ◊solution).

solution two or more substances mixed to form a single, homogeneous phase. One of the substances is the ***solvent*** and the others (***solutes***) are said to be dissolved in it.

The constituents of a solution may be solid, liquid, or gaseous. The solvent is normally the substance that is present in greatest quantity, although when one of them is a liquid this is considered to be the solvent even if it is not the major substance.

solvent substance, usually a liquid, that will dissolve another substance (see ◊solution). Although the commonest solvent is water, in popular use the term refers to low-boiling-point organic liquids that are harmful if used in a confined space. They can give rise to respiratory problems, liver damage, and neurological complaints.

Typical organic solvents are petroleum distillates (in glues), alcohols (for synthetic and natural resins such as shellac), esters (in lacquers, including nail varnish), ketones (in cellulose lacquers and resins), and chlorinated hydrocarbons (as paint stripper and dry-cleaning fluids). The fumes of some solvents have an intoxicating effect and this has given rise to the dangerous practice of solvent abuse, or 'glue-sniffing'.

sound physiological sensation received by the ear, originating in a vibration (pressure variation in the air) that communicates itself to the air, and travels in every direction, spreading out as an expanding sphere. All sound waves in air travel with a speed dependent on the temperature; under ordinary conditions, this is about 330 metres per second. The ◊pitch of the sound depends on the number of vibrations imposed on the air per second, but the speed is unaffected. The ◊loudness of a sound is dependent primarily on the amplitude of the vibration of the air.

The lowest note audible to a human being has a frequency of about 15–16 Hz (vibrations per second), and the highest one of about 20,000 Hz; the lower limit of this range varies little with the person's age, but the upper range falls steadily from adolescence onwards. Pressure waves of a frequency higher than the upper range are called ◊ultrasounds.

sound wave the longitudinal wave motion with which sound energy travels through a medium. It carries energy away from the source of the sound without carrying the material itself with it. Sound waves are mechanical; unlike electromagnetic waves, they require vibration of their medium's molecules or particles, and this is why sound cannot travel through a vacuum.

species the lowest level of classification; a distinguishable group of organisms that resemble each other or consist of a few distinctive types (varieties), and that can all interbreed to produce fertile offspring. Related species are grouped together in a genus.

All living human beings belong to the same species because they can all interbreed, even though they may differ considerably in such features as skin colour, height, head shape, and so on. Other examples of species are lions, Douglas firs, cabbage white butterflies, and sperm whales.

specific heat capacity quantity of heat required to raise unit mass (one kilogram) of a substance by one kelvin ($1K$). The unit of specific heat capacity in the SI system is the joule per kilogram per kelvin ($J\ kg^{-1}K^{-1}$).

specific latent heat the heat that changes the physical state of a unit mass (one kilogram) of a substance without causing any temperature change.

The ***specific latent heat of fusion*** of a solid substance is the heat required to change one kilogram of it from solid to liquid without any temperature change.

The ***specific latent heat of vaporization*** of a liquid substance is the heat required to change one kilogram of it from liquid to vapour without any temperature change.

spectrum (plural ***spectra***) arrangement of frequencies or wavelengths when electromagnetic radiations are separated into their constituent parts. Visible light is part of the ◊electromagnetic spectrum and most sources emit waves over a range of wavelengths that can be broken up or 'dispersed'; white light can be separated into red, orange, yellow, green, blue, indigo, and violet.

speed the rate at which an object moves. The constant speed v of an object may be calculated by dividing the distance s it has travelled by the time t taken to do so, and may be expressed as:

$$v = s/t$$

The usual units of speed are metres per second or kilometres per hour.

Speed is a scalar quantity in which direction of motion is unimportant (unlike the vector quantity ◊velocity, in which both magnitude and direction must be taken into consideration). See also ◊distance–time graph.

speed of light the speed at which light and other ◊electromagnetic waves travel through empty space. Its value is 299,792,458 metres per second. The speed of light is the highest speed possible, according to the theory of relativity, and its value is independent of the motion of its source and of the observer. It is impossible to accelerate any material body to this speed because it would require an infinite amount of energy.

speed of reaction alternative term for ◊rate of reaction.

speed of sound the speed at which sound travels through a medium, such as air or water. In air at a temperature of 0°C, the speed of sound is 331 metres per second. At higher temperatures, the speed of sound is greater; at 18°C it is 342 metres per second. It is greater in liquids and solids; for example, in water it is around 1,440 metres per second, depending on the temperature.

speed–time graph graph used to describe the motion of a body by illustrating how its speed or velocity changes with time. The gradient of the graph gives the object's acceleration: if the gradient is zero (the graph is horizontal) then the body is moving with constant speed or uniform velocity; if the gradient is constant, the body is moving with uniform acceleration. The area under the graph gives the total distance travelled by the body.

sperm or ***spermatozoon*** the male ◊gamete of animals. Each sperm cell has a head capsule containing a nucleus, a middle portion containing ◊mitochondria (which provide energy), and a long tail (flagellum). In most animals, the sperm are motile, and are propelled by the flagellum. The sperm cells of animals that carry out internal ◊fertilization are usually released, with secretions from various sex glands, in the form of a fluid called ◊semen.

spermicide any cream, jelly, pessary, or other preparation that kills the sperm cells in semen. Spermicides are used for contraceptive purposes, usually in combination with barrier methods such as the ◊condom or the diaphragm. Spermicide used alone is only 75% effective in preventing pregnancy.

sphincter ring of muscle found at various points in the alimentary canal, which contracts and relaxes to control the movement of food.

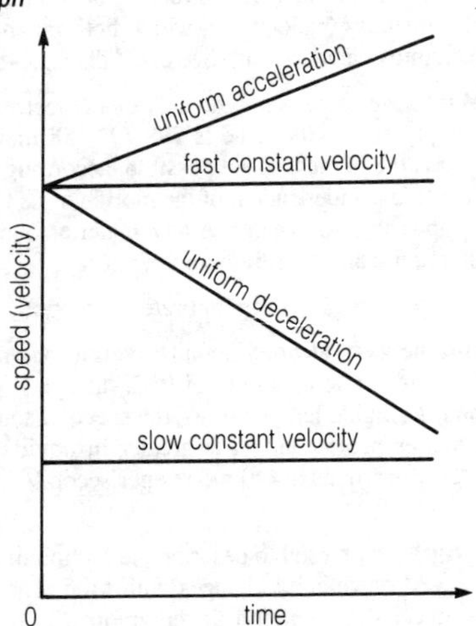

The ***pyloric sphincter***, at the base of the stomach, controls the release of the gastric contents into the duodenum. After release the sphincter contracts, closing off the stomach.

spinal cord in vertebrates, a major component of the ◊central nervous system. It is enclosed by the bones of the ◊vertebral column, and links the peripheral nervous system to the brain.

spine another name for the ◊vertebral column, the backbone of vertebrates.

spiracle in insects, the opening of a ◊trachea, through which oxygen enters the body and carbon dioxide is expelled.

spleen in vertebrates, an organ situated behind the stomach that produces ◊lymphocytes; it forms a part of the lymphatic system. The spleen also regulates the number of red blood cells in circulation by destroying old cells, and stores iron.

spongy mesophyll irregularly shaped cells in the centre of a leaf, forming a layer between the ◊palisade cells and the lower epidermis.

sperm

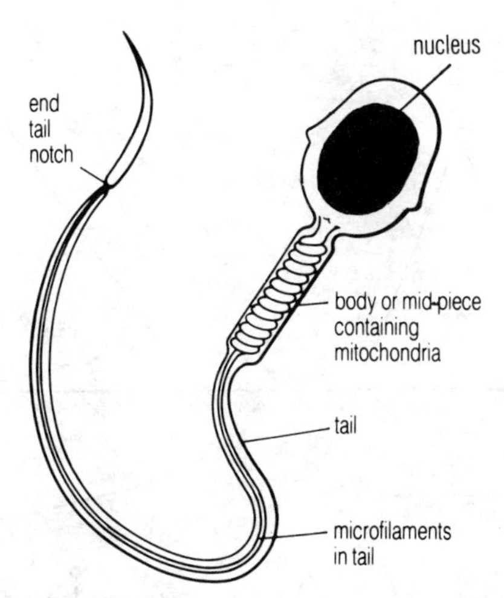

Spongy mesophyll contains chlorophyll and its role in photosynthesis is aided by an efficient gas-exchange mechanism. Air spaces between the cells are continuous with the stomata (see ◊stoma), small pores found on the undersurface of the leaf. Gases can circulate freely, with carbon dioxide entering the cells if needed and oxygen passing out of the leaf as an excretory product.

spore small reproductive or resting body, usually consisting of just one cell. Unlike a ◊gamete, it does not need to fuse with another cell to develop into a new organism. Spores are produced by the lower plants, most fungi, some bacteria, and certain protozoa. They are generally light and easily dispersed by wind movements.

spring balance instrument for measuring weight that relates the weight of an object to the extent to which it stretches or compresses a vertical spring. According to ◊Hooke's law, the extension or compression will be directly proportional to the weight, providing that the spring is not overstretched. A pointer attached to the spring indicates the weight on a scale,

stability

stable equilibrium

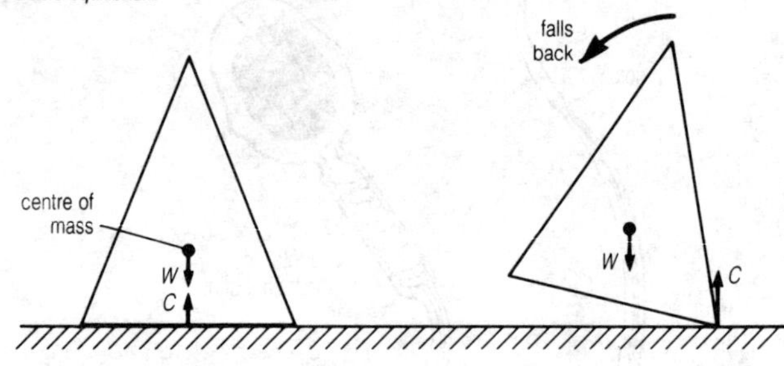

unstable equilibrium

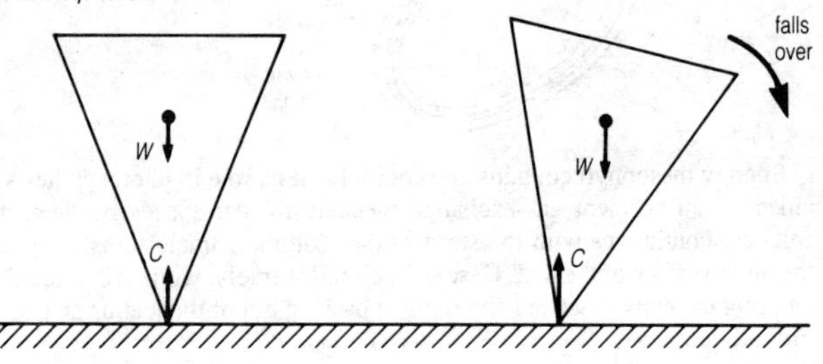

neutral equilibrium

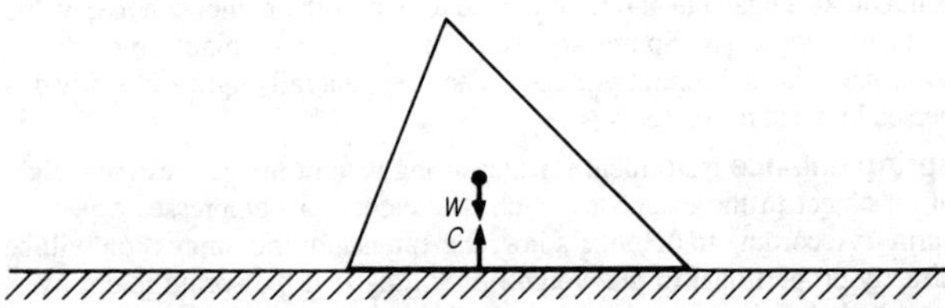

W stands for weight; *C* stands for contact force

which may be calibrated in newtons (the SI unit of force) for physics experiments, or in grams, kilograms, or pounds (units of mass) for everyday use.

stability measure of how difficult it is to move an object from a position of ◊equilibrium with respect to gravity.

An object displaced from equilibrium does not remain in its new position if its weight, acting vertically downwards through its ◊centre of mass, no longer passes through the line of action of the contact force (the force exerted by the surface on which the object is resting), acting vertically upwards through the object's new base. If the lines of action of these two opposite but equal forces do not coincide they will form a couple and create a ◊moment that will cause the object either to return to its original rest position or to topple over into another position.

An object in ***stable equilibrium*** returns to its rest position after being displaced slightly. This form of equilibrium is found in objects that are difficult to topple over; these usually possess a relatively wide base and a low centre of mass—for example, a cone resting on its flat base on a horizontal surface. When such an object is tilted slightly its centre of mass is raised and the line of action of its weight no longer coincides with that of the contact force exerted by its new, smaller base area. The moment created will tend to lower the centre of mass and so the cone will fall back to its original position.

An object in ***unstable equilibrium*** does not remain at rest if displaced, but falls into a new position; it does not return to its original rest position. Objects possessing this form of equilibrium are easily toppled and usually have a relatively small base and a high centre of mass—for example, a cone balancing on its point, or apex, on a horizontal surface. When an object such as this is given the slightest push its centre of mass is lowered and the displacement of the line of action of its weight creates a moment. The moment will tend to lower the centre of mass still further and so the object will fall on to another position.

An object in ***neutral equilibrium*** stays at rest if it is moved into a new position—neither moving back to its original position nor on any further. This form of equilibrium is found in objects that are able to roll, such as a cone resting on its curved side placed on a horizontal surface. When such an object is rolled its centre of mass remains in the same position, neither rising nor falling, and the line of action of its weight continues to coincide with the contact force; no moment is created and so its equilibrium is maintained.

stamen in flowers, the male reproductive organ. A typical stamen consists of a ***filament***, with an ***anther***, the pollen-bearing organ, at its apex.

standard temperature and pressure (STP) a standard set of conditions for experimental measurements, to enable comparisons to be made between sets of results. Standard temperature is 0°C and standard pressure 1 atmosphere (101,325 Pa).

star luminous globe of gas, mainly hydrogen and helium, which produces its own heat and light by nuclear reactions. Clusters of stars are held together by gravity to form ◊galaxies, such as our own Milky Way. Although stars shine for a very long time—many millions of years—they are not eternal, and have been found to change in appearance at different stages in their 'lives'.

Stars are born when nebulae (giant clouds of dust and gas) contract under the influence of gravity into denser bodies. As each new star contracts, the temperature and pressure in its core rises. At about 10 million°C the temperature is hot enough for a nuclear reaction to begin (the fusion of hydrogen nuclei to form helium nuclei); vast amounts of energy are released, contraction stops, and the star begins to shine. Stars at this stage are called ***main-sequence stars***—the Sun is such a star and is expected to remain at this stage for the next five billion years.

When all the hydrogen at the core of a main-sequence star has been converted into helium, the star swells to become a ***red giant***, about 100 times its previous size and with a cooler, redder surface. When, after this brief stage, the star can produce no more nuclear energy, its outer layers drift off into space and its core collapses in on itself to form a small and very dense body called a ***white dwarf***. Eventually the white dwarf fades away, leaving a non-luminous ***dark body***.

Some very large main-sequence stars do not end their lives as white dwarfs—they pass through their life cycle quickly, becoming red ***supergiants*** that eventually explode into brilliant ***supernovae***. Part of the core remaining after such an explosion may collapse to form a small 'superdense' star, which consists almost entirely of neutrons and is therefore called a ***neutron star***. Neutron stars called ***pulsars*** spin very quickly, giving off pulses of radio waves (rather as a lighthouse gives off flashes of light). If the collapsing core of the supernova is very massive it does not form a neutron star; instead it forms a ***black hole***, a region so dense that its gravity not only draws in all nearby matter but also all radiation, including its own light.

starch polysaccharide made up of monosaccharide glucose units. It is produced by plants as a carbohydrate food reserve, and is stored in the form of ***starch grains*** in the cells of roots, tubers, seeds, and fruit. Starch is therefore an important source of energy for many primary consumers (herbivores and omnivores). For humans, the main dietary sources of starch are cereals, legumes (beans and peas), potatoes, and root vegetables.

states of matter the forms (solid, liquid, or gas) in which material can exist (see ◊change of state). Whether a material is solid, liquid, or gas depends on its temperature and the pressure on it. The transition between states takes place at definite temperatures, called melting point and boiling point.

◊Kinetic theory describes how the state of a material depends on the movement and arrangement of its atoms or molecules. A hot ionized gas, or plasma, is often called the fourth state of matter, but liquid crystals, ◊colloids, and ◊glass also have a claim to this title.

state symbol symbol used in chemical equations to indicate the physical state of the substances present. The symbols are: (s) for solid, (l) for liquid, (g) for gas, and (aq) for aqueous.

static electricity ◊electric charge acquired by a body by means of electrostatic induction or friction. Rubbing different materials can produce static electricity, seen in the sparks produced on combing one's hair or removing a nylon shirt. In some processes static electricity is useful, as in paint spraying where the parts to be sprayed are charged with electricity of opposite polarity to that on the paint droplets.

steam dry, invisible gas formed by vaporizing water. The visible cloud that normally forms in the air when water is vaporized is due to minute suspended water particles. Steam is widely used in chemical and other industrial processes and for the generation of power.

steel alloy or mixture of iron and up to 1.7% carbon, sometimes with other elements, such as manganese, phosphorus, sulphur, and silicon. It has innumerable uses, including the manufacture of ships and cars, and the construction of tower-block frames and machinery of all kinds.

Steels with only small amounts of other metals are called ***mild*** or ***carbon steels***. These steels are far stronger than pure iron, with properties varying with the composition. ***Alloy steels*** include greater proportions of other metals; for example, stainless steel must contain at least 11% chromium.

Steel is produced by removing impurities, such as carbon, from raw or pig iron produced by a ◊blast furnace. The main industrial process is the ◊***basic-oxygen process***, in which pure oxygen is blown at high pressure through molten pig iron and scrap steel. The surface of the metal is disturbed by the blast and the oxidized impurities are burned out as gases or as slag. High-quality steel is made in an ***electric furnace***. A large electric current flows through electrodes in the furnace, melting a charge of scrap steel and iron. The quality of the steel produced can be controlled precisely because the temperature of the furnace can be maintained exactly and there are no combustion by-products to contaminate the steel.

stem the main supporting axis of a plant that bears the leaves, buds, and reproductive structures; it may be simple or branched. The plant stem usually grows above ground, although some grow underground, including ◊rhizomes, and ◊tubers. Stems contain a continuous vascular system that conducts water and food to and from all parts of the plant. In plants exhibiting thickening, or secondary growth, the stem may become woody, forming a main trunk, as in trees, or a number of branches from ground level, as in shrubs.

stigma in a flower, the surface at the tip of a ◊carpel that receives the ◊pollen. It often has short outgrowths, flaps, or hairs to trap pollen and may produce a sticky secretion to which the grains adhere.

stimulus any change in environmental factors, such as light, heat, or pressure, which can be detected by an organism's receptors.

stoma (plural ***stomata***) pore in the epidermis of a plant, particularly on the under surface of a leaf. Each stoma is surrounded by a pair of guard cells that are crescent shaped when the stoma is open but can collapse to an oval shape, thus closing off the opening between them. Stomata allow the exchange of carbon dioxide and oxygen (needed for ◊photosynthesis and ◊respiration) between the internal tissues of the plant and the outside atmosphere. They are also the main route by which water is lost from the plant by ◊transpiration, and they can be closed to conserve water, the movements being controlled by changes in turgidity of the guard cells.

stomach the first cavity in the gut. In mammals it is a bag of muscle situated just below the diaphragm. Food enters it from the oesophagus, is digested by the acid and ◊enzymes secreted by the stomach lining, and then passes into the duodenum. Some plant-eating mammals, such as cows and

stoma

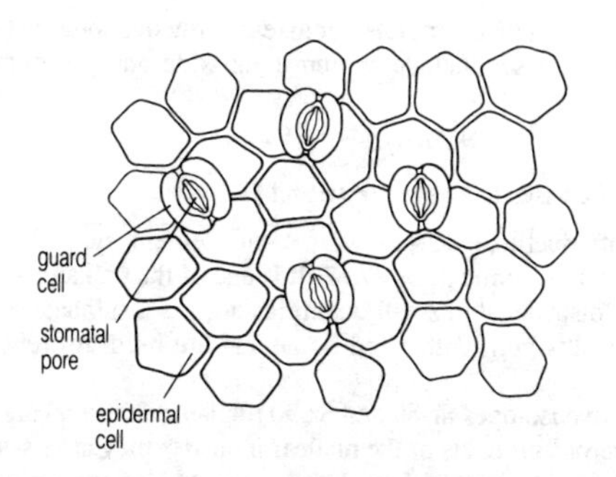

sheep, have multichambered stomachs that harbour bacteria in one of the chambers to assist in the digestion of ◊cellulose.

stopping distance the minimum distance in which a vehicle can be brought to rest in an emergency from the moment that the driver notices danger ahead. It is the sum of the ◊thinking distance and the ◊braking distance.

stress and strain in the science of materials, measures of the deforming force applied to a body (stress) and of the resulting change in its shape (strain). For a perfectly elastic material, stress is proportional to strain (see ◊Hooke's law).

striped muscle or ***striated muscle*** muscle tissue responsible for most types of movement. It is attached to the skeleton, and contracts and relaxes under voluntary nervous control.

Under the microscope it has a distinctive striped appearance, resulting from the coordinated arrangement of protein molecules within the fibres. Unlike ◊involuntary muscle, fast contraction is possible.

strong acid acid that fully or almost dissociates in aqueous solution (see ◊dissociation). Strong acids include sulphuric acid and hydrochloric acid.

$$H_2SO_{4(aq)} \rightarrow 2H^+ + SO_4^{2-}$$

$$HCl_{(aq)} \rightarrow H^+ + Cl^-$$

The pH of strong acids lies between 1 and 3.

strong base ◊alkali that completely ionizes in aqueous solution (see ◊dissociation). Strong bases include sodium hydroxide and potassium hydroxide.

$$NaOH_{(aq)} \rightarrow Na^{+} + OH^{-}$$

The pH of strong bases lies between 10 and 14.

strontium soft, ductile, pale-yellow, metallic element, symbol Sr, atomic number 38, relative atomic mass 87.62. It is one of the ◊alkaline-earth elements, widely distributed in small quantities only as a sulphate or carbonate. Strontium salts burn with a red flame and are used in fireworks and signal flames.

The radioactive isotopes Sr-89 and Sr-90 (half-life 25 years) are some of the most dangerous products of the nuclear industry; they are fission products in nuclear explosions and in the reactors of nuclear power plants. Strontium is chemically similar to calcium and deposits in bones and other tissues, where the radioactivity is damaging.

structural formula diagrammatic representation of the atoms in a molecule; see ◊organic chemistry.

style in flowers, the part of the ◊carpel bearing the ◊stigma at its tip. In some flowers it is very short or completely lacking, while in others it may be long and slender, positioning the stigma in the most effective place to receive pollen.

sublimation the conversion of a solid to vapour without passing through the liquid phase.

Some substances that do not sublime at atmospheric pressure can be made to do so at low pressures. This is the principle of freeze-drying, during which ice sublimes at low pressure.

substrate in biochemistry, a compound or mixture of compounds acted on by an enzyme.

succession in ecology, series of changes that occurs in the structure and composition of the vegetation in a given area from the time it is first colonized by plants (***primary succession***), or after it has been disturbed by fire, flood, or clearing (***secondary succession***).

If allowed to proceed undisturbed, succession leads naturally to a stable climax community (for example, oak forest or savannah grassland) that is determined by the climate and soil characteristics of the area.

succession

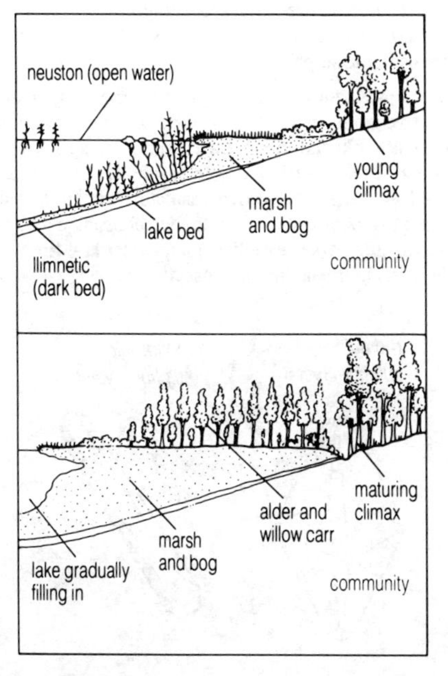

sugar soluble ◊carbohydrate, either a monosaccharide or disaccharide, usually with a sweet taste. The term is popularly used to refer to sucrose (cane or beet sugar) only.

Sugars may be described as reducing (all monosaccharides and the disaccharides maltose and lactose) or non-reducing (all other disaccharides) depending on whether they produce a positive result (change in colour) when tested with Benedict's solution (see ◊food test).

sulphate SO_4^{2-} salt or ester derived from sulphuric acid. Most sulphates are water soluble (the exceptions are lead, calcium, strontium, and barium sulphates), and require a very high temperature to decompose them.

The commonest sulphates seen in the laboratory are copper(II) sulphate ($CuSO_4$), iron(II) sulphate ($FeSO_4$), and aluminium sulphate ($Al_2(SO_4)_3$). The ion is detected in solution by using barium chloride or barium nitrate to precipitate the insoluble sulphate.

sulphur brittle, pale-yellow, non-metallic element, symbol S, atomic number 16, relative atomic mass 32.064. It occurs in three allotropic forms: two crystalline (rhombic and monoclinic) and one amorphous. It burns in air with a blue flame and a stifling odour. Insoluble in water but soluble in carbon disulphide, it is a good electrical insulator. Sulphur is widely used in the manufacture of sulphuric acid (used to treat phosphate rock to make fertilizers) and in making paper, matches, gunpowder and fireworks; in vulcanizing rubber; and in medicines and insecticides.

sulphur

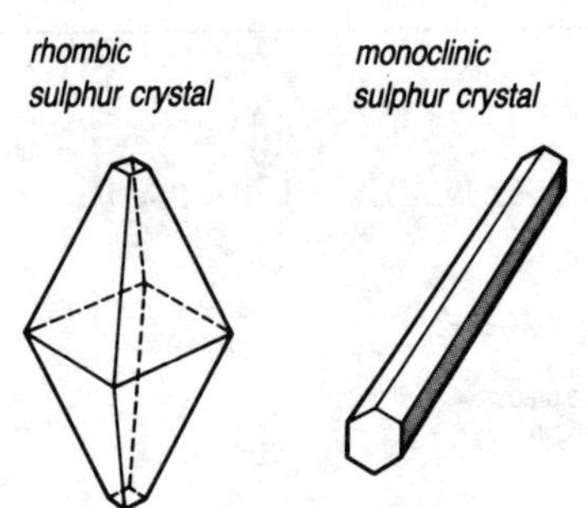

sulphur dioxide SO_2 pungent gas produced by burning sulphur in air or oxygen. It is used widely for disinfecting food vessels and equipment, and as a preservative in some food products. It occurs in industrial flue gases and is a major cause of ◊acid rain.

sulphuric acid H_2SO_4 dense, viscous, colourless liquid that is extremely corrosive, causing severe burns. Its principal properties are as follows.

affinity for water Left open to the air, the concentrated acid increases its volume as it absorbs water vapour from the atmosphere. It is used to dry gases (except ammonia).

dehydration The concentrated acid is a powerful dehydrating agent that will remove the water of crystallization from hydrated salts such as copper sulphate crystals.

$$CuSO_4.5H_2O \leftrightarrow CuSO_4 + 5H_2O$$

It will also remove the elements of water from many organic compounds, for example sucrose.

$$C_{12}H_{22}O_{11} - 11H_2O \rightarrow 12C$$

oxidation Hot, concentrated sulphuric acid will oxidize metals and some non-metals.

$$Cu + 2H_2SO_4 \rightarrow CuSO_4 + SO_2 + 2H_2O$$

$$C + 2H_2SO_4 \rightarrow CO_2 + 2SO_2 + 2H_2O$$

acidity Diluted, it acts as a strong, dibasic ◊acid, with the typical reactions of an acid.

$$H_2SO_4 + aq \leftrightarrow H^+_{(aq)} + HSO_4^-$$

Because of its chemical properties, sulphuric acid has many industrial uses. These include petrol refining and the manufacture of fertilizers, detergents, explosives, and dyes.

sulphur trioxide SO_3 colourless solid prepared by reacting sulphur dioxide and oxygen in the presence of a vanadium(V) oxide catalyst in the ◊contact process. It reacts violently with water to give sulphuric acid.

$$2SO_2 + O_2 \rightarrow 2SO_3$$

$$SO_3 + H_2O \rightarrow H_2SO_4$$

The violence of its reaction with water makes it extremely dangerous. In the contact process, it is dissolved in concentrated sulphuric acid to give oleum ($H_2S_2O_7$).

Sun the star at the centre of our Solar System—an immense ball of luminous gas consisting of 70% hydrogen and 30% helium. With a diameter of 1,392,000 km, it is about 1 million times larger than the Earth. The Sun's energy comes from the nuclear reactions that take place at its core, the fusion of hydrogen into helium. At the core the temperature is an intense 15 million °C, but this cools down to 5,800°C at the surface.

Sun

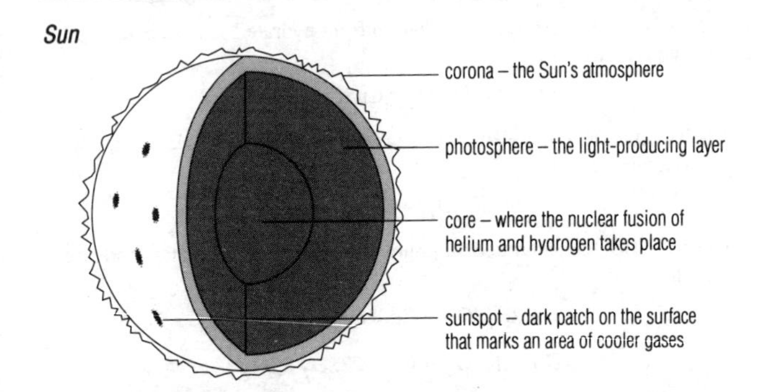

The Sun is the source of light and heat for the planets in the Solar System, determining their climate and weather conditions—without it there would be no life on Earth.

supersaturation the state of a solution that has a higher concentration of ◊solute than would normally be obtained in a ◊saturated solution.

Many solutes have a higher ◊solubility at high temperatures. If a hot saturated solution is cooled slowly, sometimes the excess solute does not come out of solution. This is an unstable situation and the introduction of a small solid particle will cause the solute to crystallize.

support the mechanism by which an organism holds itself and maintains its shape in relation to gravity.

Plants support themselves by keeping their cells turgid (see ◊turgor) and, in higher plants, by the rigidity of their vascular bundles (xylem and phloem). Large plants such as trees and shrubs undergo thickening (secondary growth) and strengthen their xylem vessels with lignin to form tough wood.

Animals tend to support themselves by means of a hard skeleton, which may be either internal (◊endoskeleton) or external (◊exoskeleton). It is important that such structures should not impede the motion of the animal, therefore most skeletons have movable joints. Exoskeletons are usually shed periodically to allow growth, while endoskeletons grow with the developing organism. Certain soft invertebrates such as earthworms make use of the pressure exerted on their body walls by their internal fluids to maintain a semi-rigid structure. Many aquatic organisms, both plant and animal, rely on the buoyancy provided by water for their support.

surface area:volume ratio the ratio of an animal's surface area (the area covered by its skin) to its total volume. This is high for small animals, but low for large animals such as elephants. The ratio is important for homeothermic (warm-blooded) animals because the amount of heat lost by the body is proportional to its surface area, whereas the amount generated is proportional to its volume. Very small birds and mammals, such as hummingbirds and shrews, lose a lot of heat and need a high intake of food to maintain their body temperature. Elephants, on the other hand, are in danger of overheating, which is why they have no fur.

surface tension the property that causes the surface of a liquid to behave as if it were covered with a weak elastic skin; this is why a needle can float on water. It is caused by the exposed surface's tendency to contract to the smallest possible area because of unequal cohesive forces between molecules at the surface. Allied phenomena include the formation of droplets, the concave profile of a meniscus, and the ◊capillary action by which water soaks into a sponge.

suspension colloidal state consisting of small solid particles dispersed in a liquid or gas (see ◊colloid).

suspensory ligament in the ◊eye, a ring of fibre supporting the lens. The suspensory ligaments are attached to the ciliary muscles, the circle of muscle mainly responsible for changing the shape of the lens during ◊accommodation. If the ligaments are put under tension, the lens becomes flatter, and therefore able to focus on objects in the far distance.

sweat gland gland within the skin of mammals that produces surface perspiration. In primates, sweat glands are distributed over the whole body, but in most other mammals they are more localized; for example, in cats and dogs, they are restricted to the feet and around the face.

switch device used to turn an electric current on and off, usually by closing or opening a circuit, or to change the route of a particular current. For example, switches are used to turn a light on or off, to select a channel on a television set, or to set the cooking time on an electric oven.

symbiosis any close relationship between two organisms of different species, and one where both partners benefit from the association. A well-known example is the pollination relationship between insects and flowers, where the insects feed on nectar and carry pollen from one flower to another.

synapse the junction between two ◊neurons (nerve cells), or between a neuron and a muscle (a neuromuscular junction), across which a nerve impulse is transmitted. The two cells involved are not in direct contact but separated by a narrow gap called the ***synaptic cleft***. The threadlike extension, or ◊axon, of the transmitting neuron has a slightly swollen terminal point, the ***synaptic knob***. This forms one half of the synaptic junction and houses membrane-bound vesicles, which contain a chemical ◊neurotransmitter. When nerve impulses reach the knob, the vesicles release the transmitter and this flows across the gap and binds itself to special receptors on the receiving cell's membrane.

synapse

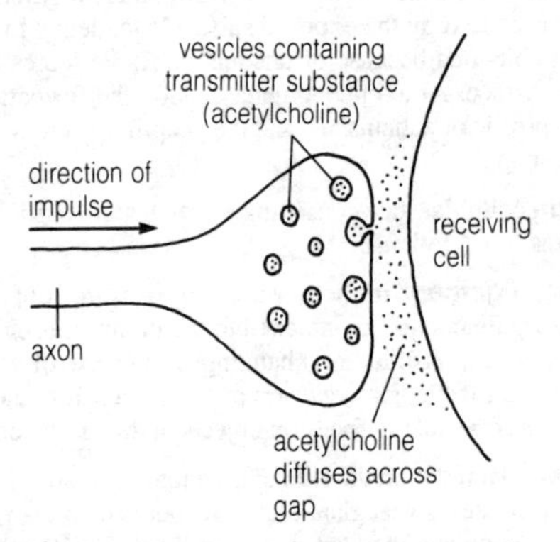

synoptic chart weather chart in which symbols are used to represent the weather conditions experienced over an area at a particular time. Information gathered from weather stations is used to draw up the charts, with lines called isobars being used to link areas of equal atmospheric pressure and other symbols being used to represent fronts of cold and warm air.

synovial fluid a viscous yellow fluid that bathes movable joints between the bones of vertebrates. It nourishes and lubricates the ◊cartilage at the end of each bone.

synoptic chart

synoptic chart of a typical depression

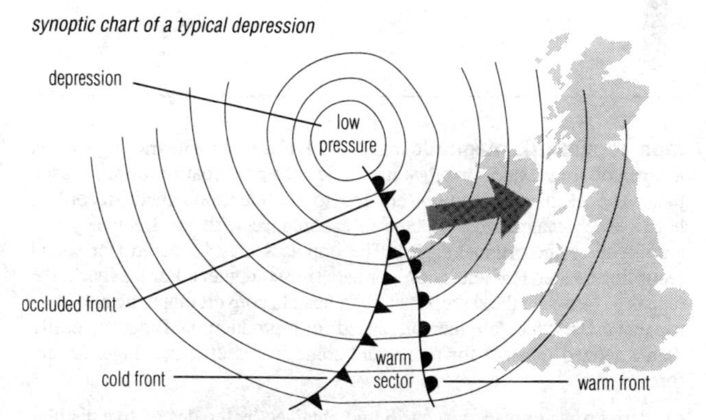

synthesis the formation of a substance or compound from simpler compounds. The synthesis of a drug can involve several stages from the initial material to the final product; the complexity of these stages is a major factor in the cost of production.

system collection of organs and tissues that work together to perform one or more coordinated functions. The blood system involves many structures —the heart, the blood, the blood vessels—all working together to transport materials around the body. Similarly, the nervous system is able to respond to a stimulus because the muscles, brain, and nerves are working as a coordinated system. Other systems include the digestive and urinary systems.

T

tape recording, magnetic method of recording electric signals on a layer of iron oxide, or other magnetic material, coating a thin plastic tape. The electrical impulses are fed to the electromagnetic recording head, which magnetizes the tape in accordance with the frequency and amplitude of the original signal. The impulses may be audio (for sound recording), video (for television), or data (for computer). For playback, the tape is passed over the same, or another, head to convert magnetic into electrical signals, which are then amplified for reproduction. Tapes are easily demagnetized (erased) for reuse, and come in cassette, cartridge, or reel form.

tar dark brown or black viscous liquid obtained by the destructive distillation of coal, shale, and wood. Tars consist of a mixture of hydrocarbons, acids, and bases. See also ◊coal tar.

taste sense that detects some of the chemical constituents of food. The human ◊tongue can distinguish only four basic tastes (sweet, sour, bitter, and salty) but it is supplemented by the nose's sense of smell. What we refer to as taste is really a composite sense made up of both taste and smell.

Teflon trade name for ◊polytetrafluoroethene (PTFE), a tough, waxlike, heat-resistant plastic used for coating non-stick cookware and in gaskets and bearings.

telecommunications communications over a distance, generally by electronic means. Today it is possible to communicate with most countries by telephone cable, or by satellite or microwave link, with over 100,000 simultaneous conversations and several television channels being carried by the latest satellites.

Integrated-services digital network (ISDN) is a system that transmits voice and image data on a single transmission line by changing them into digital signals, making videophones and high-quality fax possible; the world's first large-scale centre of ISDN began operating in Japan in 1988.

telephone instrument for communicating by voice over long distances. The transmitter (mouthpiece) consists of a carbon microphone, with a diaphragm that vibrates when a person speaks into it. The diaphragm vibrations compress grains of carbon to a greater or lesser extent, altering their resistance to an electric current passing through them. This sets up variable electrical signals, which travel along the telephone lines to the receiver of the person being called. There they cause the magnetism of an electromagnet to vary, making a diaphragm above the electromagnet vibrate and give out sound waves similar to those that entered the mouthpiece originally.

telephone

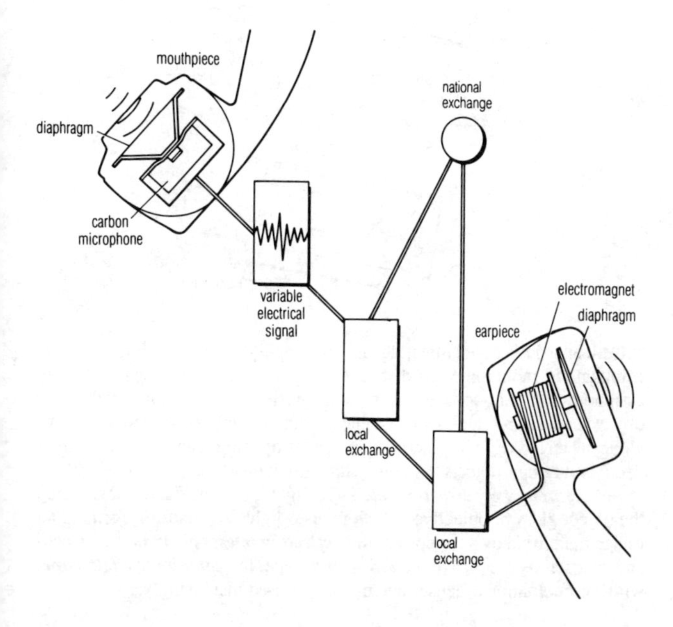

telescope

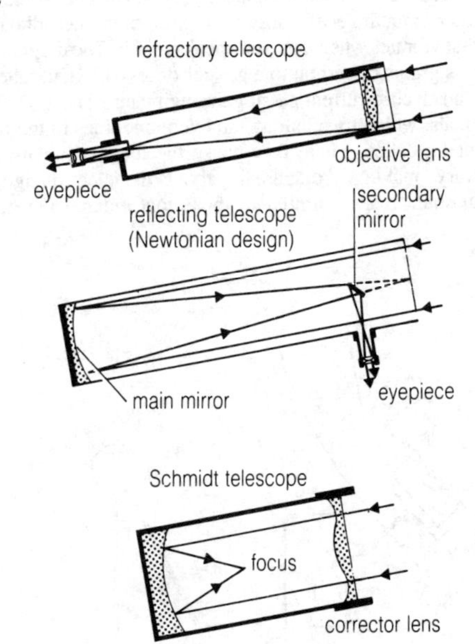

telescope device for collecting and focusing light and other forms of electromagnetic radiation from distant objects. It is a major research tool in astronomy. A telescope produces a magnified image, which makes the object seem nearer, and it shows objects fainter than can be seen by the eye alone. A telescope with a large aperture, or opening, can distinguish finer detail and fainter objects than one with a small aperture.

In a ***refractory*** or ***refracting telescope***, light is collected by a ◊lens called the object glass or objective, which focuses light down a tube, forming an image magnified by an eyepiece. In a ***reflecting telescope***, light is collected and focused by a concave mirror. A third type, the ***catadioptric telescope***, with a combination of lenses and mirrors, is used increasingly.

Large telescopes can now be placed in orbit above the distorting effects of the Earth's atmosphere. Telescopes in space have been used to study infrared, ultraviolet, and X-ray radiation that does not penetrate the atmosphere but carries much information about the births, lives, and deaths of stars and galaxies. The 2.4-m/94-in Hubble Space Telescope, launched 1990, can see the sky more clearly than can any telescope on Earth, despite an optical defect that impairs its performance.

television (TV) reproduction at a distance by radio waves of visual images. For transmission, a television camera converts the pattern of light it takes in into a pattern of electrical charges. This is scanned line by line by a beam of electrons from an electron gun, resulting in variable electrical signals that represent the visual picture. These vision signals are combined with a radio carrier wave and broadcast. The TV aerial picks up the wave and feeds it to the receiver (TV set). This separates out the vision signals, which pass to a cathode-ray tube. The vision signals control the strength of a beam of electrons from an electron gun, aimed at the screen and making it glow more or less brightly. At the same time the beam is made to scan across the screen line by line, mirroring the action of the gun in the TV camera. The result is a re-creation of the pattern of light that entered the camera. Thirty pictures are built up each second with interlaced scanning in North America (25 in Europe), with a total of 525 lines in North America and Japan (625 lines in Europe).

tellurium silver-white, weakly metallic element, symbol Te, atomic number 52, relative atomic mass 127.60. Chemically it is similar to sulphur and selenium, and it is considered as one of the sulphur group. It occurs naturally in telluride minerals, and is used in colouring glass blue–brown, in the electrolytic refining of zinc, in electronics, and as a catalyst in refining petroleum.

temperature the state of hotness or coldness of a body, and the condition that determines whether or not it will transfer heat to, or receive heat from, another body. It is measured in degrees Celsius (before 1948 called centigrade), kelvin, or degrees Fahrenheit.

temperature regulation the ability of an organism to control its body temperature.

Although some plants have evolved ways of resisting extremes of temperature, sophisticated mechanisms for maintaining the correct temperature are found only in multicellular animals. Such mechanisms may be behavioural, as when a lizard moves into the shade to cool down. Mammals and birds have internal control and are known as ***homeotherms***. These animals are insulated with fat, hair, or feathers to conserve heat produced by metabolic activities. Other adaptations allow heat to leave the body when the animal is in danger of overheating, for instance during intense activity. Such mechanisms include sweating, increased flow of blood through the skin (vasodilation), and panting.

temperature scale scale marked on a thermometer that gives a reading of temperature. The scales most widely used today are the ◊Celsius (or centigrade) scale, the ◊kelvin (or absolute) scale, and the ◊Fahrenheit scale. Each scale has at least two fixed points that are used to calibrate all those thermometers using the scale. The fixed points most commonly used are the ice point and steam point of water. On the Celsius scale these are designated 0°C and 100°C respectively, the interval between them being divided equally into 100 units or degrees.

temporary hardness hardness of water that is removed by boiling (see ◊hard water).

tendon or ***sinew*** a cord of tough, fibrous connective tissue that joins muscle to bone in vertebrates. Tendons are largely composed of the protein collagen, and because of their inelasticity are very efficient at transforming muscle power into movement.

tension reaction force set up in a body that is subjected to stress. In a stretched string or wire it exerts a pull that is equal in magnitude but opposite in direction to the stress being applied at its ends. Tension originates in the net attractive intermolecular force created when a stress causes the mean distance separating a material's molecules to become greater than the equilibrium distance. It is measured in newtons.

terminal voltage potential difference (pd) or voltage across the terminals of a power supply, such as a battery of cells. When the supply is not connected in circuit its terminal voltage is the same as its ◊electromotive force (emf); however, as soon as it begins to supply current to a circuit its terminal voltage falls because some electric potential energy is lost in driving current against the supply's own ◊internal resistance. As the current flowing in the circuit is increased the terminal voltage of the supply falls.

territory in behaviour, a fixed area from which an animal or group of animals excludes other members of the same species. Animals may hold territories for many different reasons; for example, to provide a constant food supply, to monopolize potential mates, or to ensure access to refuges or nest sites. The size of a territory depends in part on its function: some nesting and mating territories may be only a few square metres, whereas feeding territories may be as large as hundreds of square kilometres.

testa the outer coat of a seed, formed after fertilization of the ovule. It has a protective function and is usually hard and dry. In some cases the coat is adapted to aid dispersal, for example by being hairy. Humans have found uses for many types of testa, including the fibre of the cotton seed.

testis (plural ***testes***) the organ that produces ◊sperm in male animals. In vertebrates it is one of a pair of oval structures that are usually internal, but in mammals (other than elephants and marine mammals), the paired testes (or testicles) descend from the body cavity during development, to hang outside the abdomen in a sac called the ***scrotum***.

testosterone in vertebrates, hormone secreted chiefly by the testes. It promotes the development of secondary sexual characteristics in males. In animals with a breeding season, the onset of breeding behaviour is accompanied by a rise in the level of testosterone in the blood.

tetrachloromethane or ***carbon tetrachloride*** CCl_4 chlorinated organic compound that is a very efficient solvent for fats and greases. It is a toxic solvent and its use is restricted.

tetraethyl lead $Pb(C_2H_5)_4$ compound added to leaded petrol as a component of antiknock to increase the efficiency of combustion in car engines. It is a colourless liquid that is insoluble in water but soluble in organic solvents such as benzene, ethanol, and petrol.

tetrapod four-limbed vertebrate. The tetrapods include mammals, birds, reptiles, and amphibians. Birds are included because they evolved from four-legged ancestors, the forelimbs having become modified to form wings. Even snakes are tetrapods, since their lack of limbs is secondary.

thermal decomposition irreversible breakdown of a compound into simpler substances by heating it. The catalytic ◊cracking of hydrocarbons is an example.

thermal dissociation reversible breakdown of a compound into simpler substances by heating it (see ◊dissociation). The splitting of ammonium chloride into ammonia and hydrogen chloride is an example. On cooling, they recombine to form the salt.

$$NH_4Cl_{(s)} \leftrightarrow NH_{3(g)} + HCl_{(g)}$$

thermistor device in which electrical ◊resistance falls as temperature rises. The current passing through a thermistor increases rapidly as its temperature rises, and so they are used in electrical thermometers.

thermite reaction reaction between powdered aluminium and iron(III) oxide. This mixture is called ***thermite***. When heated, the aluminium reduces the iron(III) oxide to iron in a highly exothermic reaction.

$$2Al + Fe_2O_3 \rightarrow Al_2O_3 + 2Fe$$

The heat produced is enough to melt the iron produced in the reaction, so the mixture is sometimes used in welding.

thermometer instrument for measuring temperature. There are many types, designed to measure different temperature ranges to varying degrees

thermometer

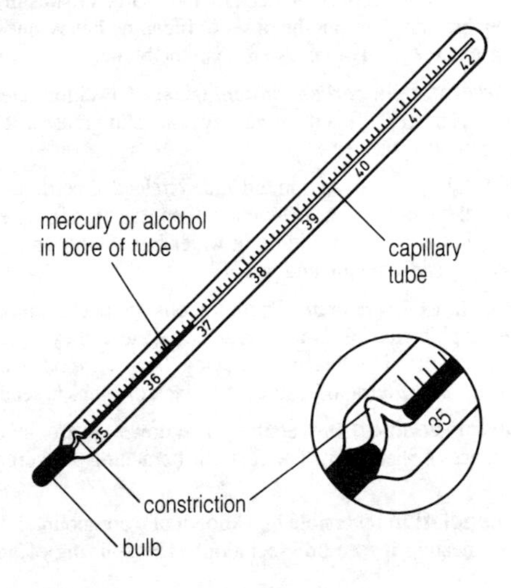

of accuracy. Each makes use of a different physical effect of temperature. See ◊temperature scale.

A ***clinical thermometer*** is used in medicine to measure body temperature. It has a limited temperature range of 35–42°C, but gives readings in steps of 0.1 or 0.2 degrees. A narrow constriction in the tube just above the bulb allows the thermometer to maintain a constant reading even after it has been removed from the patient: the mercury or alcohol is able to expand upwards into the bore of the tube as its temperature rises, but cannot, as it cools, return to the bulb unless the thermometer is shaken sharply.

thermosetting plastic or ***thermoset*** type of ◊plastic that remains rigid when set, and does not soften with heating. Thermosets have this property because the long-chain polymer molecules cross-link with each other to give a rigid structure. Examples include Bakelite, resins, melamine, and urea–formaldehyde resins.

thermosoftening plastic or ***thermoplastic*** type of ◊plastic that always softens on repeated heating. Thermoplastics include polyethene, polystyrene, nylon, and polyester.

thiamine or ***vitamin*** B_1 ◊vitamin of the B complex. Its absence from the diet causes the disease beriberi.

thinking distance distance travelled by a vehicle from the moment its driver notices danger ahead to the moment its brakes are applied. Thinking distance is directly proportional to the speed of the vehicle, and is also increased if the driver is tired or under the influence of alcohol. The total distance required to bring a vehicle to rest (its ◊stopping distance) is the sum of the thinking distance and the ◊braking distance.

thorax in tetrapod vertebrates, the part of the body containing the heart and lungs, and protected by the rib cage. It is separated from the abdomen by the muscular diaphragm in mammals.

In arthropods, such as insects, the thorax is the middle part of the body, between the head and abdomen.

throat in humans, the passage that leads from the back of the nose and mouth to the ◊trachea and ◊oesophagus. It includes the ◊pharynx and the ◊larynx, the latter being at the top of the trachea. The term is also used to mean the front part of the neck, both in humans and other vertebrates.

thunderstorm storm caused by strongly rising air currents. Lightning occurs when large amounts of electric charge built up on a cloud discharge to another cloud, or the Earth. Air in the path of this lightning ionizes and expands suddenly, creating the loud sound known as thunder.

thyroid ◊endocrine gland situated in the neck in front of the trachea. It secretes several hormones, among them thyroxin, a hormone that stimulates growth, metabolism, and other functions of the body.

tibia the anterior of the pair of bones found between the ankle and the knee. In humans, the tibia is the shinbone.

ticker-timer device used to time the motion of an object by printing dots at regular time-intervals on a length of ticker-tape attached to that object. When the timer is driven (through a transformer) by mains electricity, its printing head will vibrate with a frequency of 50Hz, printing 50 dots a second on the tape. The movement of the object pulls the ticker-tape past the printing head, and as the speed of the object increases, the distance between adjacent dots grows wider. If the marked tape is cut up into lengths that represent the distance travelled by an object in a certain time (the object's speed), an experimental speed–time graph called a ***tape chart*** can be constructed, which can then be used to measure the object's acceleration and other aspects of its motion.

tidal energy energy derived from the tides. The tides gain their potential energy from the gravitational forces acting between the Earth and the Moon. If water is trapped at a high level during high tide, perhaps by means of a barrage across an estuary, it may then be gradually released and its associated ◊gravitational potential energy exploited to drive turbines and generate electricity. Several schemes have been proposed for the Bristol Channel, but environmental concerns as well as construction costs have so far prevented any decision from being taken.

tide rise and fall of sea level due to the gravitational forces of the Moon and Sun. High water occurs at an average interval of 12 h 24 min 30 s. The highest or ***spring tides*** are at or near new and full Moon; the lowest or ***neap tides*** when the Moon is in its first or third quarter. Some seas, such as the Mediterranean, have very small tides.

timbre in music, the characteristic quality of a ◊sound. Different instruments sound different when they play a note of the same pitch because the ◊harmonics present, and their relative amplitudes, vary from instrument to instrument, and different transient sounds (sounds that last only a short time) occur at the beginning of the note.

tin soft, silver-white, malleable and somewhat ductile, metallic element, symbol Sn, atomic number 50, relative atomic mass 118.69. Tin exhibits ◊allotropy, having three forms: the familiar lustrous metallic form above 13.2°C; a brittle form above 161°C; and a grey powder form below 13.2°C (commonly called tin pest or tin disease). The metal is quite soft (slightly harder than lead) and can be rolled, pressed, or hammered into extremely thin sheets; it has a low melting point. In nature it occurs rarely as a free metal. It resists corrosion and is therefore used for coating and plating other metals.

Tin and copper smelted together form the oldest desired alloy, bronze; since the Bronze Age (3,500 BC) that alloy has been the basis of both useful and decorative materials. Tin is also alloyed with metals other than copper to make solder and pewter.

tissue a collection of cells that perform a similar function. Thus, nerve and muscle are different kinds of tissue in animals.

tissue culture process by which cells from a plant or animal are removed from the organism and grown under controlled conditions in a sterile medium containing all the necessary nutrients. Tissue culture can provide information on cell growth and differentiation, and is also used in the propagation of plants. See also ◊meristem.

titanium strong, light-weight, silver-grey, metallic element, symbol Ti, atomic number 22, relative atomic mass 47.90. The ninth most abundant element in the Earth's crust, its compounds occur in practically all igneous rocks and their sedimentary deposits. It is very strong and resistant to corrosion, and is used in building high-speed aircraft and spacecraft, and in hip-replacement joints; it is also widely used in making alloys, as it unites with almost every metal except copper and aluminium. Titanium oxide is used in high-grade white pigments.

titration technique used to find the concentration of one compound in a solution by determining how much of it will react with a known amount of another compound in solution.

One of the solutions is measured by ◊pipette into the reaction vessel. The other is added a little at a time from a ◊burette. The end-point of the reaction is determined with an ◊indicator or an electrochemical device.

tongue in tetrapod vertebrates, a muscular organ usually attached to the floor of the mouth. It has a thick root attached to a U-shaped bone, and is

titration

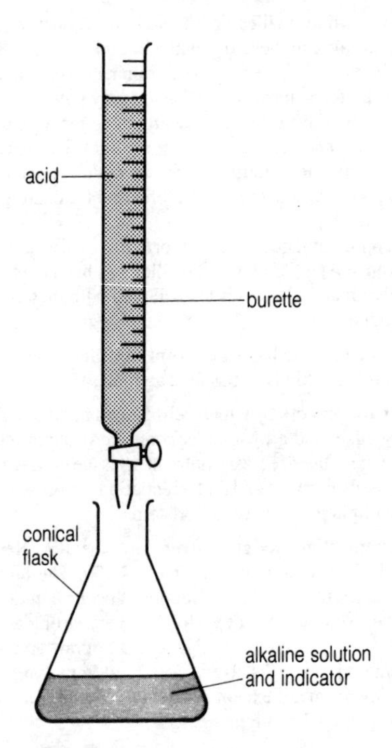

covered with a ◊mucous membrane containing nerves and 'taste buds'. It directs food to the teeth and into the throat for chewing and swallowing. In humans, it is crucial for speech; in other animals, for lapping up water and for grooming, among other functions.

tonsils in higher vertebrates, masses of lymphoid tissue situated at the back of the mouth and throat, and on the rear surface of the tongue. The tonsils contain many ◊lymphocytes and are part of the body's defence system against infection.

tooth

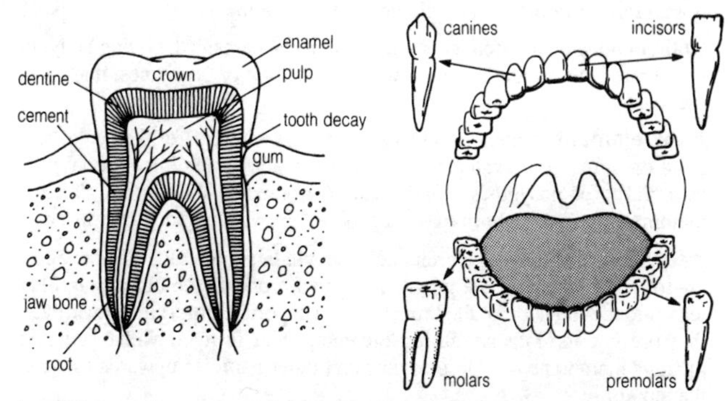

tooth in vertebrates, one of a set of hard, bonelike structures in the mouth, used for biting and chewing food, and in defence and aggression. In humans, the first set (20 milk teeth) appear from age six months to two and a half years. The permanent ◊dentition replaces these from the sixth year onwards, the wisdom teeth (third molars) sometimes not appearing until the age of 25 or 30. Adults have 32 teeth: two incisors, one canine, two premolars, and three molars on each side of each jaw. Each tooth consists of an enamel coat (hardened calcium deposits), dentine (a thick, bonelike layer), and an inner pulp cavity, housing nerves and blood vessels. Mammalian teeth have roots surrounded by cement, which fuses them into their sockets in the jawbones. The neck of the tooth is covered by the ◊gum, so that the enamel-covered crown rises out above the gum line.

touch sensation produced by specialized nerve endings (receptors) in the skin. Some respond to light pressure, others to heavy pressure. Temperature detection may also contribute to the overall sensation of touch. Many animals, such as nocturnal ones, rely on touch more than humans do. Some have specialized organs of touch that project from the body, such as whiskers or antennae.

toxic poisonous or harmful. Lead from car exhausts, asbestos, and chlorinated solvents are some examples of toxic substances that occur in the environment; generally the effects take sometime to become apparent (anything from a few hours to many years). The cumulative effects of toxic waste pose a serious threat to the ecological stability of the planet.

toxin or ***poison*** any chemical molecule that can damage the living body. In vertebrates, toxins are broken down by the action of ◊enzymes, mainly in the liver.

trace element element necessary in minute quantities for the health of a plant or animal. For example, magnesium, which occurs in chlorophyll, is essential to photosynthesis, and iodine is needed by the thyroid gland of mammals for making hormones that control growth and body chemistry.

trachea in air-breathing vertebrates, the windpipe, the tube that conducts air from the larynx in the throat to the two bronchi (see ◊bronchus) that lead into the lungs. Like the bronchi, it is supported with rings of cartilage that prevent its collapse during breathing, and is lined with a ciliated mucous membrane that propels dust and other particles upwards towards the mouth.

In insects, the tracheae are large air-filled tubes that conduct air from the spiracles, small pores in the exoskeleton, into the body. Like vertebrate tracheae, they are supported by cartilaginous rings. The tubes extend throughout the insect, occasionally opening out in to large air-filled sacs. The thinnest branches of the tubes are the ***tracheoles***, which have thin walls and are often filled with fluid. These smaller tubes are the true sites of gas exchange.

transect in ecology, an imaginary line drawn across a piece of land, along which animal and plant diversity can be sampled. In fieldwork surveys the line can be made real by stretching a string between poles, and the diversity of species sampled by using a ◊quadrat. The data obtained may then be used to show how the wildlife varies in the area under study, and perhaps related to changes in abiotic and biotic conditions. For example, the progressive change in diversity revealed by a transect survey of a beach, from the sand dunes to the high-tide mark, might be related to changing conditions of water exposure.

transformer device in which, by electromagnetic induction, an alternating current (AC) of one voltage is transformed to another voltage, without change of ◊frequency. Transformers are widely used in electrical apparatus of all kinds, and in particular in power transmission where high voltages and low currents are utilized.

A transformer has two coils, a primary for the input and a secondary for the output, wound on a common iron core. The ratio of the primary to the secondary voltages (and currents) is directly (and inversely) proportional to the number of turns in the primary and secondary coils.

If the numbers of turns in the primary and secondary coils are n_1 and n_2, the primary and secondary voltages are V_1 and V_2, and the primary and secondary currents are I_1 and I_2, then the relation between these quantities may be expressed as:

$$V_1/V_2 = I_2/I_1 = n_1/n_2$$

transfusion intravenous delivery of blood or blood products (plasma, red cells) into a patient's circulation to make up for deficiencies due to disease, injury, or surgery.

Blood transfusion was a highly risky procedure until the discovery of ◊blood groups indicated the need to check that the donated blood was compatible with that of the patient.

transistor solid-state electronic component, made of ◊semiconductor material, that can regulate a current passing through it. A transistor can act as an amplifier, oscillator, photocell, or switch, and usually operates on a very small amount of power. Transistors commonly consist of a tiny sandwich of ◊silicon or germanium, alternate layers having different electrical properties. A crystal of pure silicon or germanium would act as an insulator (non-conductor).

By introducing impurities in the form of atoms of other materials (for example, boron, arsenic, or indium) in minute amounts, the layers may be made either ***n-type*** (negative), having an excess of electrons, or ***p-type*** (positive), having a deficiency of electrons. This enables electrons to flow from one layer to another in one direction only.

Transistors have had a great impact on the electronics industry, and are now made in thousands of millions each year. They perform many of the same functions as the thermionic valve, but have the advantages of greater reliability, long life, compactness, and instantaneous action, no warming-up period being necessary. They are widely used in most electronic equipment, including portable radios and televisions, computers, satellites, and space research, and are the basis of the ◊integrated circuit (silicon chip).

transition metal any of a group of metallic elements that have incomplete inner electron shells and exhibit variable valency—for example, cobalt, copper, iron, and molybdenum. They are excellent conductors of electricity, and generally form highly coloured compounds.

translocation the movement of soluble materials through ◊vascular plants.

Roots, stems, and leaves all possess ◊vascular bundles, groups of hollow fibres that transport fluids and dissolved substances. Two types of tube exist within the bundles: ◊xylem for the upward transport of inorganic materials from root to leaf, and ◊phloem for the downward movement of organic substances formed during photosynthesis. Lower plants such as mosses lack these structures and are therefore less able to grow in dry areas.

transmission of electrical power see ◊power transmission.

transpiration the loss of water from a plant by evaporation. Most water is lost from the leaves through pores known as stomata (see ◊stoma), whose primary function is to allow ◊gas exchange between the internal plant tissues and the atmosphere. Transpiration from the leaf surfaces causes a continuous upward flow of water from the roots via the ◊xylem, which is known as the ***transpiration stream***. See also ◊xerophyte.

transverse wave wave in which the displacement of the medium's particles is at right angles to the direction of travel of the wave motion. It is characterized by its alternating crests and troughs. Simple water waves, such as the ripples produced when a stone is dropped into a pond, are transverse waves, as are the waves on a vibrating string.

All ◊electromagnetic waves have a transverse waveform; their electric and magnetic fields (rather than the particles of their medium) vibrate at right angles to their direction of travel.

transverse wave

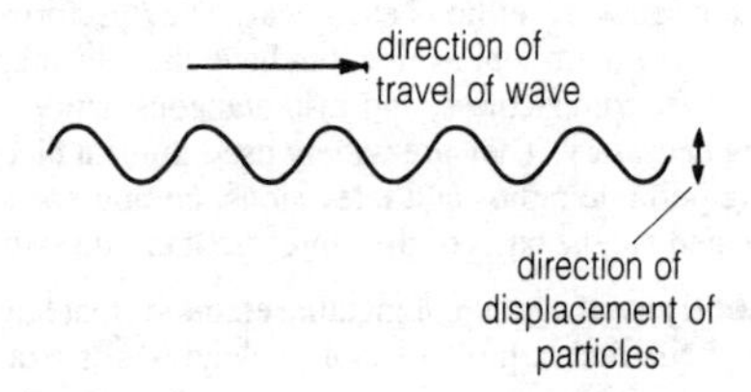

tree perennial plant with a woody stem, usually a single stem or 'trunk', made up of ◊wood, and protected by an outer layer of ◊bark. It absorbs water through a ◊root system. There is no clear dividing line between shrubs and trees, but sometimes a minimum height of 6m is used to define a tree.

tricuspid valve a flap of tissue situated on the right side of the ◊heart between the atrium and the ventricle. It prevents blood flowing backwards when the ventricle contracts.

triple bond three covalent bonds between adjacent atoms, as in the ◊alkynes ($-C\equiv C-$)

tritium radioactive isotope of hydrogen, three times as heavy as ordinary hydrogen, consisting of one proton and two neutrons. It has a half-life of 12.5 years.

trophic level in ecology, the position occupied by a species (or group of species) in a ◊food chain. The main levels are ***primary producers*** (photosynthetic plants), ***primary consumers*** (herbivores), ***secondary consumers*** (carnivores), and ***decomposers*** (bacteria and fungi).

truth table in electronics, a diagram representing the properties of a ◊logic gate.

trypsin an enzyme in the gut responsible for the digestion of protein molecules. It is secreted by the pancreas in an inactive form known as trypsinogen. Activation into working trypsin occurs only in the small intestine, owing to the action of another enzyme, enterokinase, secreted by the wall of the duodenum. Unlike the digestive enzyme pepsin, found in the stomach, trypsin does not require an acidic environment.

tuber swollen region of an underground stem or root, usually modified for storing food. The potato is a stem tuber, as shown by the presence of terminal and lateral buds, the 'eyes' of the potato. Root tubers lack these. Both types of tuber can give rise to new individuals and so provide a means of ◊vegetative reproduction.

Unlike a bulb, a tuber persists for one season only; new tubers developing on a plant in the following year are formed in different places. See also ◊rhizome.

tumour overproduction of cells in a specific area of the body, often leading to a swelling or lump. Tumours are classified as ***benign*** or ***malignant*** (see ◊cancers).

Benign tumours grow more slowly, do not invade surrounding tissues, do not spread to other parts of the body, and do not usually recur after removal.

tungsten hard, heavy, grey-white, metallic element, symbol W, atomic number 74, relative atomic mass 183.85. It occurs in the minerals wolframite, scheelite, and hubertite. It has the highest melting point of any metal (3,410°C) and is added to steel to make it harder, stronger, and more elastic; its other uses include high-speed cutting tools, electrical elements, and thermionic couplings. Its salts are used in the paint and tanning industries.

turbine engine in which steam, water, gas, or air is made to spin a rotating shaft by pushing on angled blades, like those of a fan. Turbines are among the most powerful of machines. Steam turbines are used to drive generators in power stations; water turbines spin the generators in hydroelectric power plants; and gas turbines (as jet engines) power most aircraft and drive machines in industry.

turgor the rigid condition of a plant caused by the fluid contents of a plant cell exerting a mechanical pressure against the cell wall. Turgor supports plants that do not have woody stems.

turgor

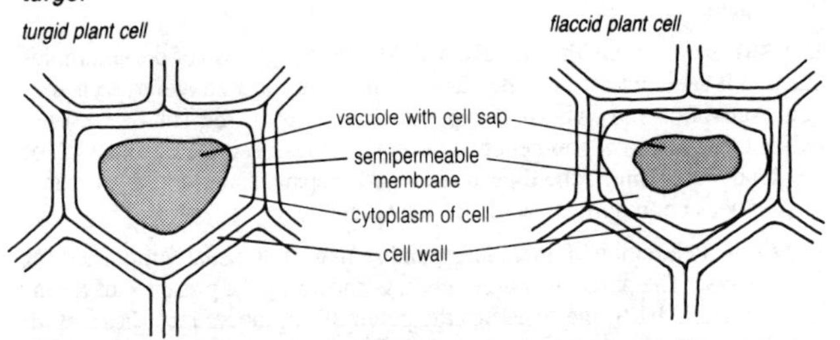

twin one of two young produced from a single pregnancy. Human twins may be genetically identical, having been formed from a single fertilized ovum (egg) that split into two cells, both of which became implanted. Non-identical twins are formed when two ova are fertilized at the same time.

tympanic membrane or ***ear drum*** membrane capable of vibrating in response to sound waves passing from the outer ◊ear. The vibrations of the membrane are transferred to the tiny bones of the inner ear, which themselves pass vibrations through to the inner ear and ◊cochlea.

ultrasound pressure waves similar in nature to sound waves but occurring at frequencies above 20,000 Hz (vibrations per second), the approximate upper limit of human hearing (15–16 Hz is the lower limit). Ultrasonics is concerned with the study and practical application of these phenomena.

ultraviolet radiation light rays invisible to the human eye, of wavelengths from about 4×10^{-7} to 5×10^{-9} metres (where the ◊X-ray range begins). Physiologically, they are extremely powerful, producing sunburn and causing the formation of vitamin D in the skin.

Ultraviolet rays are strongly germicidal and may be produced artificially by mercury vapour and arc lamps for therapeutic use. The radiation may be detected with ordinary photographic plates or films down a wavelength of about to 2×10^{-6} metres. It can also be studied by its fluorescent effect on certain materials.

umbilical cord in mammals, the connection between the ◊embryo and the ◊placenta. It has one vein and two arteries, transporting oxygen and nutrients to the developing young, and removing waste products. At birth, the connection between the young and the placenta is no longer necessary. The umbilical cord drops off or is severed, leaving a scar called the navel.

unicellular organism organism consisting of a single cell. Most are invisible without a microscope but a few, such as *Amoeba*, may be visible to the naked eye. The main groups of unicellular organisms are bacteria, protozoa, unicellular algae, and unicellular fungi, or yeasts.

unit standard quantity in relation to which other quantities are measured. There have been many systems of units. Some ancient units, such as the day, the foot, and the pound, are still in use. ◊SI units, the latest version of the metric system, are widely used in science.

universal indicator mixture of pH ◊indicators, each of which changes colour at a different pH value. The indicator is a different colour at different values of pH, ranging from red (at pH 1) to purple (at pH 13).

The pH of a substance may be found by adding a few drops of universal indicator and noting the colour, or by dipping an absorbent paper strip that has been impregnated with the indicator.

universe all of space and its contents. The study of the universe is called cosmology. The universe is thought to be between 10 and 20 billion years old. It is mostly empty space with billions of galaxies dotted around it. The most distant galaxies lie 10 billion light years or more away from the Earth, and are moving further apart as the universe expands. There are several theories as to how the universe came into being. One of these, the Big Bang theory, suggests that the universe was created in a single explosive event and has been expanding ever since.

unsaturated compound compound in which two adjacent atoms are bonded by two or more covalent bonds.

Examples are ◊alkenes and ◊alkynes, where the two adjacent atoms are both carbon. The laboratory test for unsaturated compounds is to add bromine water, which then becomes decolorized.

unsaturated solution solution that is capable of dissolving more solute than it already contains at the same temperature.

upthrust upwards force experienced by all objects that are totally or partly immersed in a fluid (liquid or gas). It acts against the weight of the object, and, according to Archimedes' principle, is always equal to the weight of the fluid displaced by that object. An object will float when the upthrust from the fluid is equal to its weight. See ◊floating.

uranium hard, lustrous, silver-white, malleable and ductile, radioactive, metallic element of the ◊actinide series, symbol U, atomic number 92, relative atomic mass 238.029. It is the most abundant radioactive element in the Earth's crust, its decay giving rise to essentially all the radioactive elements in nature; its final decay product is the stable element lead. Uranium combines readily with most elements to form compounds that are extremely poisonous. The chief ore is ◊pitchblende.

Uranium was long considered to be the element with the highest atomic number to occur in nature. The isotopes U-238 and U-235 have been used to help determine the age of the Earth. It is one of the three elements capable of ◊nuclear fission (the other two are plutonium and thorium). Uranium-238, which comprises about 99% of all naturally occurring uranium, has a half-life of 4.51×10^9 years. Because of its abundance, it is the isotope from which plutonium is produced in breeder nuclear reactors. The isotope U-235 has a half-life of 7.13×10^8 years and comprises about 0.7% of naturally occurring uranium; it is used directly as a fuel for nuclear reactors and in the manufacture of nuclear weapons.

urea $CO(NH_2)_2$ waste product formed in the mammalian liver when nitrogen compounds are broken down. It is excreted in urine. When purified, it is a white, crystalline solid. In industry it is used to make urea–formaldehyde plastics (or resins), pharmaceuticals, and fertilizers.

ureter tube connecting the kidney to the bladder. Its wall contains fibres of ◊involuntary muscle, whose contractions aid the movement of urine out of the kidney.

urethra in mammals, a tube connecting the bladder to the exterior. It carries urine, and in males, semen.

urinary system the system of organs that removes nitrogeneous waste products and excess water from the bodies of animals. In mammals, it consists of a pair of kidneys, which produce urine; ureters, which drain the kidneys; a bladder, which stores the urine before its discharge, and a urethra, through which the urine is expelled.

urine an amber-coloured fluid made by the kidneys from the blood. It contains excess water, salts, proteins, waste products in the form of urea, a pigment, and some acid.

uterus a hollow muscular organ of female mammals, located between the bladder and rectum, and connected to the Fallopian tubes above and the vagina below. The embryo develops within the uterus, and is attached to it, after implantation, via the ◊placenta and umbilical cord. The outer wall of the uterus is composed of involuntary muscle, capable of contracting powerfully when giving birth. Its lining changes during the ◊menstrual cycle.

U-value measure of a material's heat-conducting properties. It is used in the building industry to compare the efficiency of insulating products, a good insulator having a low U-value. The U-value of a material is defined as the rate at which heat is conducted through it per unit surface area per unit temperature difference between its two sides; it is measured in watts per square metre per kelvin ($W\ m^{-2}\ K^{-1}$). In equation terms:

$$\text{U-Value} = \frac{\text{rate of loss of heat}}{\text{surface area} \times \text{temperature difference}}$$

vaccine any preparation of modified viruses or bacteria that is introduced into the body to bring about the specific ◊antibody reaction that will make the body immune against a particular disease. It is may be introduced by mouth, by hypodermic syringe, or by means of a scratch on the skin surface.

vacuole fluid-filled, membrane-bound cavity inside a cell. It may be a reservoir for fluids that the cell will secrete to the outside, or be filled with excretory products or essential nutrients that the cell needs to store. In the single-celled *Amoeba*, vacuoles are the sites of digestion of engulfed food particles. Plant cells usually have a large central vacuole for storage.

vacuum in general, a region completely empty of matter; in physics, any enclosure in which the gas pressure is considerably less than atmospheric pressure (101,325 pascals).

vagina the front passage in female mammals, linking the uterus to the exterior. It admits the penis during sexual intercourse, and is the birth canal down which the fetus passes during delivery.

valency the measure of an element's ability to combine with other elements, which can be expressed as the number of atoms of hydrogen (or any other standard univalent element) capable of uniting with (or replacing) its atoms. It is the number of electrons in the outermost shell of the atom that dictates the combining ability of an element.

The elements are described as univalent, divalent, trivalent, and tetravalent when they unite with one, two, three, and four univalent atoms respectively. Some elements have ***variable valency***—for example, nitrogen and phosphorus can both possess valencies of either three or five. The valency of oxygen is two; hence the formula for water, H_2O (hydrogen being univalent).

valve a structure for controlling the direction of the flow of a fluid, in pipes or in living organisms. In humans and other vertebrates, the contractions of the beating heart cause the correct blood flow into the arteries because a series of valves prevent it from flowing back.

vanadium silver-white, malleable and ductile, metallic element, symbol V, atomic number 23, relative atomic mass 50.942. It occurs in certain iron, lead, and uranium ores and is widely distributed in small quantities in igneous and sedimentary rocks. It is used to make steel alloys, to which it adds tensile strength.

vaporization change of state of a substance from liquid to vapour. See ◊evaporation.

vapour one of the three states of matter (see also ◊solid and ◊liquid). The molecules in a vapour move randomly and are far apart, the distance between them, and therefore the volume of the vapour, being limited only by the walls of any vessel in which they might be contained. A vapour differs from a ◊gas only in that a vapour can be liquefied by increased pressure, whereas a gas cannot unless its temperature is lowered below a specific (critical) temperature; it then becomes a vapour and may be liquefied.

variable property under investigation in an experiment. All other conditions must be kept constant throughout, for example, temperature, pressure, quantities used; only the property under investigation is varied.

variation the differences between individuals of the same species, found when examining a population. Such variations may be almost unnoticeable in some cases, obvious in others, and can concern many aspects of the organism. Typically, variation in size, behaviour, biochemistry, or colouring may be investigated. The cause of the variation may be genetic (and therefore inherited), environmental, or more usually a combination of the two. The origins of genetic variation can be traced to the recombination of the genetic material during the formation of the gametes, and, more rarely, to mutation.

variegation a description of plant leaves or stems that exhibit patches of different colours. The term is usually applied to plants that show white, cream, or yellow on their leaves, caused by areas of tissue that lack the green pigment ◊chlorophyll.

vascular bundle a strand of primary conducting tissue (a 'vein') in vascular plants, consisting mainly of water-conducting tissue, primary ◊xylem, and nutrient-conducting tissue, ◊phloem. It extends from the roots to the stems and leaves. Typically the phloem is situated nearest to the epidermis and the xylem towards the centre of the bundle.

vascular bundle

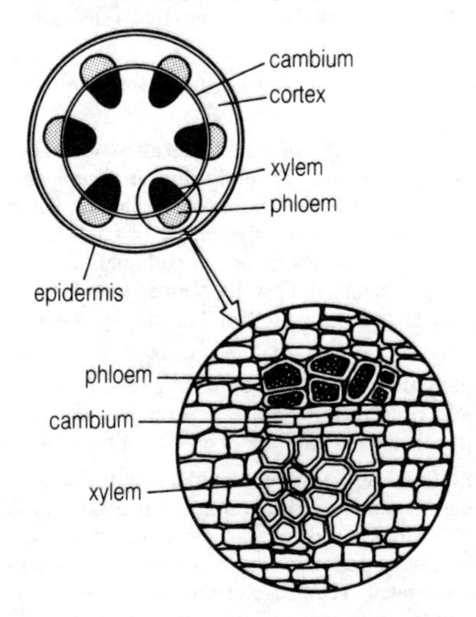

vascular plant a plant containing vascular bundles. Pteridophytes (ferns, horsetails, and club mosses), gymnosperms (conifers and cycads), and angiosperms (flowering plants) are all vascular plants.

vas deferens in male vertebrates, a tube conducting sperm from the testis to the urethra. The sperm is carried in a fluid (semen) secreted by various glands, and can be transported very rapidly when the involuntary muscle in the wall of the vas deferens undergoes rhythmic contractions, as in sexual intercourse.

vegetative reproduction ◊asexual reproduction in plants that relies not on spores but on multicellular structures formed by the parent plant. Some of the main types are ◊runners, sucker shoots produced from roots, ◊tubers, ◊bulbs, and ◊rhizomes. Vegetative reproduction has long been exploited in horticulture and agriculture, with various methods used to multiply stocks of plants; see ◊propagation of plants.

vein in animals with a circulatory system, any vessel that carries blood from the body to the heart. Veins contain valves that prevent the blood from running back when moving against gravity. They always carry deoxygenated blood, with the exception of the veins leading from the lungs to the heart in birds and mammals, which carry newly oxygenated blood.

The term is also used more loosely for any system of channels that strengthens living tissues and supplies them with nutrients—for example, leaf veins (see ◊vascular bundle), and the veins in insects' wings.

velocity the speed of an object in a given direction, or the rate of change of an object's displacement. The magnitude of the velocity v of an object travelling in a particular direction may be calculated by dividing the distance s it has travelled by the time t taken to do so, and may be expressed as:

$$v = s/t$$

The usual units of speed are metres per second or kilometres per hour. See also ◊distance–time graph.

The direction of the moving object is as important as its magnitude (or speed). If the direction of motion of a body changes, even if it is travelling at constant speed, then its velocity is also changing and it is therefore accelerating.

velocity ratio (VR) or ***distance ratio*** in a machine, the distance moved by the input force, or effort, divided by the distance moved by the output force, or load in the same time. It follows that the velocities of the effort and the load are in the same ratio. Velocity ratio has no units. See also ◊efficiency.

ventral in animals, term describing the lower surface, or the surface closest to the ground. The ventral surface of vertebrates is the surface furthest from the backbone; it faces forwards in bipedal (two-legged) vertebrates such as humans.

ventricle one of a pair of powerful pumping chambers in the lower half of the vertebrate heart. The ventricles are characterized by their thick muscular walls and their dependence on the coronary artery. A heart attack (coronary thrombosis) occurs when the coronary artery is blocked by a clot, and the ventricles are denied an adequate supply of oxygenated blood.

venule small vein, found between the capillary beds and the larger veins. It contains deoxygenated blood at low pressure.

vertebra (plural ***vertebrae***) one of the small bones that make up the ◊vertebral column. They vary in form according to position and are capable of only limited movement.

vertebral column or ***spine*** the backbone, which gives support to an animal and protects the central nervous system. It is made up of a series of bones called ***vertebrae*** (26 in most mammals), running from the skull to the tail with a central canal containing the nerve fibres of the spinal cord. In tetrapods (four-limbed vertebrates) the vertebrae show some specialization with the shape of the bones varying according to position. In the chest

vertebral column

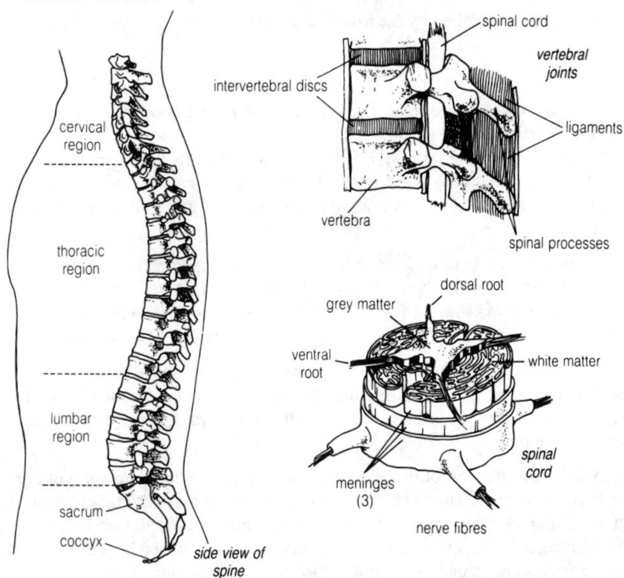

region the upper or thoracic vertebrae are shaped to form connections to the ribs. The backbone is only slightly flexible to give adequate rigidity to the animal structure.

In humans, there are seven cervical vertebrae, in the neck; 12 thoracic, in the upper trunk; five lumbar, in the lower back; the sacrum (consisting of five rudimentary vertebrae fused together, joined to the hipbones); and the coccyx (four vertebrae, fused into a tailbone).

vertebrate any animal with a backbone. The 41,000 species of vertebrates include mammals, birds, reptiles, amphibians, and fishes. They include most of the larger animals, but in terms of numbers of species are only a tiny proportion of the world's animals.

VHF (abbreviation for ***very high frequency***) referring to radio waves that have very short wavelengths. They are used for interference-free transmissions. VHF transmitters have a relatively short range because the waves cannot be reflected over the horizon like longer radio waves.

villus (plural *villi*) small fingerlike projection extending into the interior of the small intestine and increasing the absorptive area of the gut wall. Digested foods, such as sugars and amino acids, pass into the villi and are carried away by the circulating blood.

vinegar 4% solution of ethanoic (acetic) acid produced by the oxidation of alcohol, used to flavour food and as a preservative in pickles.

virus an infectious particle consisting of a core of nucleic acid (DNA or RNA) enclosed in a protein shell. Viruses are acellular and are able to function and reproduce only if they can invade a living cell to use the cell's system to replicate themselves. In the process they may disrupt or alter the host cell. Among diseases caused by viruses are chickenpox, the common cold, herpes, influenza, rabies, AIDS, and many plant diseases. Recent evidence implicates viruses in the development of some forms of cancer. ◊Antibiotic drugs cannot be used to treat viral diseases, but ◊vaccines can offer protection against infection.

vitamin any of various chemically unrelated organic compounds that are necessary in small quantities for the normal functioning of the body. Many act as coenzymes, small molecules that enable ◊enzymes to carry out their work. They are normally present in sufficient amounts in a balanced diet. Deficiency of a vitamin will normally lead to a metabolic disorder ('deficiency disease'), which can be remedied by sufficient intake of the vitamin.

Vitamins are generally classified as ***water-soluble*** (B and C) or ***fat-soluble*** (A, D, E, and K).

vitreous humour transparent jellylike substance behind the lens of the ◊eye. It gives rigidity to the spherical form of the eye and allows light to pass through to the retina.

vocal cords folds of tissue within a mammal's larynx. Air passing over them makes them vibrate, producing sounds. Muscles in the larynx change the pitch of the sound by adjusting the tension of the vocal cords.

volatile term describing a substance that readily passes from the liquid to the vapour phase.

volcano crack in the Earth's crust through which hot magma (molten rock) and gases well up. The magma becomes known as lava when it reaches the surface. A volcanic mountain, usually cone shaped with a crater on top, is formed around the opening, or vent, by the build-up of solidified lava and ashes (rock fragments).

Some volcanoes throw out lava explosively, while in others the material flows out gently over the crater rim. Most arise at plate margins (see ◊plate tectonics), where the movements of the plates generate magma or allow it to rise from the mantle beneath. There are about 600 active volcanoes on Earth. Some volcanoes may be inactive (dormant) for long periods.

volt SI unit (symbol V) of electromotive force or electric potential. A small battery has a potential of 1.5 volts; the domestic electricity supply in the UK is 240 volts (110 volts in the USA); and a high-tension transmission line may carry up to 500 kilovolts.

voltage term commonly used for ◊potential difference (pd).

voltage amplifier electronic device that increases an input signal in the form of a voltage or ◊potential difference, delivering an output signal that is larger than the input by a specified ratio.

voltmeter instrument for measuring potential difference (voltage). It has a high internal resistance (so that it passes only a small current), and is connected in parallel with the component across which potential difference is to be measured. To measure an AC (alternating-current) voltage, the circuit must usually include a rectifier (see ◊rectification).

VR abbreviation for ◊velocity ratio.

W

washing soda common name for hydrated ◊sodium carbonate. It is sometimes added to washing water to 'soften' it (see ◊hard water).

water H_2O liquid without colour, taste, or odour, an oxide of hydrogen. Water is the most abundant substance on Earth, and is essential to all forms of life. It has a unique range of properties.

Water is a reactive substance; it reacts with many metals and non-metals as well as both inorganic and organic substances.

with metals Water reacts with many metals to give oxides or hydroxides. With sodium it reacts vigorously at room temperature; with zinc it forms zinc oxide when passed over it as steam at red heat.

$$2Na + 2H_2O \rightarrow 2NaOH + H_2$$

$$Zn + H_2O \rightarrow ZnO + H_2$$

with non-metals With chlorine, water forms hydrochloric acid and chloric(I) acid at room temperature; with carbon, it forms carbon monoxide and hydrogen ('water gas') when steam is passed over white-hot carbon.

$$Cl_2 + H_2O \rightarrow HCl + HOCl$$

$$C + H_2O \rightarrow CO + H_2$$

with inorganic compounds The most common reactions of water are hydrolysis (splitting) and hydration (adding water). Anhydrous copper sulphate is hydrated by the addition of water; sodium carbonate is hydrolysed to give sodium hydrogencarbonate and sodium hydroxide.

$$CuSO_4 + 5H_2O \rightarrow CuSO_4.5H_2O$$

$$Na_2CO_3 + H_2O \rightarrow NaHCO_3 + NaOH$$

with organic compounds Hydration (for example, the conversion of ethene to ethanol) and hydrolysis (as in the splitting of long-chain carbohydrates into smaller polysaccharides) are the commonest reactions.

$$CH_2{=}CH_2 + H_2O \rightarrow CH_3CH_2OH$$

$$C_{12}H_{22}O_{11} + H_2O \rightarrow 2C_6H_{12}O_6$$

The relative molecular mass of water is 18, and its molecules are held together by intermolecular forces known as hydrogen bonds. These arise between the oxygen atom of one water molecule and the hydrogen atom of an adjacent molecule, and help explain why water is a liquid even though its molecules are very small.

Water begins to freeze solid at 0°C, and to boil at 100°C. When liquid, it is virtually incompressible; frozen, it expands by $^1/_{11}$ of its volume. At 4°C, one cubic centimetre of water has a mass of one gram, its maximum density, forming the unit of specific gravity. It has the highest known specific heat, and acts as an efficient solvent, particularly when hot. Most of the world's water is in the sea; less than 3% is fresh water.

water cycle the natural circulation of water through the biosphere. Water is lost from the Earth's surface to the atmosphere either by evaporation from the surface of lakes, rivers, and oceans or through the transpiration of plants. This atmospheric water forms clouds that condense to deposit moisture on the land and sea as rain or snow. The water that collects on land flows to the ocean in streams and rivers.

water gas fuel gas consisting of a mixture of carbon monoxide and hydrogen, made by passing steam over white-hot coke. The gas was once the chief source of hydrogen for chemical syntheses such as the ◊Haber process for making ammonia, but has been largely superseded in this and other reactions by hydrogen obtained from natural gas.

water of crystallization specific amount of water chemically bonded to a salt in its crystalline state; for example, in copper(II) sulphate, there are five moles of water per mole of copper sulphate, hence its formula is $CuSO_4.5H_2O$. This water is responsible for the colour and shape of the crystalline form. When the crystals are heated gently, the water is driven off as steam and a white powder is formed.

$$CuSO_4.5H_2O_{(s)} \rightarrow CuSO_{4(s)} + 5H_2O_{(g)}$$

watt SI unit (symbol W) of power (the rate of expenditure or consumption of energy). A light bulb may use 60, 100, or 150 watts of power; an electric heater will use several kilowatts (thousands of watts).

water cycle

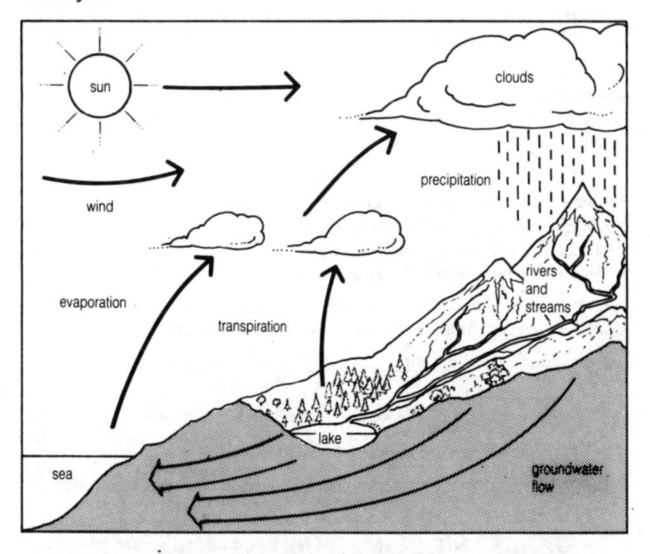

The watt is defined as the power used when one joule of work is done in one second. In electrical terms, the flow of one ampere of current through a conductor whose ends are at a potential difference of one volt uses one watt of power (watts = volts × amps).

wave disturbance travelling through a medium (or space). There are two types: in a ***longitudinal wave*** (such as a ◊sound wave) the disturbance is parallel to the wave's direction of travel; in a ***transverse wave*** (such as an ◊electromagnetic wave) it is perpendicular. The medium (such as the Earth, in the case of seismic waves) is not permanently displaced by the passage of a wave.

wave energy energy derived from that of water waves. Various schemes have been advanced since 1973, when oil prices rose dramatically and an energy shortage threatened. The best-known of these is the duck—a floating boom whose segments nod up and down with the waves. The nodding motion can be used to drive pumps and spin generators. Another device, developed in Japan, uses an oscillating water column to harness wave energy.

wave equation equation relating the speed of a wave to its frequency and wavelength. If the wave's speed is *c*, its frequency *f*, and its wavelength, then:

$$c = f\lambda$$

wavelength the distance between successive crests of a ◊wave. The wavelength of a light wave determines its colour; red light has a wavelength of about 700 nanometres, for example. The complete range of wavelengths of electromagnetic waves is called the electromagnetic ◊spectrum.

weak acid acid that only part ionizes in aqueous solution (see ◊dissociation). Weak acids include ethanoic acid and carbonic acid.

$$CH_3COOH_{(l)} + aq \leftrightarrow H^+_{(aq)} + CH_3COO^-_{(aq)}$$

$$H_2CO_{3(aq)} \leftrightarrow H^+_{(aq)} + HCO^-_{3(aq)}$$

The pH of such acids lies between 3 and 6.

weak base base that only part ionizes in aqueous solution (see ◊dissociation). Ammonia is a weak base.

$$NH_{3(g)} + H_2O_{(l)} \leftrightarrow NH^+_{4(aq)} + OH^-_{(aq)}$$

The pH of weak bases lies between 8 and 10.

weak electrolyte electrolyte that conducts electricity only moderately. Weak acids and bases are weak electrolytes.

weather the day-to-day variations in meteorological conditions, such as cloudiness, temperature and rainfall. A region's ◊climate is derived from the average weather conditions over a long period of time.

weathering process by which exposed rocks are broken down on the spot by the action of rain, frost, wind, and other elements of the weather. It differs from ◊erosion in that no movement or transportion of the broken-down material takes place.

Two main types of weathering are recognized. ***Physical weathering*** includes such effects as the splitting of rocks by the alternate freezing and thawing of water trapped in cracks, or by the alternate expansion and contraction of rocks in response to extreme changes in temperature. ***Chemical weathering*** is brought about by a chemical change in the rocks affected. The most common type is caused by rainwater that has absorbed carbon dioxide from the atmosphere and formed a weak carbonic acid. The slightly acidic rainwater is then capable of dissolving certain minerals in rocks (for example, calcium carbonate in limestone).

weedkiller another name for ◊herbicide, a chemical that kills some or all plants.

weight the force exerted on an object by ◊gravity. The weight of an object depends on its mass—the amount of material in it—and the Earth's gravitational field strength. If the mass of a body is m kilograms and the gravitational field strength is g newtons per kilogram, its weight W in newtons is given by:

$$W = mg.$$

The strength of the Earth's gravitational field strength decreases with height; consequently, an object will weigh less at the top of a mountain than at sea level. On the Moon, an object weighs only one-sixth of its weight on Earth because the pull of the Moon's gravity is one-sixth that of the Earth.

weightlessness condition in which there is no gravitational force acting on a body, either because gravitational force is cancelled out by equal and opposite acceleration, or because the body is so far outside a planet's gravitational field that no force is exerted upon it.

white blood cell or ***leucocyte*** one of a number of different cells that play a part in the body's defences and give immunity against disease. Some of these cells engulf invading microorganisms, others kill infected cells, while ◊lymphocytes produce more specific immune responses. White blood cells are colourless, with clear or granulated cytoplasm, and are capable of independent amoeboid movement. Unlike mammalia red blood cells, they possess a nucleus. Human blood contains about 11,000 leucocytes per cubic millimetre—about one to every 500 red cells . White cells are not confined to the blood, however; they also occur in the ◊lymph and elsewhere in the body's tissues.

wild type in genetics, the naturally occurring allele for a particular character that is typical of most individuals of a given species, as distinct from new alleles that arise by mutation.

wind energy energy derived from the wind. It is harnessed by sailing ships and windmills, both of which are ancient inventions, and by wind ◊turbines, aerodynamically advanced windmills that drive electricity generators when their blades are spun by the wind. Wind energy is a renewable resource that produces no direct pollution of the air; it is therefore beginning to be used to produce electricity on a large scale.

wind vane an instrument which shows the direction of the wind. It usually consists of an arrow fixed so that it can move freely in a horizontal plane as the wind blows. Wind vanes are often placed on the roofs of high buildings, where they will be exposed to the wind.

wood the hard tissue beneath the bark of many perennial plants; it is composed of water-conducting cells, or secondary ◊xylem, and gains its hardness and strength from deposits of the organic compound lignin. ***Hardwoods***, such as oak, and ***softwoods***, such as pine, have commercial value as structural material and for furniture.

The central wood in a branch or stem is known as ***heartwood*** and is generally darker and harder than the outer wood; it consists only of dead cells. As well as providing structural support, it often contains gums, tannins, or pigments which may impart a characteristic colour and increased durability. The surrounding ***sapwood*** is the functional part of the xylem that conducts water.

The ***secondary xylem*** is laid down by the vascular ◊cambium, which forms a new layer of wood annually, on the outside of the existing wood and visible as an ◊annual ring when the tree is felled.

work measure of the result of transferring energy from one system to another to cause an object to move. Work should not be confused with ◊energy (the capacity to do work, which is also measured in joules) or with ◊power (the rate of doing work, measured in joules per second).

Work is equal to the product of the force used and the distance moved by the object in the direction of that force. If the force is F newtons and the distance moved is d metres, then the work W is given by:

$$W=Fd$$

For example, the work done when a force of 10 newtons moves an object 5 metres against some sort of resistance is 50 joules (50 newton metres).

X chromosome the larger of the two ◊sex chromosomes, the smaller being the ◊Y chromosome. In female mammals, the X chromosome is paired with another X chromosome; in males, it is paired with a Y chromosome. Genes carried on the X chromosome produce the phenomenon of ◊sex linkage.

xenon colourless, odourless, gaseous element, symbol Xe, atomic number 54, relative atomic mass 131.30. It is grouped with the ◊noble gases and was long believed not to enter into reactions, but is now known to form some compounds, mostly with fluorine. It is a heavy gas present in very small quantities in the air (about one part in 20 million), and is used in bubble chambers, light bulbs, vacuum tubes, and lasers.

xerophyte a plant adapted to live in dry conditions. Common adaptations to reduce the rate of ◊transpiration include a reduction of leaf size, sometimes to spines or scales; a dense covering of hairs over the leaf to trap a layer of moist air (as in edelweiss); and permanently rolled leaves or leaves that roll up in dry weather (as in marram grass). Many desert cacti are xerophytes.

X-ray band of electromagnetic radiation in the wavelength range 10^{-11} to 10^{-9}m (between gamma rays and ultraviolet radiation; see ◊electromagnetic waves). Applications of X-rays make use of their short wavelength (such as X-ray crystallography) or their penetrating power (as in medical X-rays of internal body tissues). X-rays are dangerous and can cause cancer.

X-rays are produced when high-energy electrons from a heated filament cathode strike the surface of a target (usually made of tungsten).

xylem a tissue found in ◊vascular plants, whose main function is to transport water and dissolved mineral nutrients from the roots to other parts of the plant. In angiosperms, xylem is composed of a number of different types of cell, including the continuous conducting vessels, fibres, and thin-walled parenchyma cells.

Y chromosome the smaller of the two sex chromosomes. In male mammals it occurs paired with the other type of sex chromosome (X), which carries far more genes. The Y chromosome is the smallest of all the mammalian chromosomes and is considered to be largely inert (that is, without direct effect on the physical body).

yeast one of various single-celled fungi that form masses of minute circular or oval cells by budding. When placed in a sugar solution the cells multiply and convert the sugar into ethanol (alcohol) and carbon dioxide; see ◊anaerobic respiration. Yeasts are used as fermenting agents in baking, brewing, and the making of wine and spirits.

yolk a store of food, mostly in the form of fats and proteins, found in the ◊eggs of many animals. It provides nourishment for the growing embryo.

Z

zinc hard, brittle, bluish-white, metallic element, symbol Zn, atomic number 30, relative atomic mass 65.37. The principal ore is spalerite or zinc blende (zinc sulphide, ZnS). Zinc is little affected by air or moisture at ordinary temperatures; its chief uses are in alloys such as brass and in coating metals (for example, galvanized iron). Its compounds include zinc oxide, used in ointments (as an astringent), cosmetics, paints, glass, and printing ink.

zinc chloride $ZnCl_2$ white, crystalline compound that is deliquescent and sublimes easily. It is used as a catalyst, as a dehydrating agent, and as a flux in soldering.

zinc oxide ZnO white powder, yellow when hot, that occurs in nature as the mineral zincite. It is an amphoteric oxide and is used in paints and as an antiseptic in zinc ointment; it is the main ingredient of calamine lotion.

zoology the branch of biology concerned with the study of animals. It includes the description of present-day animals, the study of evolution of animal forms, anatomy, physiology, embryology, behaviour, and geographical distribution.

zygote an ◊ovum (egg) after ◊fertilization but before it undergoes cleavage to begin embryonic development.

Appendix I

Periodic table of the elements

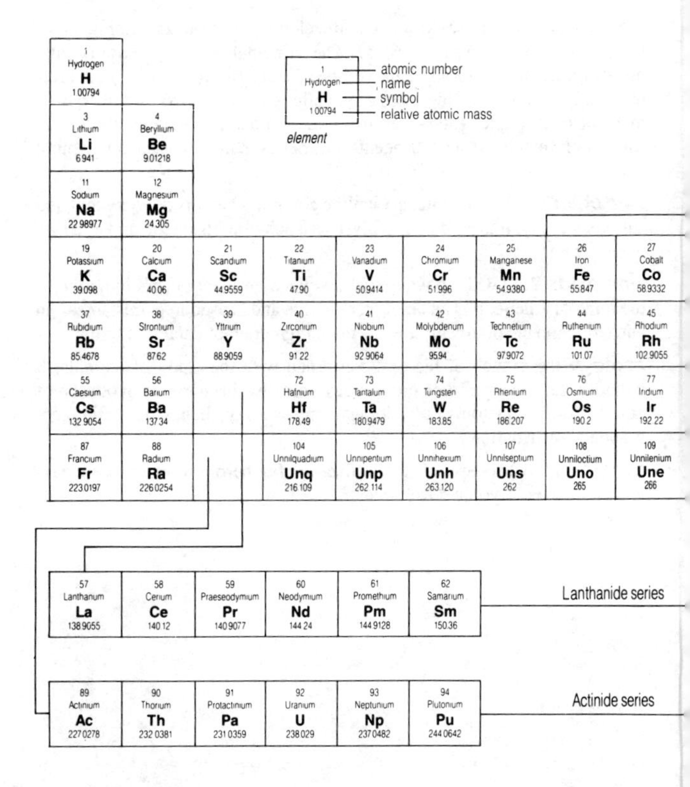

								2 Helium **He** 4 00260
			5 Boron **B** 10 81	6 Carbon **C** 12 011	7 Nitrogen **N** 14 0067	8 Oxygen **O** 15 9994	9 Fluorine **F** 18 99840	10 Neon **Ne** 20 179
			13 Aluminium **Al** 26 98154	14 Silicon **Si** 28 086	15 Phosphorus **P** 30 97376P	16 Sulphur **S** 32 06	17 Chlorine **Cl** 35 453	18 Argon **Ar** 39 948
28 Nickel **Ni** 58 70	29 Copper **Cu** 63 546	30 Zinc **Zn** 65 38	31 Gallium **Ga** 69 72	32 Germanium **Ge** 72 59	33 Arsenic **As** 74 9216	34 Selenium **Se** 78 96	35 Bromine **Br** 79 904	36 Krypton **Kr** 83 80
46 Palladium **Pd** 106 4	47 Silver **Ag** 107 868	48 Cadmium **Cd** 112 40	49 Indium **In** 114 82	50 Tin **Sn** 118 69	51 Antimony **Sb** 121 75	52 Tellurium **Te** 127 75	53 Iodine **I** 126 9045	54 Xenon **Xe** 131 30
78 Platinum **Pt** 195 09	79 Gold **Au** 196 9665	80 Mercury **Hg** 200 59	81 Thallium **Tl** 204 37	82 Lead **Pb** 207 37	83 Bismuth **Bi** 207 2	84 Polonium **Po** 210	85 Astatine **At** 211	86 Radon **Rn** 222 0176

63 Europium **Eu** 151 96	64 Gadolinium **Gd** 157 25	65 Terbium **Tb** 158 9254	66 Dysprosium **Dy** 162 50	67 Holmium **Ho** 164 9304	68 Erbium **Er** 167 26	69 Thulium **Tm** 168 9342	70 Ytterbium **Yb** 173 04	71 Lutetium **Lu** 174 97

95 Americium **Am** 243 0614	96 Curium **Cm** 247 0703	97 Berkelium **Bk** 247 0703	98 Californium **Cf** 251 0786	99 Einsteinium **Es** 252 0828	100 Fermium **Fm** 257 0951	101 Mendelevium **Md** 258.0986	012 Nobelium **No** 259 1009	103 Lawrencium **Lr** 260 1054

Appendix II

SI units and multiples

quantity	*SI unit*	*symbol*
absorbed radiation dose	gray	Gy
amount of substance	mole*	mol
electric capacitance	farad	F
electric charge	coulomb	C
electric conductance	siemens	S
electric current	ampere*	A
energy or work	joule	J
force	newton	N
frequency	hertz	Hz
illuminance	lux	lx
inductance	henry	H
length	metre*	m
luminous flux	lumen	lm
luminous intensity	candela*	cd
magnetic flux	weber	Wb
magnetic flux density	tesla	T
mass	kilogram*	kg
plane angle	radian	rad
potential difference	volt	V
power	watt	W
pressure	pascal	Pa
radiation dose equivalent	sievert	Sv
radiation exposure	roentgen	r
radioactivity	becquerel	Bq
resistance	ohm	W
solid angle	steradian	sr
sound intensity	decibel	dB
temperature	°Celsius	°C
temperature, thermodynamic	kelvin*	K
time	second*	s

*SI base unit

SI prefixes

multiple	*prefix*	*symbol*	*example*
1,000,000,000,000,000,000 (10^{18})	exa-	E	Eg(exagram)
1,000,000,000,000,000 (10^{15})	peta-	P	PJ (petajoule)
1,000,000,000,000 (10^{12})	tera-	T	TV (teravolt)
1,000,000,000 (10^{9})	giga-	G	GW (gigawatt)
1,000,000 (10^{6})	mega-	M	MHz (megahertz)
1,000 (10^{3})	kilo-	k	kg (kilogram)
100 (10^{2})	hecto-	h	hm (hectometre)
10	deca-	da-	daN (decanewton)
1/10 (10^{-1})	deci-	d	dC(decicoulomb)
1/100 (10^{-2})	centi-	c	cm(centimetre)
1/1,000 (10^{-3})	milli-	m	mA(milliampere)
1/1,000,000 (10^{-6})	micro-	m	µF (microfarad)
1/1,000,000,000 (10^{-9})	nano-	n	nm(nanometre)
1/1,000,000,000,000 (10^{-12})	pico-	p	ps (picosecond)
1/1,000,000,000,000,000 (10^{-15})	femto-	f	frad(femtoradian)
1/1,000,000,000,000,000,000 (10^{-18})	atto-	a	aT(attotesla)